Annual Conference and Expo of the Association of Avian Veterinarians (AAV 2025)

Phoenix, Arizona, USA
3-6 October 2025

ISBN: 979-8-3313-2911-2

Printed from e-media with permission by:

Curran Associates, Inc.
57 Morehouse Lane
Red Hook, NY 12571

Some format issues inherent in the e-media version may also appear in this print version.

Association of Avian Veterinarians
PO Box 9
Teaneck, NJ 07666
USA

Phone: 720-458-4111
Fax: 720-398-3496

office@aav.org

Additional copies of this publication are available from:

Curran Associates, Inc.
57 Morehouse Lane
Red Hook, NY 12571 USA
Phone: 845-758-0400
Fax: 845-758-2633
Email: curran@proceedings.com
Web: www.proceedings.com

ASSOCIATION OF AVIAN VETERINARIANS

AAV ANNUAL CONFERENCE & EXPO

2025 PROCEEDINGS

FRIDAY, OCTOBER 3 – MONDAY, OCTOBER 6, 2025
GILA RIVER RESORT & CASINO
PHOENIX, ARIZONA

ASSOCIATION OF AVIAN VETERINARIANS
AAV AVIAN MEDICINE VIRTUAL

INCLUDES **BONUS SECTION**

AAV.ORG

Introduction

We invite you to enjoy the 46th Annual AAV Conference Proceedings. Within the proceedings, you will find a wealth of information related to the variety of lectures presented at this year's conference.

AAV recognizes and thanks the following committees and staff for their time and hard work in assuring a quality program and publication:

AAV Executive Board and Conference Committee

Jennifer Graham, DVM, DABVP (Avian Practice), DABVP (Exotic Companion Mammal Practice), Dipl ACZM, *President, Conference Committee Member*
Brynn McCleery, DVM, DABVP (Avian Practice), *President-Elect, Conference Committee Member*
Stephanie Lamb, DVM, DABVP (Avian Practice), *Immediate Past President, Conference Committee Chair*
Len Donato, VMD, DABVP (Avian Practice), *Treasurer, Conference Committee Member*
Kristina Palmer, RVT, VTS (Clinical Practice-Exotic Companion Animal), *AAV Hands-On Workshop Coordinator*
Salina Locke, DVM, DABVP (Avian Practice), Cert AqV, *Education Committee Chair*
Jen Riley, DVM, *Advisory Council Representative to the AAV Executive Board*
Elizabeth Mackey, DVM, CertHbV, *Executive Director*

Veterinary Review and Editing Committee

Rachel Baden, DVM, DABVP (Avian Practice)
Kyra Berg, DVM, Dipl ACZM
Victoria Bender, DVM
Joao Brandao, LMV, MS, DipECZM (Avian)
Sue Chen, DVM, DABVP (Avian Practice)
John Chitty, BVetMed, CertZooMed, CBiol, MRSB, MRCVS
Erica Cervasio, DVM
Bob Dahlhausen, DVM
Bob Doneley, BVSc, FANZCVS
Jesse Fallon, DVM, PhD
Grace Gallentine, DVM
Jennifer Graham, DVM, DABVP (Avian Practice), DABVP (Exotic Companion Mammal Practice), Dipl ACZM
Cheryl Greenacre, DVM, DABVP (Avian Practice), DABVP (Exotic Companion Mammal Practice)
David Sanchez-Migallon Guzman, LV, MS, DipECZM (Avian), DipECZM (Small Mammal), Dipl ACZM
Rudiger Korbel, DipECZM (Avian), Prof, Dr med vet habil
Stephanie Lamb, DVM, DABVP (Avian Practice)
Isabelle Langlois, DVM, DABVP (Avian Practice)
Stephany Lewis, DVM, CWR, DABVP (Avian Practice)
Salina Locke, DVM, DABVP (Avian Practice), Cert AqV
An Martel, DVM, PhD, DipECZM
Alicia McLaughlin, DVM
Brynn McCleery, DVM, DABVP (Avian Practice)
Julianne McCready, DVM, Dipl ACZM
Bianca Murphy, DVM, DABVP (Avian Practice)
Anna Osofsky, DVM, DABVP (Avian Practice)
Olivia Petritz, DVM, Dipl ACZM
Patrick Redig, DVM, PhD
Julianne Richard, DVM, MPH, CertAqV
Jaime Samour, MVZ (Hons), PhD, DipECZM (Avian), FRSB
Nico Schoemaker, DVM, PhD, DipECZM (Small Mammal), DipECZM (Avian)
Patrick Sullivan, DVM, DABVP (Avian Practice)
Thomas N. Tully, Jr., DVM, MS, DABVP (Avian Practice), DipECZM (Avian)
Yvonne van Zeeland, DVM, MVR/MSc, PhD, DipECZM (Avian), DipECZM (Small Mammal), CPBC
Kristina Vuong, DVM, DABVP (Avian Practice)
Kenneth Welle, DVM, DABVP (Avian Practice)
Perri Wiggin, DVM

Technical Editor

Charlotte Kirk Baer, MS

Proceedings Layout and Design — BrightLizard, Inc.

Elizabeth Smith

Conference Sponsors

HARRISON'S BIRD FOODS

VMD Veterinary Molecular Diagnostics, Inc.

ZuPreem

COLORADO EXOTIC

avian & exotic veterinary care

ARIZONA EXOTIC

Proceedings Sponsor: Avian & Exotic Animal ClinPath Labs

Continuing Education Accreditation

The American Association of Veterinary State Boards (AAVSB) Registry of Approved Continuing Education (RACE) committee has reviewed and approved the 46th Annual Association of Avian Veterinarians Conference & Expo program as meeting the standards adopted by the AAVSB. Please contact the AAVSB RACE program at race@aavsb.org or 877-698-8482 if you have any questions regarding Continuing Education for this program.

Disclaimer

The material appearing in this publication comes exclusively from the authors and contributors identified in each manuscript. The techniques and procedures discussed reflect the personal knowledge and experience of the authors and contributors, and demonstrate their views of the methods that may be used for these medical procedures. The procedures demonstrated do not incorporate all known techniques, are not exclusive, and other techniques and technology may also be available. Any questions or requests for additional information concerning any of the manuscripts should be addressed directly to the authors.

AAV does not research, review, or otherwise verify any of the information contained in this publication. Opinions expressed in this publication are those of the authors and contributors and not necessarily those of AAV. AAV is not responsible for errors or for opinions expressed in this publication.

AAV expressly disclaims any warranties or guarantees, expressed or implied, and shall not be liable for damages of any kind in connection with the material, information, techniques, or procedures set forth in this publication.

Publication Terms

Conference proceedings manuscripts are accepted as full manuscripts or as scientific abstracts. Scientific abstracts are accepted if the author(s) express(es) intent to submit their work for consideration for future publication in a peer-reviewed scientific journal. Several distinguished veterinary journals, including the *Journal of the American Veterinary Medical Association* and the *American Journal of Veterinary Research* may reject work that has been previously published of 250 words or more and included additional content such as figures, tables, illustrations, and references. Scientific abstracts published in the AAV proceedings are still required to contain data and meaningful information to the reader.

For permission to reproduce or order copies of this publication, contact the AAV association office: www.aav.org, 1802 Antioch Church Road, Watkinsville, GA 30677, USA, email pubs@aav.org. Visit www.aav.org for information on AAV educational conferences, membership, and other association programs.

Table of Contents

Section 1: Friday Fundamentals Session

Section 2: Plenary Session

Section 3: Wildlife Session I

Section 4: Cardiology and Coelomic Cavity Session

Section 5: Technician Session

Section 6: Wildlife Session II

Section 7: Computed Tomography, Avian Bornavirus, and Reproductive Topics Session

Section 8: Analgesia and Anesthesia Session

Section 9: Species-Specific and Potpourri I Session

Section 10: Management, Business, and Potpourri II Session

Section 11: Wildlife and Then Some Session

Section 12: Masterclass Session I

Section 13: Masterclass Session II

A Message from Our 2025 Proceedings Sponsor

Section 1

Friday Fundamentals Session

Sue Chen, DVM, DABVP (Avian Practice);
Anna Osofsky, DVM, DABVP (Avian Practice)
Session Chairs

Setting the Stage for Success: Educating Parrot Owners During the First Veterinary Visit

Anna Osofsky, DVM, DABVP (Avian Practice)

Session #107

Affiliation: From Carrollton West Pet Hospital, Carrollton, TX 75007, USA.

Abstract: The first veterinary visit is a critical opportunity to establish a foundation for a parrot's lifelong health and well-being. This lecture will provide veterinary professionals with the information and resources they need to educate parrot owners during that inaugural visit. Topics will include essential husbandry practices, a review of the nutritional requirements of their species of bird, environmental enrichment, and the importance of preventive care. By equipping owners with this knowledge, veterinarians can enhance compliance, improve patient outcomes, and cultivate a more informed, compassionate clientele.

Introduction

Much of the information that the author provides to clients starts with key handouts available on the Association of Avian Veterinarians (AAV) website; some of these handouts are available to the public on the Bird Owner Resources page,[1] which can be printed in numerous languages, and some require AAV membership to access.[2] Veterinary technicians are an excellent resource to use as the primary client educators; the doctor can come in after this information has been reviewed and then answer any questions the bird's family has. Some families come in well-educated while others do not; occasionally, choices must be made about what to focus on during the initial visit. The author has found that few people will read much of the information provided, so it needs to be at least briefly reviewed during the visit. If you have not read these handouts and are just starting out, they are great resource for your learning as well!

The author's clinic schedules 1-hour new bird visits to provide time for educating the bird's family and answering all their questions.

The Bird Information Pack

The author provides the following handouts to every new family seen at the clinic. We call this compilation of material our bird information pack.

1. Basic care for companion parrots. (AAV Public)
2. Protecting your pet birds from household dangers. (AAV Public)
3. Signs of illness in companion birds. (AAV Public)
4. Ultraviolet lighting for companion birds. (AAV Membership required)
5. Avian chlamydiosis and psittacosis. (AAV Public)

6. Digital scales. (AAV Membership required)

7. Foraging for parrots. (AAV Public)

8. Local emergency clinics that see birds. (Must be made by your clinic for your location)

9. The QR code (AAV Membership required, under the AAV Promotional Materials, QR cards, and Posters tab) for the "How to Medicate Your Parrot" video; the video is also available on the bird owner resources page.

Additional Client Education Handouts, Provided as Needed

- Transitioning pet parrots away from a seed diet. (AAV Public). This can be a challenge for clients to carry out so it needs to be reviewed in a little more detail. Some clients arrive not knowing what pellets even are.

- How to introduce bathing to your bird. (AAV Membership required) Bathing is important for birds. However, we don't want to force them to do an activity they are frightened of, so training is sometimes needed.

- Wing trimming. (AAV Membership required) Most avian veterinarians would prefer to allow companion birds the pleasure of flight, as it is important for both cardiovascular and psychological health. However, flight is not without some degree of risk, so discussion of this important topic may be required.

- What the author calls the Scott Echols Nutrition Handout.[3] This client education handout was primarily written by Scott Echols, DVM, DABVP (Avian Practice); the author has made a few edits to update it with his approval. This handout provides a list of the most nutritious fruits, vegetables, and legumes. It also provides a list of healthy whole grains, which Echols feels should be included in the diet because the finely ground foods used in pellets can predispose birds to diarrhea and changes to their gut microbiota. Echols recommends that parrots be fed 25% of the diet as these grains, 50% of the diet as pellets, and 25% of the diet as healthy fruits and vegetables, though he leaves room for changing percentages based on species, lifestyle, and medical concerns. The grains can be purchased and mixed to be fed as part of the ration; they can also be cooked like rice, frozen in ice cube trays, and defrosted for part of the daily meal. This handout is only provided to clients that have more in-depth questions about diet and nutrition or for patients with metabolic diseases that might benefit from whole grains/fewer pellets in the diet (e.g., renal disease, chronic diarrhea).

- Baby bird information. The AAV does not recommend sale of unweaned parrots to the public. Occasionally these birds are brought into our clinics. The AAV has two public resources for this situation: (1) A video entitled Caring for Baby Parrots: Handfeeding and (2) a handout entitled Weaning Baby Parrots.[1]

Species-Specific Information

- Macaws living with powder-down producing birds. Macaws, especially blue and gold macaws (*Ara ararauna*), can develop hypersensitivity pneumonitis, a respiratory condition that can be fatal, when housed with dander-producing birds such as cockatoos, cockatiels (*Nymphicus hollandicus*), and grey parrots (*Psittacus* species). The author recommends keeping macaws in separate parts of the house from these other species. If this is not possible, a good-quality HEPA filter (or two) near the cage is recommended.

- Eclectus parrots (*Eclectus roratus*). Many avian veterinarians feel that we don't truly know the nutritional requirements of these birds; generally, it is recommended that they be offered more fresh fruit than most parrot species. The author recommends 50-60% of the diet as pellets and 40-50% as fresh fruits and vegetables. A wide variety of produce should be offered, though it can be on a rotating basis. Tropical fruits, berries, and dark orange and green produce (but no avocado!) tend to be most nutritious.

- Budgerigar (*Melopsittacus undulatus*), lovebird (*Agapornis* species), and cockatiel diets. A study performed by David Phalen, DVM, PhD but only published in proceedings showed that budgerigars appear to need less dietary calcium than other psittacine birds and, as a result, pelleted diets have the potential to cause metastatic mineralization of the kidneys.[4] He postulates that cockatiels and lovebirds may be similarly sensitive.[5] What to feed these species of birds is a topic of debate in the avian medical community, with some practitioners recommending the same diet as for other parrots, while some may find a high-quality seed diet with a mineral block or cuttlebone acceptable. The author recommends about 50% (or less) of the diet as pellets and 50% (or more) as seed with the addition of fresh veggies and fruits; a mineral block or cuttlebone should also be offered. Interestingly, budgies did not seem to ingest cuttlebone to the point of renal mineralization in Phalen's study.

Diagnostic Testing

The author recommends annual blood work for all parrots in which the author feels venipuncture can safely be performed (usually cockatiels or larger birds). Clients are advised that blood work serves two purposes: assessing overall health, since we know parrots are prey species and excellent at hiding illness, and establishing a baseline of what is normal for that individual. For example, the author has many patients that consistently have a low albumin on protein electrophoresis or a mildly elevated uric acid on the biochemical profile; knowing what is normal for the patient helps the clinician interpret blood work when the patient is sick.

Infectious disease testing should be discussed and recommended as appropriate, such as fecal direct and float for parasites, avian gastric yeast (AGY, *Macrorhabdus ornithogaster*), polyoma virus, psittacine beak and feather disease virus (PBFD, a circovirus), avian bornavirus, *Chlamydia psittaci*, and Psittacid herpesvirus testing. Though fecal parasite testing is usually negative, the author had a young quaker parrot (*Myiopsitta monachus*) patient die recently due to roundworm impaction, illustrating that this is still an important test. Clinical judgement must be applied when recommending tests. Since *Chlamydia* is zoonotic, testing for it should always be recommended. Other tests can be performed as appropriate. *Spironucleus meleagridis* infections are common in cockatiels and can be found on a fecal direct. AGY is common in budgerigars and can be identified via PCR, fecal direct, or a fecal Gram stain. Polyoma is mostly a concern if birds are less than 4 months old or are going to a breeding situation. PBFD testing is especially important if there is a new bird being added to a household with established Old World parrots.

A minimum of 30 days of quarantine should be recommended if a new bird is being added to an established flock.

Final Topics of Discussion

For clients that are bringing in their new bird, the author always tries to help them get on the right track. It is important to review that parrots have rich lives in the wild, much of which we take away from them by putting them in our homes. Thus, it is important that the family becomes their new "flock," that the bird be included in family life as much as possible, and that in addition to time with the flock, foraging and toys need to be provided for enrichment. The family should avoid allowing a pair bond to form between the bird and one of the humans in the family; aren't we lucky that we get to discuss masturbation with clients on a regular basis in avian medicine?

At the end of the visit, the author always advises that an annual exam is recommended (our computer system generates a reminder for owners, similar to preventive care visit reminders for other species) and an exam is needed right away if the bird ever appears ill.

Future Visits

Suggestions for improvements in care can be made at future visits; for example, the author commonly finds that bird parents have not followed recommendations about UV light exposure or foraging after the initial visit, so these topics can be reviewed again at the annual exam.

Husbandry Handouts for Other Avian Orders

The AAV also has husbandry handouts available for backyard poultry, ducks, rhamphastids such as toucans, pigeons, canaries and finches. Many of the handouts recommended above as part of the bird information pack apply to these species as well.

References

1. Association of Avian Veterinarians. Bird Owner Resources. Available at: https://www.aav.org/page/birdowners. Accessed March 30, 2025.

2. Association of Avian Veterinarians. Publications library. Available at: https://www.aav.org/page/publicationdownload. Accessed March 30, 2025.

3. Association of Avian Veterinarians. Publications library. Available at: https://www.aav.org/page/publicationdownload. Accessed April 30, 2025.

4. Phalen DN. The role of dietary calcium in the development of metastatic mineralization in parrots. *Proc Annu Conf Assoc Avian Vet Australasian Committee*. 2009;34-37.

5. Phalen DN. Diseases of the avian urinary system. *Vet Clin North Am Exot Anim Pract*. 2020;23:21-45.

Building Confidence in Avian Examinations: A Guide for New Practitioners

Anna Osofsky, DVM, DABVP (Avian Practice)

Session #108

Affiliation: From Carrollton West Pet Hospital, Carrollton, TX 75007, USA.

Abstract: Performing a thorough and efficient avian physical examination is a cornerstone skill for any avian veterinarian. This paper outlines a systematic approach to conducting a physical examination on a parrot, including handling techniques, key anatomic assessments, and considerations for different species. Special considerations for other commonly kept species such as chickens, pigeons, and canaries also are reviewed.

Introduction

Companion birds and backyard poultry are commonly presented to avian veterinarians for routine health assessments and medical concerns. Unlike traditional companion animals, birds require specialized handling and examination techniques to minimize stress while ensuring a comprehensive evaluation. This paper provides a step-by-step guide to conducting a thorough physical examination on an avian patient; as with physical examinations in any species, developing a consistent routine when performing the examination is essential to make sure nothing is missed.

Pre-Examination Observation

Before handling the bird, observing the parrot in its carrier or cage allows for an assessment of mentation, posture, respiration, and droppings. If the patient appears fragile, consider pre-oxygenation/provision of warmth, discussing your concerns with the owner prior to handling, and staging the examination as appropriate.

Pro tip: Adult birds should be alert in the examination room; resting with eyes closed is usually a sign of illness. Young birds (usually less than 10 months of age) are sometimes sleepy after an examination, even when well.

Restraint and Handling

When examining parrots, proper restraint is essential for both the bird's safety and the veterinary team's protection. Towels or other cloth (e.g., baby blankets or fleece blankets cut into pieces) are usually used for capture, to prevent the birds from becoming hand phobic and to protect the restrainer's hands from bites. Towels will not prevent a strong beak from biting one's hand, but they do make it hard for the bird to see the hand.

The author generally asks the owner what they think the best way to get the bird in hand is, although owners sometimes make poor choices and we end up with a bird flying around the examination room. Some owners can place the bird into an acrylic weighing container from which it can easily be captured with a towel.

Other owners can gently towel their bird prior to handing it to the veterinary team member. Some birds must be caught from the cage and some from the examination room floor. Avoid chasing the patient around and pouncing on it with the towel as this can be frightening for a prey species. When catching a bird from the examination table or cage, determine if the bird will step up onto a toweled hand and allow itself to be gently wrapped. Sometimes the author will have the bird step onto a toweled hand and then kneel slowly on the examination room floor prior to wrapping in the towel in case the bird spooks and tries to jump to the floor. Occasionally, a parrot can be examined without any restraint; this is most common with baby birds and outgoing cockatoos. Small birds brought into the clinic in large cages sometimes must be captured with the room lights off while an assistant uses a transilluminator to provide a small amount of light for the person catching the bird.

Pro tip: The author prefers to use huck towels for capturing very small parrots such as budgies and parrotlets. These towels are made of thin material and, therefore, are easy to feel the bird through and thus, avoid injuring them.

Restraint technique depends on the size of the bird. With smaller patients, a one-handed technique with the thumb and forefinger restraining the head and the palm restraining the wings can be used. With a larger bird, the hand is cuffed around the neck with the thumb and forefinger pressed against the mandible and the other hand holding the distal wings/legs/tail or allowing the patient to perch on the hand. Since birds have fixed lungs, lack a diaphragm, and use a bellows system to circulate air through their lungs, care must be taken to make sure the bird can freely move the sternum to ventilate. Care should also be taken when restraining parrots with featherless facial patches (e.g., African grey parrots and macaws); if they are restrained by the face, visible bruises are often left behind. The restraint technique described above does not require any pressure be placed on the face.

Many practitioners prefer to keep the patient in the towel for the examination. The author usually prefers to have the patient out of the towel for the exam as this makes it easier to see all parts of the patient's body. In addition, some patients, especially Amazon parrots, tend to overheat during the exam, which is made worse by a towel. Some patients are more comfortable when in a towel, in which case, the author will keep them in it.

It is essential to have staff members who are comfortable restraining parrots helping in the examination room. Poor restraint can lead to bites, injury to the patient, and loss of confidence from the pet parent. The author trains new personnel on how to restrain parrots when the patients are sedated prior to allowing awake restraint in front of an owner. The three most important things to remember are: control the head of the bird, do not compress the sternum of the bird, and if the bird is biting someone, put the bird down (safely) and it will usually let go. It is good practice to hold the bird in an upright position for examination because this makes breathing easier, especially in patients with coelomic distension.

Once the bird is in hand, it should be monitored carefully by the restrainer. If the bird is breathing too heavily, becoming too stressed, or suddenly exhibiting a decrease in muscle tone or awareness then it should be gently put down somewhere safe.

Pro tip: For initial examinations, have your technician explain prior to the examination how the bird will be restrained during the exam, because some owners get very nervous and think the bird is being choked.

Systematic Physical Examination

Always perform the physical examination in the same order. You do not have to use the order outlined here but be consistent. The author usually performs the entire examination and then returns to any areas of concern after. Since parrots are non-domesticated prey species and easily stressed, always have everything needed for the examination at the ready. Equipment needed: scale with or without a weighing container, appropriately sized towel, infant stethoscope, something to open the mouth with, sterile lubricant and a cotton tipped applicator if the cloaca needs to be examined, and a transilluminator.

Pro tip: Try your best to know whether the species you are about to examine is sexually dimorphic and, if so, at what age gender can be assessed. Nothing loses pet-parent confidence faster than asking if the owner knows if an Eclectus parrot is male or female.

Step 1. Weigh the bird

The bird can be asked to stand directly on a small digital scale, placed in a tared container on the small digital scale, or weighed on a cat scale. Slippery things can be scary, so put something on the scale or in the container to make it more comfortable for the patient, such as a small towel or a perch. The cat scale is usually needed for large cockatoos and macaws due to their size. Occasionally, a large parrot is unwilling to stand on the cat scale, in which case the travel cage can be weighed with the bird in it and out of it to calculate weight. It is helpful to compare previous weights for the patient with the current weight; owners appreciate hearing this information as well. Always weigh in grams or kilograms; this is more accurate for small patients than pounds and ounces.

Step 2. Examine the head

The eyes should be examined with a transilluminator. Evaluate the lids, cornea, conjunctiva, anterior chamber, and lens. A PLR can usually be induced but keep in mind that birds have striated muscle in the iris, so they can change pupil size at will. The fundus can also be examined with direct ophthalmoscopy, although the author usually reserves this for patients with ocular or neurologic disease. The avian retina is avascular and there is no tapetum. To provide nutrients to the retina, birds instead have the pectin, a heavily pigmented extension of the choroid. Check for blepharospasm, exophthalmia, and ocular discharge. Also check for swelling around the eyes, which most often indicates sinus disease rather than ocular disease.

The nares should be examined next. Look for debris within the nares or discharge collecting on the feathers around the cere. Some species have a very prominent opercula within the nares and other do not; pay attention so you can learn normal anatomy. Some species, such as the smaller cockatoos and Eclectus parrots, have their nares hidden by feathers, so a bit of a hunt is needed to find them.

The ears should be found and the canals checked for patency.

Pro tip: If the ear canal appears to have disappeared, ask your restrainer to reduce any tension on the skin on that side of the head as it is likely they are pulling it down and functionally obstructing the canal.

The beak should be evaluated for symmetry, normal shape and apposition of the upper and lower beak, and retained keratin, which is most often seen on the upper beak (or rhinotheca). Owners commonly ask if the ridge of flaking keratin seen on the tip of the rhinotheca is normal. New layers of keratin grow in under the old one and the tip of the beak is where the old keratin starts to flake off. A similar process can be seen happening on the lower beak (or gnathotheca) but owners seldom inquire about this.

Pro tip: Birds with normal beaks should not need beak trims.

The oral cavity should be examined. This is often one of the least popular parts of the examination for the patient. Sometimes the choanal slit and glottis can be seen while the patient vocalizes, without any inference from the examiner. Other times it is possible to get the bird to bite at the transilluminator tip for a quick peak at the choana. If these fail, the beak can be opened with tape stirrups (i.e., pieces of medical tape folded on themselves, appropriately sized for the patient), or paper clips for budgies and parrotlets. The author does not use metal speculums as these can damage the keratin and even bone of the beak when the bird bites down. Care must be taken not to hyperextend the upper beak as it can be luxated. Evaluate the choanal slit and the papillae. The slit should be clear and the papillae sharp. Evaluate mucous membrane color in patients without pigmented mucosa. Evaluate the glottis.

Pro tip: Extremely stressed grey parrots can cry red tears during the examination. This most often happens if the patient's mouth is being pried open with stirrups so the author avoids doing this in grey parrots unless it is essential (e.g., a patient with upper respiratory disease). The red is due to literal blood cells in tears and it means the patient is very stressed. This phenomenon can occasionally be seen in other parrot species. It is wise to warn pet parents ahead of time that this can happen; they are more likely to believe you if forewarned.

Pro tip: Parrots with pathologic regurgitation or vomiting rather than behavioral regurgitation will often have dried vomitus or mucus on the head feathers from shaking the material out of their mouths.

Step 3. Palpate and examine the trunk

Start with the ingluvies (i.e., crop); it is often empty and flat but can have varying amounts of ingesta. Ingesta usually palpates as firm particulate matter rather than fluid-filled, unless it is a baby bird being hand fed formula. Palpate the pectoral musculature and assess a body condition score. The author uses a 1 to 9 scale with 1 being no muscle palpable, there is just sternal bone, and 9 feeling like a Thanksgiving turkey. Hydration can also be assessed at this time by sliding the skin side to side over the pectoral musculature. In euhydrated birds, the skin should slip easily back and forth while in dehydrated birds it will stick to the muscle. Palpate the keel bone, which should be straight, not curved. Some birds, such as grey parrots, can have very prominent keel bones despite appropriate pectoral muscle mass.

Pro tip: Some birds have atrophied pectoral musculature, for a variety of reasons, including lack of opportunities for flight, and are otherwise healthy birds. This can be alarming on initial palpation because they palpate as underweight.

The only window into the coelom is caudal to the sternum. The abdominal coelom is concave in normal birds that are not laying eggs. Obese birds will sometimes have a subcutaneous fat pad overlying the coelom which can initially palpate as coelomic distension. Often, the caudal aspect of the ventriculus is palpable in the coelom. Shelled eggs can also often be palpated. The vent should be evaluated as well; there should not be droppings collecting on the feathers around the vent. In hormonally up-regulated birds, the vent often appears a bit loose and edematous.

The dorsal aspect of the trunk can be assessed next. Palpate the back then evaluate the uropygial gland at the base of the tail; the uropygial gland is the only sebaceous gland of birds. It is a bilobed gland with a tuft of feathers wicking oily material out of it.

Pro tip: Amazon parrots, pionus parrots, and hyacinth macaws lack uropygial glands, so do not waste your time trying to find one in these species.

Step 4. Auscultation

Auscultate the heart and record a beats per minute rate.

Pro tip: Bird heart rates are often very fast. The author counts for 6 seconds and adds a zero and considers this acceptably accurate. Count in groups of 10 and use fingers, if needed, to keep track of how many 10s you have counted.

Auscultate several locations over the keel. The author finds murmurs easier to hear over the more caudal aspects of the sternum. Auscultate the back, between the wings, to listen for lung and air sac sounds; normal lungs and air sacs are fairly quiet. Under the wings can also be auscultated if needed; both the heart and respiratory tract can be auscultated here. Finally, place the stethoscope on the head to listen to the sinuses for congestion; the clients always get a kick out of this since the bird looks like it is wearing a little hat.

Step 5. Assess the legs, feet, and wings

Quickly palpate the legs for symmetry. Assess grip strength of the feet. Examine the plantar surfaces for pododermatitis lesions; don't forget to look at the hocks as well as the toes. If a band is present, record the location and number. Pull out each wing to assess extension and wing withdrawal. If any remiges are trimmed, count and record them. Blood feathers can also be recorded. The cutaneous ulnar vein can be assessed at the ventral aspect of the cubital joint. It usually needs to be exposed with a small amount of alcohol on cotton. This vessel can be used to assess hydration and cardiac function. It should refill instantly after digital blanching.

Step 6. Cloacal eversion

The author performs this procedure only on New World species that are medium to large in size (e.g., Amazon parrots and macaws) or in patients presenting with cloacal disease. New World parrots are prone to chronic herpesvirus infections that can lead to papillomas anywhere in the GI tract. The cloaca and oral cavity are the only locations we can easily evaluate for these lesions. The author uses sterile lubricant and a cotton tipped applicator for this procedure in most cases. Smaller species, such as conures, would require a mini-tip culturette swab for eversion.

Pro tip: Warn the pet parent that cloacal eversion can cause a little bit of fresh blood in the next dropping. Again, forewarning builds more trust than explanation after the fact.

Step 7. Return the bird to the carrier, owner, or weighing container, depending on next steps

Watch how well the patient recovers from restraint. One to two minutes of heavy breathing, sometimes with the wings held away from the body, can be normal. Prolonged or profound heavy breathing can be a sign of disease.

Optional Step 8. Assess ambulation

In patients that may have orthopedic or neurologic disease, watching them walk around can be helpful. The bird might fall off the table onto the floor and further injure itself, thus the author puts a large towel onto the floor and has the bird walk around on that. Limping and/or ataxia can often be discovered this way. In addition, the bird may be willing to step up onto the veterinarian's hand in this situation, allowing grip strength and symmetry to be assessed. If the patient is small and flighted, gait can be assessed in a large weighing container or the carrying cage.

Pro tip: If the patient is biting onto something you'd prefer it not to bite (e.g., your stethoscope, your technician's scrub) trying to pry the beak off seldom works. Providing an alternative to bite often does work. Tongue depressors are made of non-toxic birch wood that is often fun for the parrot to bite and can serve as a great distractor. But remember, if the bird is biting part of a person, put the bird down in or on a safe place, such as the examination room table, as soon as possible.

Plumage and skin

Plumage and skin should be assessed over the course of the entire examination. Normal head feathers but abnormal body feathers suggest feather damaging behavior. Parrots have featherless tracks (apteric tracts) in many places on their bodies so it is normal to be able to part the feathers and see skin. You should not be able to see skin in a bird without moving feathers to the side. Retained feather sheaths on head feathers is normal in birds that don't have a conspecific to groom their heads for them. Numerous broken tails feathers (rectrices) are very common in juvenile cockatiels as they seem prone to falling and breaking them. The author does not consider this a sign of illness if the bird is otherwise healthy. Similarly, juvenile birds commonly have stress bars on the rectrices; this should resolve when new feathers grow in.

Special Considerations for the Chicken Examination

If you are used to parrot examinations, chickens are a lovely breath of fresh air. They are less stressed by handling and don't have a hookbill with which to bite you. Chickens can be picked up, with your hands gently compressing the wings to the body so they don't flap, and placed onto the examination table. A towel or other non-slip surface should be placed on the table for them to stand on. They will also readily stand on the cat scale to be weighed. Minimal restraint is required; the restrainer can simply gently hold the chicken in place on the table.

Chickens are not parrots and palpate quite differently in many ways. Experience is the best teacher for becoming comfortable with what is normal for chickens versus parrots. One difference is the comb; in breeds that do not have a pigmented comb, it is red and can be used to assess the capillary refill time (should be less than 2 seconds). A pale comb can indicate anemia. A dark comb (that is normally red) can indicate poor oxygenation or cardiovascular disease.

The shape of the sternum is narrower than in parrots, which makes chickens feel thin when one is used to palpating parrot pectoral musculature. In laying hens, the coelom is normally convex rather than concave due to the large reproductive tract. Healthy hens do not have droppings collecting on the plumage ventral to the vent. This finding usually indicates pathologic coelomic distension. The vent of laying hens is quite relaxed and loose compared to a parrot's vent. The author finds chicken hearts quieter on auscultation compared to parrot hearts.

Because chickens live outside, they commonly have lice and mites, especially when they are ill. Examine the feathers and skin under the wings, around the uropygial gland, and around the vent for these parasites.

The chicken's feet need to be picked up off the table for examination of the plantar surfaces since pododermatitis is common in this species. The feet can be evaluated either by picking one foot up at a time while the restrainer supports the bird or the restrainer can pick the entire bird up by the trunk, compressing the wings to the body, and tilting the body slightly back.

Cloacal palpation can be performed in chickens using a gloved finger and sterile lubricant. The finger is slowly introduced into the vent and through the length of the cloaca. Applying gentle digital pressure ventrally while the examiner's other hand pushes up on the external ventral coelom can allow for palpation of an egg or any caudal coelomic masses.

Pro tip: Chickens dislike having their oral cavities examined so the author saves this for last. Chicken mouths can be opened using one's fingers rather than any sort of speculum. Their choanal papillae are arrayed in rows across the choana rather than lining the edge of the choanal slit.

Pro tip: Chickens produce two types of droppings: regular and cecal. The regular droppings are usually fairly dry and solid with a small pile of urates on top; they can be brown or green in color. The cecal droppings are brown, moist, and malodorous. Save droppings for a fecal direct and float since intestinal parasites are common in birds living outside!

Special Considerations for the Pigeon and Dove Examination

Similar to chickens, doves and pigeons are very cooperative patients. They can usually be captured with bare hands and no need for towels. Restraint for most of the examination can be done using the ice cream cone technique, with the tail, legs and pelvis, and caudal aspects of the wings held in one hand and the rest of the bird not being restrained at all. Hand position will need to change for the examiner to look at the legs, feet, vent, and uropygial gland. The mouth can be opened with fingers rather than a speculum. These are usually fun patients, although diamond doves are small and can be a bit high stress.

Special Considerations for the Canary and Finch Examination

Canaries and finches are not dangerous but they can be frightening little patients. They are easily stressed and should be given a break if restraint is more than 2-3 minutes. As with the parrot examination, have all your supplies at the ready to minimize length of restraint. Since these birds are so small, some sort of magnification can aid greatly in the examination. The author uses 3.5x surgical magnification loupes.

Canaries and finches can be captured with a paper towel or huck towel and then removed from the towel for the examination. The trachea can be transilluminated to look for tracheal mites. The abdominal wall can also be evaluated for hepatomegaly, ascites, and intestinal dilation, which is visible through the skin. Evaluating the trachea and the abdomen may require wetting the feathers with alcohol on a cotton ball but care must be taken not to cause hypothermia. The color of the pectoral musculature can be assessed. If it is pale, the patient is likely anemic.

Conclusions

A well-structured physical examination is essential for avian health assessments. By employing proper restraint techniques and a systematic approach, veterinarians can maximize diagnostic yield while minimizing stress for the patient. Learning what is normal for each species takes time and experience, but everyone can get there in the end. Once a clinician is confident in what constitutes normal, the physical examination is the most valuable diagnostic tool available in our arsenal.

How to Prevent Behavior and Welfare Problems in Parrots: 40 Years and Counting

Jan Hooimeijer, DVM, RPBC

Session #113

Affiliation: From Consultancy Practice for Birds, Schoolstraat 7A, 7941 CA Meppel, The Netherlands.

Abstract: Preventing behavior and welfare problems in parrots requires acknowledging their needs as non-domesticated, highly intelligent animals with advanced cognitive abilities and social behaviors. Proper housing, nutrition, enrichment, and care are essential, along with opportunities to spend time outside the cage and outdoors. Climate control must be adapted to the species. It is important not to underestimate parrots' sensitivity to human body language and attitudes, or to overlook the significance of lack of privacy as a serious stress factor. Ensuring that parrots are approached and treated with respect—thereby fostering mutual respect and trust—is essential. When parrots are not tame or do not feel comfortable in the presence of their caretaker, behavior and welfare issues are predictable. A 5-step behavioral protocol can be used to reduce fear and stress, foster mutual respect and mutual trust, and shape desired behavior by acknowledging parrots as intelligent prey animals. Definitions of normal, undesired, desired, enforced, and displacement behavior provide tools to understand and manage parrot behavior as well as our behavior. Displacement behavior can be used to prevent and resolve conflicts. Wing clipping may provide more freedom outside the cage and outdoors, promoting social interaction, while acknowledging the need for parrots to be part of the family as the most important form of enrichment. Having a parrot on the shoulder can result in insecure behavior.

Introduction

Preventing behavior and welfare problems begins with optimizing housing, feeding, and caring for parrots in a way that fulfills their needs in captivity. It means creating an environment in which parrots feel comfortable and experience being approached with respect for their intelligence, cognitive abilities, and social skills as non-domesticated prey animals.

Welfare Problems in Parrots Require Acknowledging Their Basic Needs

It is important to acknowledge that malnutrition is a major cause of insecurity, which in turn leads to health, behavior, and welfare problems. In the author's experience, the Association of Avian Veterinarians (AAV) was founded in 1980, and still, after 45 years of AAV, malnutrition remains the leading cause of health and welfare issues in parrots. While proper balanced certified organic nutrition is available, it is not yet universally acknowledged as the standard of care. In my experience most of the birds in The Netherlands are still eating seed mixtures that are nutrient deficient and, as most of the pelleted diets, not certified organic. Malnutrition can be considered as one of the most common captivity-related health and welfare problems. The serious consequences of malnutrition were documented as long ago as 1979. As noted by Prof. Zwart, vitamin A deficiency in parrots is one of the best-known and most common causes of serious health problems.[1,2] Health problems are a major reason for insecure behavior and welfare problems.[3-5] We must also acknowledge that when dealing with vitamin A deficiency caused by malnutrition, we are usually dealing with a combination of deficiencies.

Lack of sunlight, lack of privacy, and lack of positive social relationships all contribute to behavioral and welfare issues, and ultimately to health problems. Climate control is essential to prevent health and welfare problems. For instance, having a budgerigar from the Australian desert is very different from having an Amazon parrot from the tropical rainforests of South America. Temperature and humidity must be adjusted based on the species kept.[6] As established in poultry, optimal ventilation is crucial to prevent respiratory and welfare problems.[7]

In general, there is still a lack of awareness that parrots are non-domesticated prey animals being kept under unnatural conditions in captivity. These highly intelligent birds possess remarkable cognitive abilities. Without understanding their normal behavior, and while underestimating their intelligence, a wide variety of behavioral problems are likely to occur, depending on the species, the individual bird, and its circumstances.[8]

Parrot behavior is influenced by several factors, including hereditary. In nature, behavior is based on innate characteristics and shaped by observational learning, experience, intelligence, and cognitive processing. Behavior is ultimately driven by survival, both individually and as a species. To understand behavior, we must study ethology—the science of animal behavior—by observing how animals act in the wild. In doing so, we also learn about our own behavior acknowledging that we are part of the same animal kingdom.

Parrots are anatomically and behaviorally adapted to life as prey animals. A defining anatomic trait is the placement of their eyes on the sides of their heads, which allows them to observe their entire environment. In contrast, predators such as dogs, cats, owls, and humans have front-facing eyes, enabling binocular vision. Because of this, parrots may interpret our body language and attitude as threatening. Insecurity, stemming from a fear of predation, is a normal and essential aspect of their behavioral repertoire. Humans who smile, show their teeth, or speak animatedly may be perceived as predators. Failing to understand the impact of our predator-like characteristics can lead parrots to develop insecure or defensive behaviors.

Parrots are intelligent birds with exceptional cognitive skills. Studies show that parrots understand human posture, intention, and attitude. They learn by observing other birds, animals, and humans in their environment. Parrots evaluate and process information, draw inferences, and behave accordingly. In many ways, interacting with parrots is like interacting with children; they are eager to demonstrate what they've learned. The Model/Rival technique developed by Irene Pepperberg exemplifies this intelligence and the power of observational learning in parrots.[8,9] On a personal note, the author offers an anthropomorphic reflection: From the parrots' perspective, it must be frustrating to be admired only for their beauty or cuteness, rather than being appreciated for their intelligence, cognitive abilities, and social skills. Showing respect for parrots and acknowledging their evolutionary history spanning over 60 million years is fundamental to prevent behavior and welfare problems.

Definitions of Various Behaviors as Tools to Understand Animal and Human Behavior

These definitions have been developed by the author, with the exception of the definition of displacement behavior. The definitions serve as tools to understand and prevent behavior and welfare problems in parrots, other animals, and humans.

Normal behavior

Normal behavior refers to species-typical, individual behavior that is not intended to manipulate the caretaker or elicit a specific response. It is behavior that does not aim to create a problem. Examples include eating, drinking, grooming, playing, sitting in the cage, or perching on a hand. Normal behavior is not motivated by the desire to get a treat or reward or any attention/response. In the experience of the author, rewarding normal behavior may lead the bird to develop additional behaviors intended to attract attention from the caretaker. These new patterns, if aimed at prompting a reaction, can be classified as undesired behavior. For this reason, withholding reinforcement for normal behavior can be an effective way to prevent the development of undesired behavior.

Undesired behavior

Undesired behavior is intended to manipulate the caretaker's behavior, elicit a response, or even create a problem. This may include screaming, talking, biting, or feather picking. Any human response such as fear, frustration, anger, concern, or even laughter can be interpreted by the bird as a reward, reinforcing the undesired behavior. Consequently, acknowledging that a bird's behavior is problematic can actually increase its occurrence. Not rewarding undesired behavior is important to prevent it from becoming entrenched and escalating into habitual or maladaptive behavior patterns.

Desired behavior

Desired behavior is created and shaped by the caretaker, acting as a positive role model in a non-intimidating manner, as if the parrot is not there. Desired behavior is not the result of a direct command or request, nor is it behavior motivated by the expectation of a treat. Desired behavior is not the result of fear for punishment. Examples include the parrot showing attention and curiosity toward positive activities by the caretaker, such as talking about novel toys or interacting with its surroundings in a constructive way. Observational learning can be considered as desired behavior. Rewarding desired behaviors in a non-intimidating way is showing respect for the bird. This approach fosters self-esteem, mutual respect, and trust, while simultaneously reducing the likelihood of undesired behavior. In effect, reinforcing desired behavior prevents undesired behavior.

Enforced behavior

Enforced behavior arises from coercion, commands, or external motivation such as treats or fear of punishment. It may also be the result of social pressure or force. In humans, financial incentives often play a similar role in enforcing behavior. Enforced behavior may also be the outcome of fear of punishment. Rewarding enforced behavior is the opposite of rewarding desired behavior. It diminishes autonomy and can undermine mutual respect and mutual trust, particularly in intelligent, socially complex animals like parrots.

Displacement behavior

Tinbergen, one of the founders of ethology, described displacement behavior in gulls during conflict situations involving territorial behavior in breeding colonies. When a gull must choose between fighting or fleeing but neither option is viable, it may instead engage in seemingly unrelated actions such as preening, pulling at grass as if gathering nesting materials, or looking away from a competitor.[10-13] This behavior serves to de-escalate tension. In some instances, a freeze response may also occur.[14] Displacement behavior represents an alternative to the typical fight-or-flight response when those options are not possible, such as when a bird is confined in a cage. Displacement behavior is described as the result of two contradicting instincts in a particular conflict situation. Similarly, parrots in both natural and captive environments may exhibit displacement behavior in response to conflict or stress. While these behaviors may appear unrelated or out of order, they often lead to a shift in the social dynamic for instance, triggering displacement behavior in another parrot and de-escalating the conflict situation. Typical examples of displacement behavior in parrots include nail-biting, scratching, grooming, wing-flapping, tail-shaking, head-shaking, or tilting the head to look upward. While grooming and preening are normally associated with comfort and well-being, they can also emerge as displacement behaviors during conflict situations.[15] If caretakers respond to displacement behavior in ways that are perceived as rewarding, such as offering attention or comfort, this may inadvertently reinforce the behavior. Over time, this can transform a normal stress response into undesired behavior. Displacement behavior is most frequently observed when parrots, as prey animals, are approached by humans in an intimidating way. For example, face-forward with forward-facing eyes, which mimics the typical gaze of a predator, such as a dog, cat, owl, or bird of prey. It is also common in parrots exhibiting territorial behavior, particularly when perched atop or inside their cages. In these situations, it is essential to understand how parrots perceive and process sensory information. Research suggests that parrots engage in complex information processing, meaning they do not merely react reflexively to stimuli. They interpret their environment and choose how to respond.[9,16,17] Failure to recognize displacement behavior, or responding to it inappropriately, can lead to the development of abnormal repetitive behaviors, also known as maladaptive behaviors, such as feather destructive behavior. These may persist even after the original conflict trigger has disappeared. The expression of displacement behavior likely has a genetic component, which may explain why some species are more prone to developing feather-destructive tendencies.[8] In addition, the way displacement behavior is displayed is influenced by environmental factors and early learning experiences. For this reason, hand-rearing baby parrots is considered problematic; young parrots must observe their parents to learn normal behavioral patterns, including appropriate displacement behaviors in conflict situations.[20,21]

How to manage displacement behavior

It is vital to recognize and validate displacement behavior as normal behavior in a perceived conflict situation. The bird is coping with stress in the only way available. Any direct response in that moment, whether positive, negative, or reassuring, may inadvertently reinforce the behavior and contribute to maladaptive outcomes. The most effective approach is to show displacement behavior, being in the presence of the parrot, avoiding reacting, as if the parrot is not there. By doing so, the caretaker neither escalates the conflict nor rewards the displacement behavior. Recognizing and appropriately navigating displacement behavior is a cornerstone of respectful, informed avian care. To prevent undesired behavior, it is crucial not to reward displacement behavior, intentionally or unintentionally. Preventing undesired behavior begins with awareness, followed by showing displacement via a non-reactive presence.

Preventing Behavior and Welfare Problems by Implementing the 5-Step Behavior Protocol

The 5-step protocol has been developed as a tool to create mutual respect and mutual trust. The most important aspect of working with parrots is the attitude of the person caring for them. A respectful, relaxed, positive and informed approach forms the foundation of showing respect. The following 5-step protocol is designed for veterinarians and caregivers and can be applied to any parrot (animal or child) including individuals already exhibiting behavior or welfare-related concerns.

Step 1. Become a role model, demonstrate being happy and comfortable, while having no negative intentions in the presence of the parrot.

Begin by acknowledging that parrots are prey animals. Avoid direct eye contact, which can be perceived as threatening. Instead, turn your body slightly or show your back. Allow the bird to observe you from a safe distance, as though it were an audience member watching a play. Relax your posture and engage playfully with nearby objects. Toss and catch a parrot toy, fiddle with a pen, tear up paper, or gently rub a towel on your face. These actions should not be directed at the parrot but performed neutrally, as if the bird is not present. Parrots are more likely to feel at ease with people who appear comfortable around birds and with themselves. Your behavior, whether confident or nervous, communicates crucial information to the parrot. By playing a role and creating a positive, interesting, and predictable environment, you are creating desired behavior.

Step 2. Reward the desired behavior created by Step 1.

Through your attitude in Step 1, you have allowed the parrot to feel comfortable, safe, and curious. The bird's attentive observation of you is a reason to reward it, by telling the bird that it is the most beautiful creature on this planet or expressing appreciation for its presence. By rewarding the bird in a non-intimidating posture, you strengthen the parrot's sense of safety and connection.

Step 3. Become a teacher, acknowledging and rewarding the intelligence of the parrot.

Acknowledge and respect the parrot's intelligence by enriching its environment with verbal descriptions of what it can see: birds in trees, colorful toys, the texture of objects, and other items. Describe even its own body parts, like the beak or tail, to help build associative learning. Speak as though the parrot is not present. This technique, based on observational learning skills, allows parrots to absorb information without pressure. Research by Pepperberg has shown that parrots may eventually request labels for new items in their environment.[17] Reward this curiosity gently and avoid a direct gaze, which may be experienced as confrontational.

Step 4. Allow the parrot to learn that it may participate and explore.

Offer the bird the opportunity to engage by showing that it may touch your pen, toy, towel, or paper. It is not uncommon for parrots to reach out gently with their tongue being motivated to touch. Encourage exploration by offering small, non-threatening items and praising each positive action. You may also offer your hand, tell the bird that it may step up. This should not be framed as a command or a way to earn a treat. Instead, it creates an internal motivation to step-up that builds confidence and trust.

Step 5. Build resilience and acceptance through mutual respect and trust.

At this stage, mutual respect and trust have typically been established. The bird feels safe in your presence and is more open to new experiences. If the bird shows fear of an object like a towel, or a setting such as the examination room, do not reassure it. Reassurance in that moment can function as a reward, reinforcing the fear response. Instead, begin again at Step 1, being happy and relaxed, and follow the protocol. Soon, the bird will recognize that the situation poses no threat. This process fosters self-esteem, enabling the bird to confidently face new situations such as a clinical examination, blood draws, grooming, wing clipping, gentle toweling, and more.[8] In the experience of the author, it is amazing to see that birds cooperate during handling or procedures like nail clipping or grooming the beak.

Preventing, Dealing With, and Redirecting Undesired Behavior and Welfare Problems

Preventing and dealing with undesired behavior and welfare problems requires understanding the causes of undesired behavior, seeking a diagnosis. One must endeavor to understand the reason for the behavior as well as the consequences of that behavior. This includes evaluating the environment, the responses the parrot experience and natural instincts of parrots as intelligent, social prey animals. We need to learn to look through the eyes of the parrot to understand its behavior. It is important to consider that in many cases, undesired behavior in captivity originates as the result of rewarding normal behavior. It is important to consider that our behavior, attitude, and mindset can be important factors contributing to behavior and welfare problems.

When dealing with undesired behavior, it is crucial to understand that any response, whether positive, negative, emotional, or even subconscious, can serve as a reward based on the highly developed cognitive abilities of parrots. Even ignoring a behavior can reinforce undesired behavior, because ignoring is experienced as a negative response. Ignoring is the opposite of showing displacement behavior. The key is to avoid unintentionally rewarding undesired behavior.

Preventing behavior and welfare problems begins by ensuring optimal health and welfare by providing certified organic pelleted food, access to sunlight and fresh air, opportunities for exercise, and an enriched environment that encourages the bird's intelligence and natural behaviors. This includes toys and food that stimulate foraging.

Providing privacy is also vital. Using branches with leaves in and around the cage creates hiding spots and a sense of security. Positive social interactions with other birds, animals, and humans, such as participating in walks, bike rides, picnics, or family visits with appropriate safety considerations, can prevent both undesired behavior and welfare problems.

Natural climbing opportunities like offering flexible branches of various diameters to activate and support muscle development and balance, should be provided for exercise. In nature parrots are exploring in the trees, looking for food and even sleeping on flexible branches. These branches also satisfy the bird's instinct to chew and explore.

Understanding a bird's attachment to its cage is important. The cage may be seen as a nesting site, triggering territorial or reproductive behavior.[22] The only time in nature when parrots spend day and night in one spot is during the breeding season in and around their nesting site. To prevent the cage from becoming a trigger for hormonal activity, it is important to have the bird outside the cage and outdoors as much as possible. It is important to provide a separate sleeping cage in a different room, where the parrot can sleep 10-12 hours in complete darkness. In nature, birds do not sleep in the same location where they forage, drink, or spend their time during the day.

Recognize that birds feel more insecure when they are perching in a higher position. It is, in the experience of the author, a common reason for insecurity and aggressive, defensive behavior that looks like dominant behavior when parrots are allowed to sit on the shoulder or when the parrot is sitting on the top of the cage. Owners are amazed when the parrot, that was sitting on the shoulder, feels comfortable sitting on my hand without the intention to climb to my shoulder. Holding the bird on your hand, similar to the way falconers have worked with birds of prey for thousands of years, fosters a better dynamic. The relationship between falconers and their birds of prey can be considered as based on mutual respect and mutual trust.

Redirecting undesired behavior can be accomplished by using displacement behavior. This creates a shift, redirecting the bird's attention away from undesired behavior toward curiosity, making the bird feel comfortable. By using displacement behavior, you do not reward the undesired behavior and instead you create desired behavior. According to the author, this has been successfully applied both in the wild and in clinical practice. The first step of the 5-step behavior protocol is basically showing displacement behavior. This way you create a reason to reward the bird at Step 2 for the created desired behavior; following it up with Step 3 and 4, will prevent or redirect undesired into desired behavior within seconds. This method enables redirection, builds trust, acknowledges and reinforces the parrot's cognitive abilities and intelligence, and is the opposite of encouraging maladaptive responses.[8]

Wing Clipping to Prevent Behavior and Welfare Problems

In nature, flying is essential for the survival of parrots, with the exception of the kakapo (*Strigops habroptilus*). Flying enables parrots to find food, water, and a safe place to sleep. It is also a primary means of escaping predators; flying in flocks provides additional protection. For parrots in captivity, particularly those kept as companion birds, flight is no longer essential for survival. Most parrots are kept under extremely unnatural and bird-unfriendly conditions. They are often confined to small cages or aviaries, primarily because they can fly. Under these conditions, they are unable to fly and spend most of their lives behind bars sitting on stiff horizontal perches. This situation can be compared to a dog living in a small kennel or chained outdoors. Both parrots and dogs subjected to such confinement are prone to predictable behavior and welfare problems. As avian veterinarians, we have a responsibility to safeguard the health and welfare of parrots kept under these extreme and unnatural conditions. A critical aspect of welfare is social integration of parrots as flock animals; they benefit greatly from being considered as part of the family flock. Spending time outside the cage, and especially outdoors with their caretaker or human flock, represents a major enrichment in their lives. Within the Consultancy Practice for Birds, wing clipping may be recommended for safety reasons, but more importantly, it is recommended because it enables parrots to experience more freedom under unnatural circumstances. It allows them to be outside of the cage and to engage in social interactions, both indoors and during daily outdoor activities with their caretaker. In the author's experience, wing clipping, when performed using a parrot-friendly attitude and integrated into behavioral consultations, can help prevent behavior issues and solve welfare problems. When parrots experience being part of the family, the likelihood that they will end up in parrot rescue facilities is significantly reduced. The question of whether wing clipping is ethical seems obvious. Since I was a young boy, I have been a birdwatcher. Watching birds fly in nature has always been fascinating. As an avian veterinarian, I have seen parrots in small cages and small aviaries in which they cannot fly, because they can fly. Most parrots are kept as prisoners, behind bars and caretakers as their guards. Those parrots are not part of the family and do not experience any freedom at all compared with the dog or the children in the same family. Many parrots that are kept as companion birds are hardly or even never outside. Sunlight can be considered as one of the basic rights of prisoners and many parrots don't even get that right. It was reason to start organizing parrot walks and parrot picnics in 2000 for my clients to emphasize that parrots must be part of the family, inside as well as outdoors. The first question to ask is whether it is ethical to have parrots as non-domesticated animals in captivity under extreme unnatural circumstances in which they cannot fly and they are cared for by caretakers who are not aware of their basic needs. I don't think that is ethical. It is the responsibility of avian veterinarians to look for ways to improve the health and welfare of parrots that are kept in captivity under bird-unfriendly circumstances. In my experience wing clipping can be a major tool to prevent behavior problems and to enhance the welfare of companion parrots.[8]

Hand Rearing Baby Parrots

As discussed during the roundtable discussion at the AAV conference in New Orleans in 2011, the necessity for hand-rearing baby parrots in the past was the result of management failures concerning housing, nutrition, and care, complemented with a lack of knowledge about the natural breeding behavior of parrots. It started as a necessary evil. Over the years, hand-rearing baby parrots became a commercial activity because of the experience that females produced more eggs, meaning more babies to sell.

The Dutch Parrot Foundation, supported by the Consultancy Practice for Birds, has successfully campaigned in The Netherlands against the hand-rearing of baby parrots. The outcome has been that since July 1, 2014, it is illegal to separate a baby parrot from the parents until the baby is fully weaned (i.e., able to eat completely on its own). The new legislation puts parrots on the existing list with dogs, cats, rabbits, chimpanzees, and other species for which the timing of separation from parents has long been regulated. During the past 10 years it has become obvious that less and less young parrots are offered because of the typical behavior and welfare problems, related to hand rearing.

The regulation is based on data from the past 20 years showing the negative consequences of separating baby parrots from their parents. A range of behavioral and welfare issues have been recognized as consequences, particularly when separation occurs before or during the first imprinting phase. Some behavior problems emerge months or even years later, especially when the bird becomes reproductively active. Insecure behavior, phobic reactions, biting, feather picking, and self-mutilation are well-documented outcomes.[19,20] Early enrichment is essential in maintaining healthy interactions later in life.[21]

Conclusions

Parrots are among the most charismatic, intelligent, and socially complex animals on the planet. When provided with a suitable living environment, they can enjoy a remarkably long life. Understanding how to prevent behavior and welfare problems is inseparable from understanding how to address and resolve them. Prevention begins with acknowledging that when parrots are not tame or do not feel safe in the presence of their caretaker, behavior and welfare issues are likely to arise. As caregivers and avian veterinarians, we must learn to view the world through the eyes of parrots as prey animals, navigating a world not designed for them. This requires us to recognize that we, as humans, possess the characteristics of predators, and must consciously adjust both our body language and our mindset. We prevent behavior and welfare problems by meeting parrots' fundamental needs: providing complete and balanced certified organic nutrition. Malnutrition remains a leading cause of health, behavior, and welfare concerns. Offer meaningful enrichment, including access to privacy and ensuring time spent outside of the cage and outdoors, and integrating the parrot into the social life of the family. In the author's experience, this is the most profound enrichment of all. The implementation of the formulated definitions of behavior and the 5-step behavior protocol offers practical and effective tools to build mutual respect and mutual trust between parrots and their caregivers. These tools foster relationships that prevent behavior problems and safeguard welfare. Preventing problems of health, behavior, welfare, and actively contributing to the conservation of parrots in the wild, are core responsibilities of the avian veterinary profession. Parrots' behavior patterns provide feedback on the quality of care they receive, just as students reflect the effectiveness of their teachers. In this way, parrots have been the author's greatest teachers, offering correction, challenge, and insight. For this, the author is deeply grateful.

If Not Us, Who? If Not Now, When?

Acknowledgments: The author thanks Harrison's Bird Foods, The Netherlands, for their generous sponsorship of parrot events throughout the country. Special thanks go to the staff of the Consultancy Practice for Birds, whose ongoing support and dedication have made this work possible over many years. I am grateful for the role models who have shaped my thinking and practice. Foremost among them: my grandfather; Prof. Niko Tinbergen, Prof. P. Zwart, Dr. Jane Goodall, Dr. Greg and Linda Harrison, Prof. Helga Gerlach, Dr. James Harris and Dr. Fern VanSant. I extend my sincere appreciation to Dr. Irene Pepperberg, with whom I have had the honor of collaborating from 2014 to 2024, educating veterinary students around the world in the fields of avian behavior, intelligence, cognition, and welfare. Finally, I am grateful for an anonymous sponsor who made it possible for me to travel, participate at conferences and continue to educate veterinary students about prevention, behavior and welfare.

References

1. Zwart P, Schreurs WHP, Dorrestein GM. Vitamin A deficiency in parrots. *21st Int Symp Erkrk Zootiere Mulhouse.* 1979;47-52.

2. Dorrestein GM, Zwart P. Practical aspects of vitamin A deficiency in parrots and cockatoos. *Proc Voorjaarsdagen.* 1981;61-62.

3. Harrison GJ, Lightfoot T. *Clinical Avian Medicine.* Vol 1-2. Palm Beach, FL: Spix Publishing; 2006.

4. Zwart P, Samour J. The avian respiratory system and its noninfectious ailments: a review. *J Exot Pet Med.* 2021;37:39-50.

5. Schmidt RE, Reavill DR, Phalen DN. *Pathology of Pet and Aviary Birds.* New York, NY: Wiley; 2015.

6. Van Sant F. Seeing the rainforest through the trees. *Proc Annu Conf Assoc Avian Vet.* 1993:228-232.

7. Greenacre CB, Morishita TY. *Backyard Poultry Medicine and Surgery.* Hoboken, NJ: John Wiley & Sons; 2015.

8. Pepperberg I, Hooimeijer J. Studying the behavior of captive parrots: respecting their cognitive abilities creates mutual respect and mutual trust. *Proc Assoc Avian Vet.* 2022;48-56.

9. Pepperberg IM. The African grey: how cognitive processing might affect allospecific vocal learning. In: Balda RP, Pepperberg IM, Kamil AC, eds. *Animal Cognition in Nature.* London, England: Academic Press;1998:381-410.

10. Tinbergen N. *Spieden en Speuren in de Vrije Natuur.* Amsterdam, The Netherlands: Uitgeverij Ploegsma;1959.

11. Tinbergen N. *Curious Naturalists.* London, England: Country Life;1958.

12. Kruuk H. *Niko's Nature: The Life of Niko Tinbergen and His Science of Animal Behaviour.* New York, NY: Oxford University Press; 2003.

13. Burkhardt RW Jr. *Patterns of Behavior: Konrad Lorenz, Niko Tinbergen and the Founding of Ethology.* Chicago, IL: University of Chicago Press; 2005.

14. Bracha HS. Freeze, flight, fight, fright, faint: adaptationist perspectives on the acute stress response spectrum. *CNS Spectrums.* 2004;9:679-685.

15. Spruijt BM, van Hooff JARAM, Gispen WH. Ethology and neurobiology of grooming behavior. *Physiol Rev.* 1992;72:825-852.

16. Zucca P. Mind of the avian patient: cognition and welfare. *Proc Annu Conf Euro Assoc Avian Vet.* 2007; 357-365.

17. Pepperberg IM. *The Alex Studies.* Cambridge, MA: Harvard University Press;1999.

18. Seibert LM. Feather-picking disorder in pet birds. In: Luescher AU, ed. *Manual of Parrot Behavior.* Oxford, England: Blackwell; 2006:255-265.

19. Orosz S. Diagnostic workup of suspected behavioral problems. In: Luescher AU, ed. *Manual of Parrot Behavior.* Oxford, England: Blackwell; 2006:233-245.

20. Schmid R, Doherr MG, Steiger A. The influence of the breeding method on the behavior of adult African grey parrots (*Psittacus erithacus*). *Appl Anim Behav Sci.* 2006;98(3-4):293-307.

21. Fox RA, Millam JR. The effect of early environment on neophobia in orange-winged Amazon parrots (*Amazona amazonica*). *Appl Anim Behav Sci.* 2004;89:117-129.

22. Van Sant F. Problem sexual behaviors of companion parrots. In: Luescher AU, ed. *Manual of Parrot Behaviour.* Oxford, England: Blackwell Publishing; 2006:233-245.

23. Hooimeijer J. A practical behavioral protocol for dealing with parrots. *Proc Annu Conf Assoc Avian Vet.* 2003;177-181.

Ten Tips for Poultry in Ten Minutes

Marli Lintner, DVM

Session #144

Affiliation: From Avian Medical Center, Lake Oswego, OR 97035, USA.

Abstract: Sometimes the most helpful practice tips do not include drug doses or expensive new medical equipment. This is especially true in backyard poultry practice where much is known, but little relates to actual circumstances. Problems encountered in backyard poultry may be different than those seen in commercial flocks. This session presents ten concise, experienced-based tips for improving poultry health, particularly in non-commercial settings.

Tip 1. Prevent Bumblefoot

Bumble foot treatments are varied and numerous, but it is best to prevent the problem. The two strongest indicators of how frequent this issue will occur in a flock come down to roosting height and hardness of the coop floor. Question owners on their morning routine. Do the hens come crashing off the roost in a mad dash to be the first outside? A heavy hen dropping 2-4 feet onto a hard slick floor will bruise the pads of her feet. Repeated hard landings result in scabs, which get thicker and cause more pressure and eventually this leads to infection. Lower coop perches to less than 2 feet, pad coop floor with non-slip mats and apply ample bedding on top. The iconic chicken ladder is a disaster in the wet Northwest climate because it is very slick. Going up is manageable for the bird, but coming down the hen slips to the first slat where it either jams digit 3, resulting in swollen joints or bruises the pads on its feet. A combination of lowering the angle and covering with a better traction surface (no slats) works well.

Tip 2. Treat Bumblefoot

When bumblefoot occurs, the goal should be to keep the pressure off the pad, not wrapping over it and adding to the pressure. A doughnut wrap with stiff layers of foam, shaped like the foot and cut out in the center to take weight off the pad, works well. Cover with a conforming material like Vet-wrap in a natural standing position. It is important to have a stiff base to maintain the normal shape of the foot and not wrap in tightly against the toe webbing. This is critical when wrapping duck feet to avoid pressure necrosis and loss of webbing. Leave in place no longer than 10-14 days.

Tip 3. Ensure Water Access for Ducks

When hospitalizing ducks, water dishes must be deep enough to allow the ducks to submerge their bills up past their nares. A full bath every 48-72 hours is also important. Without bathing, their nares fill up with thick, dark, rubbery debris. This rapidly progresses to clogged tear ducts and foamy runny eyes. While it may look like a sinus infection, if left long enough it can progress to one. A good bath in the tub clears it up quickly.

Tip 4. Provide Relief from Heat for Chickens

Hot weather is common and chickens do not tolerate heat well. Chickens do not sweat and depend on their respiratory systems for cooling. Heavy-bodied, heavy-feathered and obese birds with even a slight amount of exertion can quickly spike dangerous and even fatal body temperatures. Ambient temperatures over 80 °F put birds at risk. These are acute emergencies. Getting the bird out of the sun and having it sit on a wet towel over an ice pack uses the countercurrent blood flow in the legs to drop body temperatures quickly. Monitor the bird closely.

Tip 5. Understand the Purpose of the Preen Gland

The preen gland may be a bird's most misunderstood piece of anatomy. The purpose of this structure is not to oil the feathers but to clean the plumage that the bird can't reach with its beak. The gland fills its two lobes with a soft, water-soluble waxy material, which the bird preens out onto a small papillae or brush-like feather. This waxed papillae is then used like a dusting cloth to wipe debris off the plumage. The bird rubs its head, face and neck vigorously. The bird then cleans off the papillae, resets it and continues cleaning. This results in a clean head with no grease involved.

Tip 6. Do Cloacal Palpations with Caution

Don't try this at home and convince your clients not to do it. Seeking diagnostic information quickly and using my long thin fingers, I began doing cloacal palpations on hens in active lay. With care, practice, and a good water-soluble lubricant this can be a valuable diagnostic tool. It is possible to assess fluid buildup, egg impactions, yolk consolidation, renal enlargement, tumors, uroliths, etc. but use caution. It is possible to inadvertently collapse eggs, cause tears and bleeding. Non laying hens or bantam breeds may have vents too tight and small for this procedure.

Tip 7. Consider Causes and Misperceptions for Soft-Shelled Eggs

Soft-shelled eggs can be routinely seen for a brief period at the beginning and ending of a laying cycle as the system fires up or shuts down. With current diets and free range available rarely is calcium deficiency a factor. The shell gland, being the last section of the oviduct before the urodeum, is at risk of ascending bacterial infections. Repeat or chronic infections can result in damage to the shell gleaned such that it is rendered unable to properly form a shelled egg. Hens do not run out of follicles and can and will continue to lay eggs, albeit not as many, on into their teens. As they age muscular control of the oviduct degrades. Older hens literally may not have the ability to hold on to the egg in the oviduct long enough to put a healthy normal shell in place. Once again, it is not a calcium issue and additional dietary supplements are not effective.

Tip 8. Be Skeptical of Dietary Fads and Nutritional Pitfalls

Dietary fads come and go. This should impress on us that products touted are often not effective long term; they can fall out of favor or reveal harmful consequences. Be skeptical of new super foods with magical powers. Dark green leafy vegetables are one such item. Kale, spinach, collards and dandelion are too rich, high in oxalates, and variable in their nutritional content. A first cutting on freshly fertilized fields under perfect growing conditions is too "hot" for a chicken feed. Chickens fed large amounts over time can suffer severe and even fatal renal complications.

Tip 9. Assess Diet and Nutrition

There are additional considerations regarding dietary fads and information on the internet. On our routine client intake sheets, owners are savvy enough to know the correct answers in the dietary sections. It is important to question, probe and obtain dietary history from other angles. It might reveal the reality that their chickens are actually fed eggs (cooked and raw), cheese, yogurt, huge amounts of dried mealworms, unstable oil supplements, perishable cooked products, shrimp, bacon, dog and cat food, feeder goldfish, rancid whole grain mixes and their body weight in kale. All of these can cause upset digestive and renal systems, not to mention obesity. Some items can cause dramatic changes to urinary pH. Cloacal irritations, uroliths and vent gleet can be the result of high urine pH. Common sense is important in nutrition.

Tip 10. Keep Birds Clean

Although it is low tech, veterinary management needs to include keeping the chicken vent and cloaca clean. Fecal material can build up for a variety of reasons. Diarrhea and loose feces can naturally occur with diets high in moisture and fiber like fruit and greens. Also, a fat belly will hang out further and catch fecal matter. Fluffier feathers can attract droppings. Arthritic hens can't squat as well to defecate cleanly and saggy cloacas can leak. Leaving fecal balls in place can lead to skin infections but also gives a prime area for maggot infestation. Dry cleaning is more effective than trying to wash out. Suture scissors work well to cut feather shafts close to the skin. Leave a clear path 1- to 2-inches wide and as long as necessary. Avoid plucking because it is painful and the plumage will grow back quickly.

Conclusions

Backyard poultry present several unique challenges distinct from commercial situations. These issues can include preventing and treating bumblefoot, ensuring adequate access to water and relief from heat, addressing soft-shelled eggs, maintaining adequate diets, and keeping birds clean. Practical, low-tech interventions often yield effective results.

Physicality of the Psittacine Digestive Tract

Marli Lintner, DVM

Session #137

Affiliation: From Avian Medical Center, Lake Oswego, OR 97035, USA.

Abstract: In the avian world, the psittacine digestive tract is a comparatively short and very dynamic system. Parrots come equipped with a powerful beak for cracking and crushing, a crop for food storage, a ventriculus for grinding, and an intestinal tract with a transit time potentially as short as a few hours. All these features are designed to enable them to eat the food found in their natural environment. These features should inform us of the diets we offer, especially the physical features of that diet, the treatment methods we employ and the medications we use, to keep the psittacine digestive tract healthy.

Beak

If you have ever been bitten by a parrot you know the pressure generated can be up to 500-700 psi for a large macaw. It is impressive and painful. This same beak is sensitive enough to meticulously peel the skin off a grape, preen feathers back into place and feed tiny nestlings. Its main function however is to crack, crush and break into things. Nuts, seeds, grains etc. are opened, the contents expertly pried out and swallowed relatively whole. The beak grows almost a quarter of an inch a month and gets most of its wear and maintenance not during food gathering but by the bird itself. At times of rest parrots can be heard quietly grinding their beaks from side to side and wearing down the surfaces. When malocclusions do occur, the bird's continued attempts at wear often only serve to reinforce this malalignment. Great care should be taken when performing beak trims to preserve and/or return the beak to its proper alignment.

Salivary glands are located primarily at the base of the tongue. Smaller glands can be found in the roof of the mouth around the choanal slit. These glands are sensitive to Vitamin A deficiencies leading to hyperkeratosis and plugged ducts. The saliva produced lubricates the ingested food to help its progression down the cervical esophagus.

The choanal slit is open to the sinuses. Forcing food or medication orally faster than the patient is willing to swallow can result in the patient using their tongue to push the product up into the slit and out through the sinuses and nares.

Crop

The crop, sitting at the base of the neck, is a simple expansion of the esophagus. While it is a storage area for food items, it is also very moist and warm with a temperature of around 103 degrees.[1] Digestion does not occur here, but under certain conditions fermentation/spoilage can. It can be observed that crop contents, when produced by sexual regurgitation or even when vomiting from carsickness, have very little if any odor and appear, while moist with saliva, pretty much in the same condition as when ingested. But that is when eating a raw and primarily grain diet. In the author's experience, the more cooked and processed and therefore unstable the diet, the greater the chance of bacterial growth within the crop. Regurgitated perishable material can thus smell sour.

Cytology smears of crop contents reveal higher bacterial counts with more perishable food items. This moist warm environment is a perfect incubator for bacterial and yeast growth. While impacted crops can be caused by foreign material and infection, crop stasis is more likely to be caused by blockages or motility issues further down the digestive tract.

The cervical esophagus and crop have no functional sphincter thus overfilling the crop when gavaging can result in reflux and aspiration. Weak recumbent patients should be propped up 45 degrees to avoid reflux.

Crop contractions bring food down into the thoracic esophagus, while peristaltic contractions bring food into the proventriculus. These contractions appear to be controlled by the ventriculus. When relaxed and empty the ventriculus signals the crop to pass food along. Thus, a full ventriculus will lead to a full crop.

Proventriculus

Digestion starts in the proventriculus with the excretion of hydrochloric acid and pepsinogen. The pH in the psittacine stomach is acidic but not dramatically so, typically in the 4.2 to 4.6 range. Glands produce a sizable quantity of mucus, possibly to lubricate and facilitate moving large pieces of food into the ventriculus and to protect the mucosa from damage. Excessive mucus production by the proventricular glands can form dense ropey balls of mucus that the ventriculus is unable to pass and must eventually be vomited up. The proventriculus is a common source of the hemorrhage resulting in melena. Foreign body trauma (especially fiber ball irritation), ulcers, Avian Gastric yeast infections, toxins, and cancerous processes can cause acute bleeding episodes.

Ventriculus

Ingesta are passed into the ventriculus where mastication occurs. Thick and thin muscular layers work in a complicated dance to effectively grind food into very small particles. But the basic premise is rather simple. Large inlet, small outlet ensures ingesta is held onto until ground finely enough to escape the pyloric sphincter and move into the small intestine. This design is what enables grit, or other hard foreign objects, to remain in the gizzard almost indefinitely. Offering grit to psittacine birds has not been recommended for years. The diets we feed do not appear to require it. Grit regurgitated and fed by parents to nestlings may remain in their gizzards throughout the course of their lifetimes. The ventriculus may reflux food back into the proventriculus as needed for additional chemical digestion. The thick koilin wall of the ventriculus protects the organ from physical damage, acid and enzymes. While this lining is usually brown in color it can be stained green from bile refluxed from the small intestine.

Pyloric sphincter irritation or malfunction can lead to whole food items being passed through the digestive tract. This is problematic for the patient as no nutritional value will be derived, the small intestine will be irritated, and if large food pieces are passing through the sphincter it will have difficulty regaining its function.

While the psittacine ventriculus is well adapted to grind hard items, it is not designed for long fibrous materials. Fabric, hair, jute, grass, carpet, etc. will be balled up and retained for indefinite periods of time. Synthetic fibers may take years to pass. These fiber balls are not only irritating but can cause reoccurring digestive tract infections, slowing of ingesta, obstructions and ensuing regurgitation.

Like any muscle, exercise is important. Raw hard vegetables seem to require some effort, but nuts and grains are what this organ was designed to process.

Intestine

As food moves into the duodenal loop more digestion occurs as bile and pancreatic enzymes are released. Amylase, lipase, proteolytic enzymes along with sodium bicarbonate are excreted via ducts from the pancreas into the ascending duodenal loop. Absorption begins to take place. Motility via peristaltic contractions moves ingesta thru in a relatively short period of time. A barium series on a normal psittacine may clear in as little as two hours.

Psittacine birds do not have ceca. In other avian species this is the organ where bacterial fermentation or digestion occurs. As food passes through the psittacine digestive tract relatively quickly it is an open question as to what extent bacterial flora plays a role in their digestion.

Cloaca

The end of the system adds the urinary and reproductive tracts to the mix. Three chambers, the coprodeum, urodeum, and finally the proctodeum, are contained here. Urates, through the ureters, empty into the urodeum. This is mixed with feces but then retrograded back into the coprodeum and potentially back into the colon, where water is reabsorbed. The resulting fecal produced is a mix of feces and urates and often very telling of the diet recently consumed.

Frank blood in a fecal sample is most often from the renal or reproductive traits. Unless a cloacal issue, such as a papilloma, blood from the digestive tract will appear as melena.

Conclusions

The nutritional requirements of the various species we work with are obviously important to the care and management we give them. The physical requirements have been a later development. The equine species is a good example. We now see that continuous grazing is important for utilizing gastric acid production in the equine stomach. That bulk fiber is important for the large intestine. Even consuming food with the head down in a normal grazing posture is beneficial. And that confining horses and feeding a concentrated diet has detrimental effects on the digestive system. Now of course the equine and psittacine digestive tracts are extremely different but analyzing the psittacine digestive tract in the same fashion can give us some valuable insights. The beak is built for cracking and breaking hard objects into pieces just small enough to be consumed. The crop serves as a storage area to allow a bird to quickly consume a large meal and digest it later when safely roosted in the trees. But it is a moist, warm environment designed for non-perishable food material. The gizzard is a phenomenal muscle that grinds hard food items into a fine slurry to be quickly digested with enzymatic action from bile production and the pancreas. Without the benefit of ceca, bacterial fermentation is limited. With one opening for all three systems, the health of the lower colon and cloaca is vitally important.

So much of our knowledge of the avian digestive tract has been gained from commercial poultry production but rarely, in a commercial situation, does a hen's life expectancy exceed 18 months. In my practice we routinely see cockatiels living into their late twenties and early thirties. With longer life spans, the physical effects of dietary choices become more apparent.

References

1. Denbow DM. Gastrointestinal anatomy and physiology. In: Scanes CG, ed. *Sturkie's Avian Physiology.* Sixth Ed. St. Louis, MO: Elsevier; 2015:337-366.

Three-Dimensional Digital Anatomy of the Avian Lower Respiratory System: Structural Diversity and the Impact on Diagnosis, Obesity, Disease, and Surgical Planning

Emma R Schachner, PhD,
Stephanie L. Baumgart, PhD,
Aracely Martinez, PhD,
Amanda Pineda, BS,
Alexis Slack, BS,
Emily Clements, BS,
M. Scott Echols, DVM, DABVP (Avian Practice)

Session #172

Affiliation: From Department of Physiological Sciences, College of Veterinary Medicine, University of Florida, Gainesville, FL 32608, USA (Schachner, Baumgart, Martinez, Pineda, Slack, Clements), Department of Cell Biology and Anatomy, School of Medicine, Louisiana State University Health Sciences Center, New Orleans LA 70112, USA (Martinez), Medical Center for Birds, Oakley, CA 94561, USA (Echols).

Abstract: The avian lower respiratory system is composed of an immobilized volume-constant gas-exchanging lung, unidirectionally ventilated by a series of compliant air sacs, and highly variable diverticula that in most taxa pneumatize the postcranial skeleton.[1-6] The avian respiratory system is difficult to study in three dimensions; however, innovations in imaging and scientific rendering programs have facilitated the visualization of detailed 3D digital anatomic models that allow for mapping pulmonary tissues alongside the skeletal system in both live, sedated birds and deceased specimens for study.[7] In this study, 3D digital anatomic models of poultry, companion birds, as well as wildlife species ranging from the chicken to the red-tailed hawk, were segmented in the scientific visualization program Avizo (Thermo Fisher Scientific). These data illustrate the relationship between the gas-exchanging lung, air sacs, and diverticula in association with the skeleton and adjacent organs in 3D, and intraspecific variation in the hawks and parrots. The models demonstrate the impact of obesity on the respiratory tissues and how intracoelomic fat results in a reduced total lung volume, near collapse of the caudal air sacs (i.e., caudal thoracic and abdominal), and the resulting expansion of the cranial air sacs (i.e., cervical, interclavicular, and cranial thoracic) into the cervical and axillary regions. The results show how intraspecific and interspecific variation in air sac morphology can impact surgical planning. Lastly, the data demonstrate how knowledge of pulmonary diverticula morphology and pneumatization patterns is critical for catheterization and orthopedic surgery of the limbs.

References

1. Duncker HR. The lung air sac system of birds. A contribution to the functional anatomy of the respiratory apparatus. *Ergebnisse der Anatomie und Entwicklungsgeschichte.* 1971;45:1-171.

2. Maina JN. Pivotal debates and controversies on the structure and function of the avian respiratory system: setting the record straight. *Biol Rev.* 2017;92:1475-1504.

3. Maina JN. *Current Perspectives on the Functional Design of the Avian Respiratory System. Zoological Monographs 8.* Cham, CH: Springer Nature; 2023:191-268.

4. Martinez A, Diaz Jr RE, Grand Pre CA, et al. The lungs of the finch: three-dimensional pulmonary anatomy of the zebra finch (*Taeniopygia castanotis*). *Philos Trans R Soc B.* 2025;380:20230420.

5. Moore AJ, Schachner ER. When the lung invades: a review of avian postcranial skeletal pneumaticity. *Philos Trans R Soc B.* 2025;380:rstb.2023.0427.

6. Schachner ER, Moore AJ. Unidirectional airflow, air sacs, and the horizontal septum: what does it take to make a bird lung? *Philos Trans R Soc B.* 2025;380:20230420.

7. Schachner ER, Lawson AB, Martinez A, et al. Perspectives on lung visualization: three-dimensional anatomical modeling of computed and micro-computed tomographic data in comparative evolutionary morphology and medicine with applications for COVID-19. *Anat Rec.* 2025;308:1118-1143.

Preventing Common Causes of Respiratory Problems in Parrots

Jan Hooimeijer, DVM, RPBC

Session #152

Affiliation: From Consultancy Practice for Birds, Schoolstraat 7A, 7941 CA Meppel, The Netherlands.

Abstract: A variety of noninfectious ailments can be considered common causes of respiratory issues in parrots. A thorough questionnaire combined with a detailed physical examination are the most important tools for establishing a diagnosis. A combination of factors is responsible for the development of respiratory problems. In daily practice, the most frequent causes remain malnutrition, inappropriate climate conditions and stress factors. There is a need to focus on prevention by establishing tailored management protocols for different species to ensure the health and welfare of parrots in captivity.

Introduction

In 2014, I was asked to chair a roundtable on prevention at the Association of Avian Veterinarians (AAV) annual conference in New Orleans. In my experience, prevention is always about preventing health and welfare problems. I conducted a survey, reviewing the AAV and European Association of Avian Veterinarians conferences I attended between 1987 and 2014, looking for the word prevention in titles or key words. I also examined various textbooks in the field of avian medicine. The outcome was that the word prevention was hardly mentioned. In the literature, the topic of prevention is not a priority. In my experience as an avian veterinarian, respiratory problems are the most common health issues in parrots. It is not unusual to discover respiratory issues during an appointment originally made for other reasons. As a veterinary student, I learned from Prof. P. Zwart, who was then head of the division of pathology dealing with birds and exotic animals, about various management failures as causes of non-infectious ailments and death. Prof. Zwart motivated and supported me as my mentor to become an avian veterinarian. He can be considered one of the founders of avian and exotic animal medicine. During the ExoticsCon conference in Boston in 2023, Prof. Zwart received a lifetime achievement award from the Association of Reptile and Amphibian Veterinarians for his contributions to reptile medicine. At the age of 91, he co-authored an article in 2021 with Dr. Jaime Samour, titled The Avian Respiratory System and Its Non-infectious Ailments: A Review, published in the *Journal of Exotic Pet Medicine*.[1] Prof. P. Zwart passed away in January 2024, at the age of 94. In a way, this presentation is a tribute to his contributions.

Anatomy and Physiology of the Avian Respiratory System

The avian respiratory system is described as a dual system, consisting of lungs for the exchange of gases and a number of air sacs that function as ventilators.[2,3] The air sacs fill more than 30% of the volume of the bird's coelomic cavity.[4] The avian respiratory system is considered the most complex respiratory system in the vertebrate world.[5,6] The O_2/CO_2 exchanging surface, relative to the size of the bird, is much larger compared to that of mammals. The evolution of the avian respiratory system achieved its most efficient state, enabling birds to sustain flight. Birds can fly at high altitudes, efficiently utilizing low levels of oxygen to support their high metabolic rate.[7,8]

A Variety of Noninfectious Ailments Cause Respiratory Problems

Malnutrition

As documented by Prof. Zwart in 1979, vitamin A deficiency in parrots is one of the best-known and most common causes of serious health problems.[1,9] Vitamin A plays a role in both the innate immune system and in adaptive immunity. It provides and enhances defense against multiple diseases.[9,10] Vitamin A deficiency makes birds more vulnerable to infectious diseases, such as aspergillosis. It also causes visual and reproductive problems.[11]

Vitamin A deficiency remains a major cause of multiple health issues. One of the most obvious outcomes is epithelial squamous metaplasia which is manifested as hyperkeratosis of the oral cavity, conjunctiva, nasal lacrimal duct, upper alimentary tract and respiratory tract. In African greys massive keratin rhinoliths can be observed, distorting the nares and nasal sinus.[4] The nasal passages, tear and salivary glands, uropygial gland, and adjacent epithelia can also be affected. In African greys, there is a close relationship between the vitamin A content of the liver and the extent of the lesions.[1] In cockatiels, it has been reported that this species is more sensitive to diets containing excess vitamin A than to diets that are deficient.[12,13]

In addition to squamous metaplasia changes in the nares, the epithelium surrounding the eyes and the tear gland, we may observe that the papillae at the choana in the oral cavity are blunted or edematous.[14] The glands on the caudo-dorsal and lateral surface of the tongue may appear whitish and bulging.

As we can imagine, the epithelium of the mucous membranes throughout the respiratory tract is affected when dealing with vitamin A deficiency caused by malnutrition. This makes malnutrition a very common cause of respiratory problems.

We must acknowledge that when we are dealing with vitamin A deficiency caused by malnutrition, we are also dealing with a combination of deficiencies. An important outcome of malnutrition is skin abnormalities. Clinical signs of vitamin A deficiency may also involve the integumentary system and can manifest as overgrowth of the beak and nails. Hyperkeratosis may occur with the skin becoming dry and flaky, and we can expect feather abnormalities and molting problems associated with malnutrition.[12]

Treating only the vitamin A deficiency does not resolve the broader health issues associated with malnutrition, including deficiencies in other vitamins, trace elements, and amino acids. The most commonly fed seeds generally lack 32 essential ingredients, including vitamins, minerals, trace minerals, pigments, protein, fiber, vitamin precursors, and omega-3 fatty acids.[11]

When evaluating birds in daily practice, we observe a variety of clinical signs of malnutrition in birds fed different commercial diets. In the opinion of the author, the birds themselves show us whether the provided nutrition is sufficient. The birds are the experts.

Evaluating the quality of the feathers and identifying molting disorders is helpful in determining how long the bird has already been experiencing health problems that have gone unnoticed by the owner. It is not unusual to conclude that birds showing respiratory problems are also experiencing feather and molting disorders—sometimes long before the owner makes an appointment for what seems like a sudden respiratory issue. In my experience, it is not uncommon for feather and molting disorders to have existed for years before the owner seeks help for respiratory problems.

Evaluating the diet and advising the owner to provide reliable and organic nutrition is the first and most important measure to both prevent and solve respiratory problems.

Obesity

Obesity can be considered another long-term outcome of malnutrition and a cause of respiratory problems. In my experience, during autopsies dealing with fatty livers and excessive fat accumulation in the coelomic cavity we can notice that the fat can also compromises the respiratory system by exerting pressure on the air sacs and the heart. Certain species are more vulnerable to obesity—Amazons, rose-breasted cockatoos, budgerigars (especially English budgerigars), lories, and eclectus parrots are examples.[11] I have never seen an African grey suffering from obesity. Providing reliable, organic nutrition is the most important approach to prevent and address obesity. In my experience, offering a high-calorie complete diet can lead to weight loss while simultaneously supporting muscle development. We often observe fat loss accompanied by muscle gain, which may result in an increase in overall weight due to improved muscle mass.

Lack of exercise

Parrots in nature are forced to be active to survive. Flying—and, for many psittacine species, climbing—are essential activities that ensure muscle strength, endurance, and overall viability. Parrots are also socially active. During the breeding season, they engage in a variety of behaviors, including courtship rituals. As prey animals, they must be constantly prepared to escape from predators. Another key natural behavior is perching on flexible, moving branches, which activates their muscles to maintain balance—even while sleeping. In the author's opinion, not providing flexible perches is a highly underestimated factor when keeping parrots in captivity. Lack of exercise, as in humans, is a cause of respiratory problems—especially during physical exertion or stressful situations. Creating opportunities for parrots to exercise helps prevent and resolve health issues, including those related to the respiratory system.[15,16]

Lack of sunlight

It is not uncommon in daily practice to encounter parrots suffering from a lack of sunlight. Some parrots are never taken outside, or only occasionally during the summer when the sun is shining. Owners are often afraid of cooler weather and may not realize that parrots in the wild experience vastly different climate conditions. For instance, in Africa and Australia, it can be very cold at night and extremely hot during the day, with a large temperature difference between night and day.[15] The negative consequences of a lack of sunlight are frequently underestimated. Sunlight plays a critical role in enabling birds to produce vitamin D. Vitamin D deficiency in African greys is a known cause of serious bone disorders, such as rickets. While preening, birds spread the oil produced by the uropygial gland over their feathers. The sebaceous material contains vitamin D precursors, which are converted to be the active form of vitamin D3 when exposed to ultraviolet light. During preening, the active form of vitamin D3 is ingested. African greys and other species, like macaws, have a patch of unfeathered skin around the eye to allow direct sunlight to reach the skin. African greys are known to forage on the ground in open spaces—very different from species like amazons, which typically dwell in the canopy and do not have a uropygial gland. Beyond the physical health implications, we must also acknowledge that outdoor exposure is a significant form of enrichment—no different than for children.[16] In human medicine in the Netherlands, we are familiar with the typical winter depression that affects people between October and April. A lack of sunlight can lead to serious mental health issues. Taking parrots outside as often as possible can therefore be considered a major preventive measure against both health and welfare problems.

Climate control

Climate control involves regulating temperature, humidity, and ventilation. It is crucial for the health and welfare of parrots—as well as other animals and humans. As a veterinary student, I was taught the critical importance of climate control in poultry.[17] In poultry farming, it is well known that proper climate regulation is essential for preventing respiratory problems. As described by Dr. Fern Van Sant in her 1993 AAV conference presentation Seeing the Rainforest Through the Trees, we must consider the natural living conditions of parrots to understand their needs in captivity.[15,16] Budgerigars live in the desert environments of Australia, while amazons inhabit the tropical rainforests of South America—species with vastly different needs when it comes to temperature and humidity. Keeping a budgerigar in a centrally heated and air-conditioned home is not the same as keeping an amazon under the same conditions. It's no surprise that many respiratory problems are observed in Amazons and other birds kept as companion animals. Inappropriate climate conditions can be considered a major cause of respiratory illness. In general, outdoor climate conditions can be far better than those indoors—another reason to recommend taking parrots outside as much as possible. Another major aspect of climate control is ventilation. Ventilation prevents the circulation of fine dust and supplies fresh air. It also helps control humidity. In the wild, parrots—particularly rainforest species—are not exposed to circulating fine dust. Birds are adapted to the climate conditions of their natural habitats.[15]

Aerogenous toxins

Due to the efficiency of the avian respiratory system, parrots are highly vulnerable to airborne toxins. The famous example of canaries falling from their perches due to toxic gases in mines serves as a historical reminder—they saved miners' lives by acting as early warning systems. Toxic gases and airborne particulates in the environment can have destructive effects on the respiratory tract.[1,18] Especially dangerous are combinations of low humidity and circulating fine dust, strong fumes, cigarette smoke, or smoke from burning wood. These are all important risk factors.[1,3,12] I will never forget the case of an Amazon parrot presented with a sudden onset of severe respiratory distress. The bird entered the examination room gasping for air, sitting on the shoulder of its female owner. I immediately smelled strong fumes coming from the owner. When I asked when she had last been to the hairdresser, she explained that her daughter—a hairdresser—had done her hair the day before. The respiratory symptoms had started that same day.

Similar issues occur with fumes from scented candles and cleaning products used near the bird or its cage. Products containing ammonia or bleach are particularly problematic and can trigger respiratory distress. The most well-known and serious respiratory condition caused by an aerogenous toxin is exposure to polytetrafluoroethylene (PTFE). PTFE is found in various common home appliances with non-stick surfaces. When heated to 536°F (280°C), PTFE releases toxic fumes that can lead to sudden death. Air fresheners can also cause serious respiratory issues, as can candy cooking flavorings.[3] It is best to avoid exposing birds to any aerosolized substances to prevent respiratory problems.[1,3]

Egg binding

In practice, I have seen parrots present with respiratory problems, only to discover that the true cause was egg binding. In some cases, the owner could not believe the diagnosis, as the bird was assumed to be male and even had a male name. Diagnostic imaging (radiographs or ultrasound) may be necessary to confirm the presence of an egg to the owner. This serves as a reminder that respiratory symptoms are not always linked directly to the respiratory system itself. Other noninfectious causes of respiratory problems include internal tumors, cysts, hydrops, ascites, and hepatomegaly.[11]

Cardiac disease

Over the years, cardiac diseases, including atherosclerosis, have become typical captivity-related conditions in parrots and a common cause of a wide variety of clinical symptoms, including respiratory problems. Besides the difference between birds and humans, it makes sense to consider that atherosclerosis in birds can result from malnutrition, exposure to cigarette smoke, obesity, high blood pressure, lack of exercise, and a combination of stress factors. I have learned from Prof. Peer Zwart to look for the similarities despite the differences between species. Malnutrition, lack of sunlight, lack and serious behavioral problems can all be considered stressors that may contribute to the development of cardiac disease. Lack of privacy is another well-known cause of stress in both humans and prey animals. Parrots, as prey animals, may experience significant stress when approached by humans who unknowingly display predator-like behaviors.[16,19,20] It is normal for parrots to have a higher blood pressure than mammals. This pressure increases during stressful moments or events.[12] Providing reliable, organic nutrition, along with creating an environment and social setting in which the parrot feels comfortable, appears to be an important preventive strategy.

Stress

Stress creates insecurity and negatively affects the immune system, making parrots more vulnerable to infectious diseases such as aspergillosis. In baby parrots, hand-rearing must be considered a significant stress factor. It is known to cause behavioral and welfare problems.[21] Stress increases blood pressure and respiratory rate.[11] Parrots in captivity face a variety of stressors. Living in a cage, unable to perform natural behaviors, can be a continuous source of stress and reduced welfare.[16] Not being acknowledged as prey animals, not acknowledging the need for privacy—and being approached or handled in ways that contradict their natural instincts—can be a serious stress factor.[22,23]

Conclusions

Noninfectious ailments can be considered the most common causes of respiratory issues in parrots. In my experience, a thorough questionnaire combined with a detailed physical examination are the most important tools for establishing a diagnosis. In general, a combination of factors is responsible for the development of respiratory problems. In daily practice, the most frequent causes remain malnutrition, inappropriate climate conditions and a variety of stress factors. These findings strongly support the need to focus on prevention by establishing tailored management protocols for different species to ensure the health and welfare of parrots in captivity.

References

1. Zwart P, Samour J. The avian respiratory system and its noninfectious ailments: a review. *J Exot Pet Med.* 2021;37:39-50.

2. West JB, Watson RR, Fu Z. Major differences in the pulmonary circulation between birds and mammals. *Respir Physiol Neurobiol.* 2008;157(2-3):382-390. doi: 10.1016/j.resp.2006.12.005.

3. Maina JN. Functional morphology of the avian respiratory system, the lung-air sac system: efficiency built on complexity. *Ostrich.* 2008;79(2):117-132.

4. Schmidt RE, Reavill DR, Phalen DN. *Pathology of Pet and Aviary Birds.* New York, NY: Wiley; 2015.

5. Maina JN, West JB, Orgeig S, et al. Recent advances into understanding some aspects of the structure and function of mammalian and avian lungs. *Physiol Biochem Zool.* 2010;83(5):792-807. doi: 10.1086/652244.

6. Sawad AA, Udah DA. Morphological and histopathological study of air sacs (sacci pneumatici) in Japanese quail (*Coturnix coturnix japonica*). *Mirror Res Vet Sci Anim.* 2012;1(1):50-56.

7. Maina JN. The design of the avian respiratory system: development, morphology and function. *J Ornithol.* 2015. doi: 10.1007/s10336-015-1263-9.

8. Maina JN. Pivotal debates and controversies on the structure and function of the avian respiratory system: setting the record straight. *Biol Rev Camb Philos Soc.* 2017;92(3):1475-1504. doi: 10.1111/brv.12292.

9. Zwart P, SchreursWHP, Dorrestein GM. Vitamin A deficiency in parrots. *21st Int Symp Erkrk Zootiere Mulhouse.* 1979;21:47-52.

10. Klika E, Scheuermann DW, De Groodt-Lasseel MH, Switka BA. Pulmonary macrophages in birds (barn owl, *Tyto alba*), domestic fowl (*Gallus gallus domesticus*), quail (*Coturnix coturnix*), and pigeons (*Columba livia*). *Anat Rec.* 1996;246(1):87-97. doi: 10.1002/(SICI)1097-0185(199609)246:1<87::AID-AR10>3.0.CO;2-0.

11. Harrison GJ, Lightfoot T. *Clinical Avian Medicine.* Vol 1-2. Palm Beach, FL: Spix Publishing; 2006.

12. Ritchie BW, Harrison GJ, Harrison L. *Avian Medicine: Principles and Application.* Lake Worth, FL: Wingers Publishing; 1994.

13. Zwart P, Schreurs WHP, Dorrestein GM. Vitamin A deficiency in parrots. *21st Int Symp Erkrk Zootiere Mulhouse.* 1979;47-52.

14. Zwijnenberg RJG, Zwart P. Squamous metaplasia in the salivary glands of canaries (a case report). *Vet Q.* 1994;16(1):60-61.

15. Van Sant F. Seeing the rainforest through the trees. *Proc Annu Conf Assoc Avian Vet.* 1993:228-232.

16. Pepperberg I, Hooimeijer J. Studying the behaviour of captive parrots: respecting their cognitive abilities creates mutual respect and mutual trust. *Proc ExoticsCon.* 2022;48-56.

17. Greenacre CB, Morishita TY. *Backyard Poultry Medicine and Surgery: A Guide for Veterinary Practitioners.* Hoboken, NJ: John Wiley & Sons; 2015.

18. Forbes NA. Avian respiratory disease. *Vet Q.* 1998;20(Suppl 1):S67. doi: 10.1080/01652176.1998.10807420.

19. Hooimeijer J. Dealing with behaviour problems of parrots. *Proc DVG.* 2008.

20. Engebretson M. The welfare and suitability of parrots as companion animals: a review. *Anim Welf.* 2006;15:263-276.

21. Schmidt R. The influence of the breeding method on the behavior of adult African grey parrots. Dissertation zur Erlangung der Doktorwürde der Vetsuisse-Fakultät der Universität Bern; 2004.

22. Hooimeijer J. How to create mutual respect and mutual trust in the examination room. *Proc ExoticsCon.* 2022;550-559.

23. Hooimeijer J. A practical behavioural protocol for dealing with parrots. *Proc Annu Conf Assoc Avian Vet.* 2003;177-181.

Contrafreeloading in Parrots: Applying Cognitive Abilities in Daily Practice

Irene Pepperberg, SB, MA, PhD

Session #138

Affiliation: From Boston University, 30 Curry Circle, Swampscott, MA 01907, USA.

Abstract: In nature, animals expend energy foraging and competing with others to access resources while avoiding predation. Hence, most animals aim to forage optimally to obtain the highest caloric intake in the shortest time with the least effort. In captivity, we observe contrafreeloading, the choice to expend excess energy by foregoing freely available food to extract food from a puzzle box. Colleagues and I tested grey parrots,[1] kea,[2] and umbrella cockatoos[3] on contrafreeloading tasks differing in ecologic relevance. On one task, more kea contrafreeloaded than grey parrots and did so slightly more often, but kea completely ignored a second task (i.e., shelled versus unshelled nuts). Cockatoos tested on the second task also exceeded the grey parrots but results were affected by dominant or subordinate behavior and other individual differences. A subsequent study (unpublished) on grey parrots examined specific effects of task or task familiarity on contrafreeloading. Results suggest contrafreeloading occurs if birds consider tasks as play, not work, which would influence design of enrichment programs. Acknowledging parrots' intelligence, individual differences, and playfulness can ensure their welfare.

References

1. Smith GE, Greene D, Hartsfield LA, Pepperberg IM. Initial evidence for eliciting contrafreeloading in grey parrots (*Psittacus erithacus*) via the opportunity for playful foraging. *J Comp Psychol.* 2021;135(4): 516-533.

2. Smith GE, Bastos APM, Chodorow M, et al. Contrafreeloading in kea (*Nestor notabilis*) in comparison to grey parrots (*Psittacus erithacus*). *Sci Rep.* 2022;12(1):17415. doi: 10.1038/s41598-022-21370-6.

3. Caroll A, Pepperberg IM. Contrafreeloading in umbrella cockatoos (*Cacatua alba*): further evaluation of the play hypothesis. *J Comp Psychol.* 2025;139(2):84-97.

Section 2

Plenary Session

Jennifer Graham, DVM, DABVP (Avian Practice),
DABVP (Exotic Companion Mammal Practice),
Dipl ACZM, *AAV President*
Session Chair

Avian Influenza H5N1: A Zoonotic Threat in the One Health Era

Joseph F. Annelli, DVM, MS

Session #101

Affiliation: From National Association of Federal Veterinarians, 1300 Pennsylvania Avenue, Northwest, Suite 190, #330, Washington, DC 20004, USA.

Abstract: Avian influenza H5N1, initially identified from an isolated human outbreak in Hong Kong in 1997, has evolved into a global zoonotic threat, impacting veterinary, public, and environmental health. Highly pathogenic avian influenza (HPAI) H5N1 primarily affects avian species but has demonstrated its potential for cross-species transmission, including to mammals and humans, resulting in sporadic yet severe infections. This article comprehensively discusses epidemiology, transmission dynamics, public health concerns, and the pivotal role of veterinarians in surveillance and response measures. The significance of biosecurity, early detection, and the One Health approach in mitigating risks is underscored. The recent surge in cases in the U.S. poultry industry and isolated human infections serve as a stark reminder of the need for vigilant monitoring, particularly among veterinary professionals who may be exposed to potentially infected animals. The emergence of cats as a potential intermediate host and the growing evidence of symptomatic infections among healthcare workers further emphasizes the complexity of controlling this virus. H5N1 serves as a clear illustration of the interconnectedness of wildlife, domestic animals, and humans, reinforcing the crucial role of private practitioners as public health sentinels.

Introduction

Initially identified in humans in Hong Kong in 1997, highly pathogenic avian influenza (HPAI) H5N1 has since manifested as an instance of zoonotic spillover. The virus continues to circulate in wild bird populations, predominantly waterfowl, globally. Occasionally, it infects domestic poultry and spills over into mammals, including humans.

By 2006, the virus had disseminated across Asia, Europe, and Africa. In 2023 and 2024, the United States experienced substantial poultry losses attributed to H5N1, and recent media reports have highlighted the first confirmed human infection in the United States was in Colorado.[1] This patient had no known contact with animals, raising concerns in public health circles.

Veterinarians in public practice are instrumental in monitoring, identifying, and responding to poultry outbreaks. Veterinarians in private practice play a pivotal role in H5N1 surveillance and response efforts in humans. As per CDC guidelines,[2] practitioners handling cats exposed to H5N1 must adhere to strict precautions. Infected animals can shed the virus through saliva, feces, and respiratory secretions, posing risks to humans.

The One Health approach, which integrates human, animal, and environmental health, is crucial for managing H5N1. Surveillance, biosecurity, interagency coordination, and public awareness are essential components of this approach. As H5N1 evolves, so must our response, with veterinarians at the forefront of emerging infectious disease control. The H5N1 virus has a long history of causing outbreaks in both poultry and wild birds, with occasional transmission to humans. The virus continues to evolve and mutate, posing a potential threat to human health.

The Centers for Disease Control and Prevention (CDC) and other health organizations monitor the situation and provide updates as information becomes available. The World Health Organization (WHO) has documented 878 cases of HPAI H5N1 infection in humans and 458 (52%) fatalities in 23 countries from 2003 to July 14, 2023 (Fig 1).[3,4]

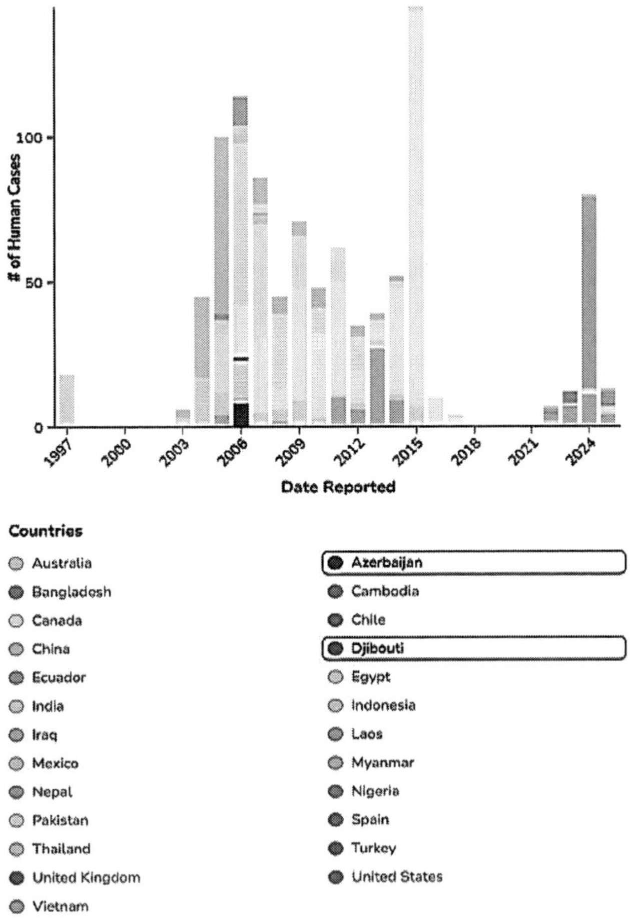

Figure 1. Past reported global human cases with highly pathogenic avian influenza A(H5N1) (HPAI H5N1) by country, 1997-2025.[4]

Global HPAI H5N1 Timeline: Emergence and Initial Outbreaks[5]

1996: First identification

In 1996 the HPAI H5N1 virus was first identified in farmed geese in Guangdong province, China.

1997: First zoonotic transmission

In 1997 outbreaks of HPAI H5N1 were detected in poultry in Hong Kong, leading to 18 human infections with six deaths, marking the first known instances of zoonotic transmission (i.e., animal to human). The virus caused high mortality in chickens on three farms in Hong Kong.

2003-2005: Re-emergence and global spread

In 2003 the H5N1 virus re-emerged in China and several other countries, causing widespread poultry outbreaks across Asia. From 2003-2005 the virus spread to poultry in Africa, the Middle East, and Europe, with wild birds playing a role in transmission. From 2003-2006 several countries reported sporadic human infections with HPAI H5N1.

2006-2019: Continued circulation and monitoring

The virus continued to circulate in both wild birds and poultry, with sporadic human infections reported from various countries.

2016-2017: New strain

A new HPAI H5N8 strain emerged and spread through eastern Asia, with multiple incursions into European countries.

2017-2018: New reassortant

A new reassortant H5N6 derived from the 2016/2017 H5N8 viruses was detected in the UK, The Netherlands, and Germany.

2020-2024: Recent outbreaks and concerns

In 2022 the first detection of influenza A(H5N1) among domestic mammals in the United States was reported. Also in 2022 the first human case of avian influenza A(H5) in the United States was reported in a poultry farm worker in Colorado (CDC, 2022). In addition, the first detection of influenza A(H5N1) among wild mammals in Canada was reported.

In 2023 a large number of sea lions died of bird flu along the coasts of Peru, Chile, Argentina, Uruguay, and Brazil.

In 2024 the current outbreak began in late March with the detection of H5N1 bird flu in dairy cows.

In 2025 (on January 6, 2025) the CDC reported the first H5 bird flu death in the United States in Louisiana.[6]

As of May 2025, avian influenza A(H5N1) has been detected in domestic cats across the United States, raising concerns among public health and veterinary communities. Since 2022, at least 130 cases of H5N1 have been confirmed in domestic cats across 23 U.S. states. These cases include both indoor and outdoor cats, as well as feral and barn cats. Infections have been reported in multiple states, including Colorado, California, South Dakota, Oregon, Kansas, Louisiana, Iowa, Minnesota, Kansas, Pennsylvania, and Illinois. Cats can contract H5N1 through various means, such as consuming infected birds, exposure to contaminated environments, and ingesting raw or unpasteurized animal products. Notably, outbreaks have been linked to the consumption of raw milk from infected dairy cows and contaminated raw pet food.[7] Infected cats often exhibit severe symptoms, including lethargy, respiratory distress, neurological issues, and loss of appetite. The disease frequently results in death, and there is currently no specific antiviral treatment for pets; care is supportive.

The CDC assesses the current risk to the public as low, while the risk to individuals with direct exposure to infected animals is considered moderate to high. To date, no confirmed cases of cat-to-human transmission have been reported in the United States, but the potential for such transmission, particularly with prolonged, unprotected exposure, cannot be dismissed.[8]

The CDC and the U.S. Department of Agriculture (USDA) are actively monitoring the situation, with efforts focused on surveillance, biosecurity measures, and public education to prevent further spread. Recent funding allocations aim to bolster these initiatives.

Recommendations for Cat Owners

Preventive measures

Pet owners are strongly advised to confine their cats indoors to minimize exposure to infected avian species and contaminated environments. Feeding cats raw or unpasteurized animal products should be strictly avoided. Regular veterinary check-ups and prompt attention to any indications of illness are paramount.

Recommendations for Pet Parrot Owners

Confirmed cases of avian influenza H5N1 in pet parrots in the United States

Confirmed cases of avian influenza H5N1 in pet parrots in the United States are rare.[9] One notable incident occurred in Washtenaw County, Michigan, where pet parrots in a household succumbed to the virus. The Michigan Department of Agriculture and Rural Development reported that these birds passed away shortly after H5N1 was detected in the area.[9] Officials emphasized that, although it is uncommon for pet birds to contract avian influenza, infection can occur through indirect exposure, such as contaminated food, cage materials, or clothing that has come into contact with wild birds.

Beyond this case

There are no widespread reports of H5N1 affecting pet parrots in the United States; however, the virus has been detected in various bird species and mammals, including domestic cats, across multiple states. This underscores the significance of vigilant biosecurity measures for all pet bird owners.

Preventive measures

Prevent contact with wild birds. Ensure that pet parrots are kept indoors or in enclosed outdoor aviaries to minimize exposure to wild birds.

Maintain cleanliness. Regularly clean and disinfect cages, feeding utensils, and any items that come into contact with your pet parrot. birds.

Monitor health. Be vigilant for signs of illness in your parrots, such as lethargy, loss of appetite, or respiratory distress, and consult a veterinarian promptly if symptoms arise.

Practice good hygiene. Wash hands thoroughly before and after handling birds and avoid wearing clothing or shoes indoors that have been exposed to outdoor environments frequented by wild birds.

Human Cases

As bird migration declines in late spring, the number of flock infections also declines. As of mid-May 2025, influenza activity in the United States has reached low levels, according to the CDC.[10] The percentage of human outpatient visits for influenza-like illness (ILI) decreased slightly from 2.1% to 1.9% over the past week of May 18, 2025 (Fig 2). The seasonal human influenza test positivity dropped to 2.9% from 3.6%, and hospitalizations declined from 2,336 to 2,008. No states are reporting moderate, high, or very high ILI activity as old May 2025.

Despite the overall decline, the CDC reported one new pediatric seasonal H1N1 flu-related death. This death brought the human cases of H1N1 influenza total for the 2024-2025 season to 227, the highest since the 2009-2010 season, which saw 288 pediatric deaths. The recent death occurred in January and was attributed to the H1N1 strain.[11]

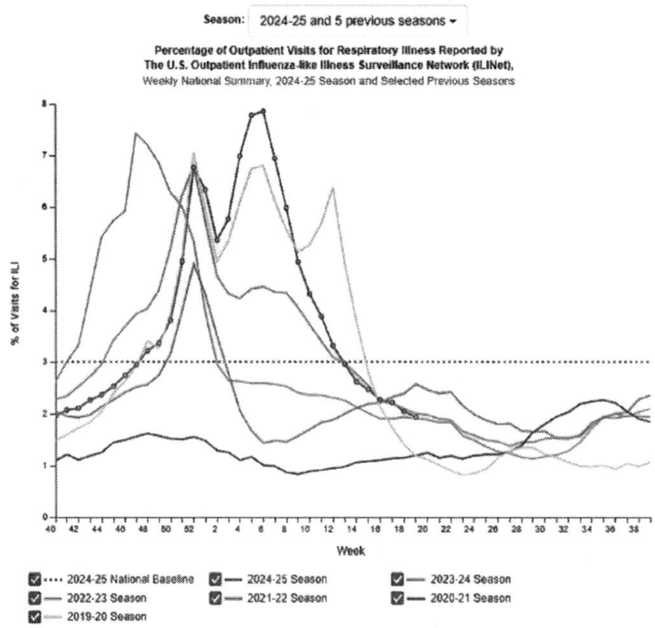

Figure 2. Seasonal influenza-like illnesses (ILI) in humans.

Professional Guidance

For personalized advice, consult with veterinarians. Stay informed through official channels such as the CDC and the USDA for updates on the situation.

Summary

HPAI H5N1 and its multiple clades have been around since 1996 and they will likely not be going away. H5N1 remains a serious and evolving zoonotic threat. Effective management requires vigilance, cross-sector cooperation, and strict biosecurity. Veterinarians play a critical role in early detection and response.

References

1. Centers for Disease Control and Prevention. 2022. First human H5N1 bird flu case in U.S. identified in Colorado. Available at: https://www.cdc.gov/media/releases/2022/s0428-avian-flu.html. Accessed June 20, 2025.

2. Centers for Disease Control and Prevention. 2024. Health alert network update: Precautions for veterinarians treating pets exposed to HPAI H5N1. Available at: https://emergency.cdc.gov/han/2024/han00498.asp. Accessed June 20, 2025.

3. World Health Organization. 2023. Cumulative number of confirmed human cases for avian influenza A(H5N1) reported to WHO, 2003-2023. Available at: https://www.who.int/docs/default-source/influenza/h5n1/h5n1-cumulative-table.pdf. Accessed June 20, 2025.

4. Centers for Disease Control and Prevention. 2025. Past reported global human cases with highly pathogenic avian influenza A(H5N1) (HPAI H5N1) by country, 1997-2025. U.S. Department of Health and Human Services. Available at: https://www.cdc.gov/bird-flu/php/avian-flu-summary/chart-epi-curve-ah5n1.html. Accessed June 20, 2025.

5. World Health Organization. 2006. H5N1 avian influenza: Timeline of major events. Available at: https://www.who.int/csr/disease/avian_influenza/timeline.pdf. Accessed June 20, 2025.

6. Centers for Disease Control and Prevention. 2025. CDC reports first H5 bird flu death in U.S. Available at: https://www.cdc.gov/media/releases/2025/s0106-h5n1-death.html. Accessed June 20, 2025.

7. Oregon Health Authority. 2024. Investigation into raw pet food linked to H5N1 in domestic cats. Available at: https://agr.wa.gov/about-wsda/news-and-media-relations/news-releases?article=42075. Accessed June 20, 2025.

8. Rabalski Ł, Milewska A, Pohlmann A, et al. 2023. Emergence and potential transmission route of avian influenza A (H5N1) virus in domestic cats in Poland, June 2023. *Eurosurveillance.* 2023;28(31):2300390. doi:10.2807/1560-7917.ES.2023.28.31.2300390

9. Nissan J. 2022. Pet parrot in Washtenaw County dies of highly pathogenic avian influenza as 'severe' outbreak spreads. Available at: https://www.fox2detroit.com/news/pet-parrot-in-washtenaw-county-dies-of-high-pathogenic-avian-influenza-as-severe-outbreak-spreads. Accessed June 20, 2025.

10. Center for Infectious Disease Research and Policy. 2025. Flu activity in US has reached low levels. Available at: https://www.cidrap.umn.edu/influenza-general/flu-activity-us-has-reached-low-levels. Accessed June 20, 2025.

11. Centers for Disease Control and Prevention. 2025. Preliminary estimated flu disease burden, 2024-2025 flu season. U.S. Department of Health and Human Services. Available at: https://www.cdc.gov/flu-burden/php/data-vis/2024-2025.html. Accessed June 20, 2025.

A Retrospective Study of Disease in Passerine Birds: 6300 Cases

Lauren V. Powers, DVM, DABVP (Avian Practice), and
Michael M. Garner, DVM, DACVP

Session #154

Affiliation: From the Avian and Exotic Pet Service, Carolina Veterinary Specialists, 12117 Statesville Rd, Huntersville, NC 28078, USA (Powers), Northwest ZooPath, 654 West Main St, Monroe, WA 98272, USA (Garner).

Abstract: The order Passeriformes is the largest and most globally diverse avian order,[1] including at least 6704 species, more than 60% of known avian species. Passerine birds are kept in managed zoological collections, as pets, and presented as wildlife.[1-3] The purpose of this descriptive retrospective archival study was to tabulate and describe diseases of passerine birds submitted to a zoologic pathology service (Northwest ZooPath) from 1995 to 2024. Diseases, disease categories, and body systems affected were tabulated for necropsies and biopsies. Causes of death and reasons for euthanasia were evaluated. A total of 6283 passerine birds were included in the study, spanning 364 species in 46 families. The most frequent species were Gouldian finches (*Erythrura gouldiae*), Taveta golden weavers (*Ploceus castaneiceps*), and canaries (*Serinus canaria domestica*). Five species were listed as endangered, four as critically endangered (eg. Bali mynah, *Leucopsar rothschildi*), and one as currently extinct (bridled white-eye, *Zosterops conspicillatus*) on the International Union for Conservation of Nature Red List. The most frequent disease categories were infectious or inflammatory diseases, trauma, and nutritional diseases (including inanition). The most common systems affected were systemic or multiple systems affected, respiratory tract, and alimentary tract. For birds with unassisted death, the most common immediate causes of death were pulmonary shock, inanition, and bacterial sepsis. Of the 6283 birds, 572 (9.1%) were euthanized. The most common categories for euthanasia were infectious or inflammatory, trauma, and degenerative conditions. Iatrogenic causes of death were 123/6283 (2.0%) of cases. Results of this study can be used to improve management conditions and welfare decisions for passerine birds under managed care.

References

1. Wahle A, Leiss L, Marcordes B, et al. Passerine birds in zoos: a global approach on distribution patterns and conservation breeding of threatened birds in zoological institutions. *J Zoo Aquar Res*. 2024:23-35.

2. Garner MM, Juan-Sallés C, Burns, RE, et al. Passeriformes, Trochiliformes, Coraciiformes, Caprimulgiformes, Apodiformes, and Piciformes. In: Terio KA, McAloose D, St. Leger J, eds. *Pathology of Wildlife and Zoo Animals*. 2nd Ed. Cambridge, MA: Academic Press; 2025.

3. Flach EJ, Dodhia HS, Guthrie A, et al. Systemic isosporiasis (atoxoplasmosis) in passerine birds at the Zoological Society of London, London Zoo. *J Zoo Wildl Med*. 2022:70-82.

Comparison of Protein Electrophoresis in Non-Laying and Laying Domestic Chickens (*Gallus gallus domesticus*)

Jared S. Zion, DVM,
Ashleigh W. Newman, VMD, DACVP,
Nicole Belcher,
Ricardo de Matos, LMV, MSc, DABVP (Avian Practice),
DipECZM (Avian), DipECZM (Small Mammal)

Session #131

Affiliation: From Arizona Exotic Animal Hospital, Phoenix, AZ 85024, USA (Zion), Cornell University College of Veterinary Medicine 602 Tower Road, Ithaca, NY 14853, USA (Newman, Belcher, de Matos).

Abstract: Protein electrophoresis (EPH) is a valuable diagnostic tool, especially in avian species where measurement of plasma proteins is considered inaccurate by routine biochemical analyzers.[1] Veterinarians caring for poultry, including the increasingly popular backyard flock, must be knowledgeable about differentiating normal physiology from pathology. Genetic selection has resulted in modern domestic hens laying an average of ~250-300 eggs/year in commercial production systems.[2] With these laying advances, reproductive pathology has become exceedingly common in poultry when presented to veterinary hospitals.[3] While EPH provides the most accurate determination of plasma proteins in avian patients, unique monoclonal-appearing gammopathies have been reported with egg-laying status, yet this remains not well characterized.[1,4] In addition, the clinical significance has not been well investigated, leading to challenges differentiating physiology from pathology. In this study, capillary-zone EPH results were compared between healthy egg-laying and non-laying hens. Twenty-four healthy hens divided equally between red star and Cornish cross breeds (36 and 14 months old, respectively) were separated into egg-laying or non-laying groups based on reported egg-laying status and coelomic ultrasonography. Comprehensive blood work analysis and capillary-zone EPH using heparinized plasma were performed. Preliminary results suggest observable differences in egg-laying hen total protein, prealbumin, alpha-2 globulins compared to non-layers. No monoclonal-like peaks were observed. Finalized statistical analysis are conveyed in the presentation. The overall use of EPH in non-traditional species is reviewed with further discussion in birds. The results of this study will help avian veterinarians better distinguish differences in EPH patterns between laying and non-laying healthy birds to reference in clinical practice.

References

1. Cray C. Protein electrophoresis of non-traditional species: a review. *Vet Clin Pathol.* 2021;50(4):478-494.
2. Qanbari S, Rubin C-J, Maqbool K, et al. Genetics of adaptation in modern chicken. *PLOS Genet.* 2019;15(4):1-21.
3. Greenacre CB. Reproductive diseases of the backyard hen. *J Exot Pet Med.* 2015;24(2):164-171.
4. Werner LL, Reavill DR. The diagnostic utility of serum protein electrophoresis. *Vet Clin North Am Exot Anim Pract.* 1999;2(3):651-662.

Pharmacokinetics of Maropitant Citrate After Intravenous and Subcutaneous Administration in Orange-Winged Amazon Parrots (*Amazona amazonica*)

Ariella Darvish, DVM,
David Sanchez-Migallon Guzman, LV, MS, DipECZM (Avian), DipECZM (Small Mammal), Dipl ACZM,
Hugues Beaufrère, DVM, PhD, Dipl ACZM, DABVP (Avian Practice), DipECZM (Avian),
Heather K. Knych DVM, PhD, DACVCP,
Olivia A. Petritz DVM, Dipl ACZM

Session #157

Affiliation: From William R. Pritchard Veterinary Medical Teaching Hospital, University California Davis, Davis, CA 95616, USA (Darvish), Department of Medicine and Epidemiology, School of Veterinary Medicine, University of California-Davis, Davis, CA 95616, USA (Guzman, Beaufrère), K. L. Maddy Equine Analytical Chemistry Laboratory, University of California-Davis, Davis, CA 95616, USA (Knych), Department of Clinical Sciences, College of Veterinary Medicine, North Carolina State University, Raleigh, NC (Petritz).

Abstract: Maropitant citrate, a neurokinin-1 receptor antagonist, is commonly used in dogs and cats to treat vomiting, nausea, motion sickness, and has shown promising analgesic effects.[1-4] Despite its empirical use in psittacine birds, pharmacokinetic data are lacking. To date, Rhode Island Red chickens are the only avian species with published maropitant pharmacokinetic data.[5] This study aimed to evaluate maropitant pharmacokinetics in orange-winged Amazon parrots (*Amazona amazonica*) following a single 1 mg/kg dose administered intravenously (IV) and subcutaneously (SC). Eight adult parrots received both treatments with an 8-week washout period. Plasma concentrations were measured via liquid chromatography with tandem mass spectrometry using blood sampled collected over 36 hours, and pharmacokinetic parameters were calculated using a noncompartmental model. After SC administration, the mean maximum plasma concentration was 130.9 ± 24.6 ng/ml at 0.5 ± 0 hours, with a terminal half-life of 6.67 hours and SC bioavailability of 85%. Plasma concentrations were negligible at 24 hours for both routes. Plasma concentrations were higher than those shown to be therapeutic in dogs for at least 30 minutes to 1 hour. The current empiric dose of 1 mg/kg and dosing interval of every 12 hours might be insufficient to achieve systemic effects comparable to those observed in other species. These results highlight the need for further research to establish optimal dosing regimens for clinical use of maropitant in psittacine birds.

References

1. de la Puente-Redondo VA, Siedek EM, Benchaoui HA, et al. Anti-emetic efficacy of maropitant (Cerenia™) in the treatment of ongoing emesis caused by a wide range of underlying clinical aetiologies in canine patients in Europe. *J Small Anim Pract.* 2007;48:93-98.

2. Benchaoui HA, Cox SR, Schneider RP, et al. The pharmacokinetics of maropitant, a novel neurokinin type-1 receptor antagonist, in dogs. *J Vet Pharmacol Ther.* 2007;30:336-344.

3. Hickman MA, Cox SR, Mahabir S, et al. Safety, pharmacokinetics and use of the novel NK-1 receptor antagonist maropitant (CereniaTM) for the prevention of emesis and motion sickness in cats. *J Vet Pharmacol Ther.* 2008;31:220-229.

4. Bozkurt G, Kaya F and Yildiz M. Does maropitant provide more effective perioperative pain management than meloxicam in bitches undergoing ovariohysterectomy? The first report on the comparison of visceral algesia-analgesia for ovariohysterectomy. *Res Vet Sci.* 2024;169:105179.

5. Mones AB, Petritz OA, Knych HK, et al. Pharmacokinetics of maropitant citrate in Rhode Island Red chickens (*Gallus gallus domesticus*) following subcutaneous administration. *J Vet Pharmacol Ther.* 2022;45:495-500.

Use of ^{18}F-Sodium Fluoride Positron Emission Tomography for the Diagnosis of Atherosclerosis in Amazon Parrots (*Amazona* spp.)

Lucyanne Megan, DVM,
Kelsey Brust, DVM, Dipl ACVR,
Mathieu Spriet, DVM, MS, Dipl ACVR, Dipl ECVDI,
Dipl ACVR (Equine Diagnostic Imaging),
Pedro Ruivo, DVM, MS, DACVP,
Marcel Gómez-Ponce, DVM,
Hugues Beaufrère, DVM, PhD, Dipl ACZM, DABVP (Avian Practice),
DipECZM (Avian)

Session #118

Affiliation: From Veterinary Medical Teaching Hospital (Megan), Department of Medicine and Epidemiology (Beaufrère), Graduate Group in Integrative Pathobiology (Gómez-Ponce, Beaufrère), Department of Surgical and Radiological Sciences (Brust, Spriet), Comparative Pathology Laboratory (Ruivo), School of Veterinary Medicine, University of California, Davis, CA 95616, USA.

Abstract: In this feasibility proof-of-concept study, ^{18}F-sodium fluoride micro positron emission tomography (^{18}F-NaF-μPET) scans and micro computed tomography (μCT) scans were obtained on 10 Hispaniolan (*Amazona ventralis*) and eight orange-winged Amazon parrots (*Amazona amazonica*). A different μPET-scan protocol was used for each species. Images were reviewed by three observers for arterial calcification and radiotracer uptake indicative of atherosclerosis. A blood lipid profile was also performed. A subset of Hispaniolan Amazon parrots was euthanized after completion of the study, and the heart and arteries were collected for histopathology. Atherosclerotic lesions were detected on both μCT and μPET scans in various arteries. The presence of streak artefacts and increased noise on μCT scan limited the ability to detect high attenuation areas of arteries and obtain observer consensus except for one case of severe arterial calcification. The μPET scan was overall more sensitive, had higher interobserver agreement in lesion scoring, and detected a high number of lesions in both species, including in coronary arteries. On histopathology on five Amazon parrots, the majority (5/6) of clinically relevant atherosclerotic lesions and some pre-atheromatous lesions (1/3) were identified on μPET scan. No association between atherosclerosis detection on μPET scan and the lipid profile was observed.

This study established a safe and effective protocol for ^{18}F-NaF molecular imaging of psittacine atherosclerosis, which proved to be more sensitive than CT scan. The ^{18}F-NaF PET scan allows for a more reliable and earlier diagnosis of atherosclerotic lesions in Amazon parrots.

Effect of 4.7-mg Deslorelin Acetate Implant on Blood Lipids and Steroid Hormones in Cockatiels

Mariana Sosa Higareda, MVZ,
David Sanchez-Migallon Guzman, LV, MS, DipECZM (Avian), DipECZM (Small Mammal), Dipl ACZM,
Marcel Gómez-Ponce, DVM,
Rachel Hirota,
Hugues Beaufrère, DVM, PhD, Dipl ACZM, DABVP (Avian Practice), DipECZM (Avian)

Session #119

Affiliation: From Department of Medicine and Epidemiology (Guzman, Beaufrère), the Graduate Group in Integrative Pathobiology (Gómez-Ponce, Beaufrère), School of Veterinary Medicine (Hirota), University of California, Davis, CA 95616, USA.

Abstract: Deslorelin acetate is a gonadotropin-releasing hormone agonist frequently used for reproductive suppression in birds. Female psittacine birds frequently undergo vitellogenesis with resulting high blood lipids, which may predispose to several lipid disorders such as atherosclerosis.1 This study evaluated the effects of deslorelin implants on blood lipids and steroid hormones in cockatiels (*Nymphicus hollandicus*). Sixteen female cockatiels were randomly divided into two groups assigned to either receive a 4.7-mg deslorelin implant subcutaneously or a sham procedure. Plasma samples were collected at baseline and at 1, 2, 4, and 6 months post-implantation. Lipid and lipoprotein profiling was performed using high-resolution polyacrylamide gel electrophoresis, and steroid hormones were analyzed via liquid chromatography-tandem mass spectrometry. Deslorelin significantly reduced plasma triglycerides, total cholesterol, and intermediate-density lipoproteins (IDL-C) at 1-month post-implantation compared to controls with some effects persisting up to 2 months for IDL-C. Two out of eight cockatiels were found to be hyperlipidemic (triglycerides > 300 mg/dl) in the deslorelin group over the 6-month treatment periods compared to seven of eight cockatiels in the control group. Progesterone concentrations were also lower in the deslorelin group, suggesting a potential mechanism for lipid reduction. Estradiol levels decreased transiently but did not reach statistical significance. These findings highlight deslorelin's potential to modulate lipid metabolism in female cockatiels and suggest a temporal effect lasting 1-2 months. This study provides novel insights into the metabolic effects of deslorelin beyond its reproductive role and may have implications for managing hyperlipidemia in avian species.

References

1. Beaufrère H, Ammersbach M, Reavill DR, et al. Prevalence of and risk factors associated with atherosclerosis in psittacine birds. J Am Vet Med Assoc. 2013;242(12):1696-1704. doi:10.2460/javma.242.12.1696.

Use of Glimepiride for Regulation of Diabetes Mellitus in a Cockatiel (*Nymphicus hollandicus*)

Zoë van der Plaats, DVM, MSc, and
Yvonne van Zeeland, DVM, MVR, PhD, DipECZM (Avian), DipECZM (Small Mammal), CPBC

Session #188

Affiliation: From Department of Clinical Sciences of Companion Animals, Division of Zoological Medicine and Surgery of Companion Animals, Ophthalmology Section at Utrecht University, 3584 CM Utrecht, The Netherlands.

Abstract: Diabetes mellitus (DM) is relatively uncommon in psittacine species, although cockatiels and budgerigars are seemingly overrepresented.[1] While insulin is the therapeutic mainstay for management of DM, some owners are reluctant to administer daily injections. Hence, oral anti-diabetic therapies are desirable. Here, we describe a 6-year-old female cockatiel (*Nymphicus hollandicus*), diagnosed with DM of unknown origin based on the history, clinical signs (i.e., polyuria/polydipsia, weakness, weight loss), and persistent hyperglycemia and glucosuria, which was treated successfully with oral glimepiride, a second-generation sulfonylurea, for 2 years until the bird eventually died due to other causes (i.e., suspected pneumonia). Following diagnosis, glimepiride was gradually titrated to effect, whereby a dose of 0.01 mg/kg PO q8-12h resulted in stable blood glucose (BG) concentrations (346-351 mg/dL), minimal glucosuria, and resolution of the clinical signs. Based on dipstick findings and BG-monitoring, glimepiride appeared to reach its peak effectiveness at 4 hours after the initial dose, with a recurrence of clinical signs noted in the successive hours, indicating a seemingly short half-life. Attempts to increase the dose to 0.03 mg/kg PO q12h repeatedly resulted in hypoglycemic events, suggestive of a narrow therapeutic index. This case demonstrates that glimepiride can be considered for long-term treatment of DM in avian patients, although further investigations into its pharmacokinetic properties, efficacy, and safety are warranted. In addition, patients with primary (type I) DM may not respond to this treatment, since sulfonylurea drugs exert their effect by stimulating insulin production and release by pancreatic beta cells.

References

1. Orosz SE, Monks D, de Matos R. Clinical endocrinology of the protein hormones. In: Speer BL, ed. *Current Therapy in Avian Medicine and Surgery.* St. Louis, MO: Elsevier; 2016:378-399.

Thermal Antinociceptive Effects After Oral Administration of a Cannabidiol and Cannabidiolic Acid-Rich Hemp Extract in Orange-Winged Amazon Parrots (*Amazona amazonica*)

David Sanchez-Migallon Guzman, LV, MS, DipECZM (Avian), DipECZM (Small Mammal), Dipl ACZM,
Lexi Durant,
Hugues Beaufrère, Dr med vet, PhD, DABVP (Avian Practice), DipECZM (Avian), Dipl ACZM

Session #122

Affiliation: From School of Veterinary Medicine (Durant), Department of Medicine and Epidemiology, School of Veterinary Medicine (Guzman, Beaufrère), University of California Davis, Davis, CA 95616, USA.

Abstract: Cannabinoids have been studied for their antinociceptive, anti-inflammatory, antipsychotic and antineoplastic effects in mammals,[1-4] making them attractive therapeutic options in birds with chronic pain. Recently, a study in orange-winged Amazon parrots evaluated the pharmacokinetics of a cannabidiol (CBD)/cannabidiolic acid (CBDA)-rich hemp (*Cannabis sativa*) extract (Ellevet, South Portland, ME).[5] The goal of the current study was to determine the thermal antinociceptive effects of the same commercial hemp extract containing CBD, CBDA, Δ -tetrahydrocannabinol 9 (THC), tetrahydrocannabinolic acid, cannabigerol, cannabigerolic acid and cannabichromene (33.5, 32.8, 1.38, 1.09, 0.74. 0.75, 1.46 mg/ml respectively) in the same species. Eight orange-winged Amazon parrots received PO hemp extract (30/29.5 and 15/14.75 mg/kg based on CBD/CBDA concentrations) or control suspension in a masked, within-subjects, complete crossover study. Agitation-sedation scores were assessed prior to each thermal challenge. Foot withdrawal response to increasing thermal stimulus was determined prior to treatment administration and at 0.75, 1.5, 3 and 6 hours post-administration. Statistical analysis was performed with linear mixed modeling on withdrawal temperature delta values and ordinal logit mixed model in agitation-sedation scores. Oral administration of hemp extract induced thermal antinociception based on a dose of 30/29.5 mg/kg CBD/CBDA. Mean withdrawal temperatures increased at the 0.75-hour time point ($P = 0.007$). There was no effect of treatment in the agitation-sedation score ($P > 0.05$). These results support the use of this formulation for pain management in this species. Further studies including higher doses are needed to fully determine the analgesic and adverse effects of CBD/CBDA-rich hemp extract in orange-winged Amazon parrots and other species.

Acknowledgments: This study was supported by the Harrison Avian Research & Conservation Collaborative and the and The Richard M. Schubot Parrot Wellness and Welfare Program, School of Veterinary Medicine, University of California-Davis.

References

1. Go YY, Kim SR, Kim DY, Chae SW, Song JJ. Cannabidiol enhances cytotoxicity of anti-cancer drugs in human head and neck squamous cell carcinoma. *Sci Rep*. 2020;10(1):20622.

2. Hartsel JA, Boyar K, Pham A, Silver RJ, Makriyannis A. Cannabis in veterinary medicine: cannabinoid therapies for animals. In: Gupta RC, Srivastava A, Lall R, eds. *Nutraceuticals in Veterinary Medicine*. Cham, Switzerland: Springer; 2019:121-155.

3. Ryan D, McKemie DS, Kass PH, Puschner B, Knych HK. Pharmacokinetics and effects on arachidonic acid metabolism of low doses of cannabidiol following oral administration to horses. *Drug Test Anal.* 2021;13(7):1305-1317.

4. Rock EM, Limebeer CL, Parker LA. Effect of cannabidiolic acid and Δ9-tetrahydrocannabinol on carrageenan-induced hyperalgesia and edema in a rodent model of inflammatory pain. *Psychopharmacology (Berl).* 2018;235(11):3259-3271.

5. Sosa-Higareda M, Guzman DS, Knych H, et al. Twice-daily oral administration of a cannabidiol and cannabidiolic acid-rich hemp extract was well tolerated in orange-winged Amazon parrots (*Amazona amazonica*) and has a favorable pharmacokinetic profile. *Am J Vet Res.* 2023:1-11.

Section 3

Wildlife Session I

Jen Riley, DVM
Session Chair

Legal Responsibilities of Veterinarians for Wild Birds in the United States

Stephany Lewis, DVM, CWR, DABVP (Avian Practice)

Session #115

Affiliation: From Department of Veterinary Clinical Medicine and Wildlife Medical Clinic, University of Illinois College of Veterinary Medicine, 1008 W. Hazelwood Dr., Urbana, IL 61802, USA.

Abstract: Veterinarians who treat native birds in the United States may include those working in wildlife rehabilitation centers, zoos, government, research, academia, and private practice. Regardless of circumstances, all veterinarians who wish to treat native birds must be aware of the laws and regulations that govern this work. This session discusses the many regulations that apply to veterinary work with releasable and non-releasable native wild birds in the United States.

Introduction

Veterinarians who treat native birds in the United States may include those working in wildlife rehabilitation centers, zoos, government agencies, research, academia, or private practice.[1] Private practitioners may be asked to see a native bird patient by a client or other public member, wildlife rehabilitator, animal control officer, conservation officer, nature center, or zoo. Regardless of the circumstance, it is imperative that all engagement with that patient be done in a legal and ethical manner. Veterinarians must ensure that they are abiding by all state and federal regulations to protect themselves and their businesses from potential legal implications, as well as to ensure that minimum standards for animal welfare are met. These proceedings are meant to guide practitioners and are not legal advice.

Definition and Role of Wildlife Rehabilitation

Native wild birds may be seen in the settings of wildlife rehabilitation or in permanently captive collections. Before specific discussion of wildlife regulations as they apply to native birds, it is important to understand the definition of wildlife rehabilitation, its role, and some of the ethical considerations surrounding it.

The National Wildlife Rehabilitator's Association (NWRA) defines wildlife rehabilitation as "the treatment and temporary care of injured, diseased, and displaced indigenous wildlife and the subsequent return of healthy animals to appropriate habitats in the wild."[2] Wildlife rehabilitation is the practice of providing responsible and professional care for sick, injured, and orphaned wild animals with the goal of releasing fully recovered animals back to their native environment. Most animals presenting to wildlife rehabilitation centers have been negatively impacted by anthropogenic activities[2-4]. Many animals have experienced vehicle strikes, window strikes, or domestic cat attacks. They may have ingested lead, pesticides, or trash. Marine wildlife often ingest or become entangled in fishing lines, hooks, and lures, or contaminated by toxicants such as crude oil. Young animals are often inadvertently "kidnapped" by well-meaning citizens, or their nests impacted by human activity. As the human population continues to grow and urban sprawl extends its reach, the human-wildlife interface is expanding, and with it so has the demand for wildlife treatment facilities. In fact, most wildlife rehabilitation centers are experiencing increased caseloads each year.

Wildlife rehabilitators provide several important services within their communities. Removing wild animals from public possession not only improves animal welfare but can have a positive impact on public health. Without the existence of licensed wildlife treatment centers, many benevolent but untrained citizens would attempt to care for wild animals in their own homes, potentially leading to public injuries, transmission of zoonotic diseases, or transmission of infectious diseases to and from domestic animals. Wildlife centers also have a vital role in educating the community about the importance of conservation and how to peacefully coexist with wildlife.

Wildlife centers can have significant impacts on conservation efforts when working with endangered and threatened species or engaging in large-scale rescue efforts within a localized area, such as during an oil spill response. The prevailing paradigm in the scientific community has considered rehabilitated and released animals as insignificant to conservation efforts; however, some evidence now exists indicating rehabilitated birds of prey may serve as mitigation for at least some anthropogenic mortality and have a measurable impact on population size.[3]

Wildlife rehabilitation centers can also contain a huge wealth of data that can contribute to biological and veterinary research. Wildlife rehabilitators and wildlife veterinarians are often the first to identify infectious disease outbreaks or toxins affecting wild populations and can alert governmental and regulating authorities. They can serve as influential voices for their patients while advocating for laws or public policy changes that affect wildlife and the environment.

Wildlife rehabilitators can be large, professionally run non-profit organizations, or individuals caring for animals out of their own homes. They can also be affiliated with academic institutions. All rehabilitators must have a sponsoring veterinarian on record to oversee the medical care of the animals treated. This can take the form of a full-time paid staff veterinarian, a part-time contract veterinarian, or a volunteer veterinarian working out of their own clinic.

Ethics of wildlife rehabilitation

Wildlife rehabilitation in the United States has the potential to improve the animal welfare of an estimated hundreds of thousands of animals each year. Due to very stringent release criteria and the advanced nature of many of the illnesses and injuries seen in the rehabilitation setting, approximately 70% of animals presenting to rehab centers are not released.[4] The humane euthanasia of animals unsuitable for release arguably makes the greatest impact on individual animal welfare. Without this service, animals with severe conditions would experience prolonged suffering and eventual death in the wild or in the hands of the public.

Release criteria for wildlife and the wildlife rehabilitator's code of ethics can be found in the *Minimum Standards for Wildlife Rehabilitation*, published by the NWRA and International Wildlife Rehabilitation Council (IWRC).[2] The reader is highly encouraged to refer to that text for additional information, along with other crucial husbandry and care standards. The review article by Willette et al., "Interrupted Lives: Welfare Consideration in Wildlife Rehabilitation" also contains critical details and references regarding welfare and ethical considerations in wildlife rehabilitation.[4]

Government Agencies Regulating Native Wild Birds

The U.S. Fish and Wildlife Service (USFWS), housed in the U.S. Department of Interior, is the primary federal agency responsible for wildlife management.[1] Avian practitioners may also interact with the U.S. Department of Agriculture (USDA). Sea turtles, seals, sea lions, and cetaceans are regulated by the National Marine Fisheries Service under the National Oceanic and Atmospheric Administration, while terrestrial mammals, non-migratory birds such as game birds and non-native species, and reptiles outside of sea turtles are regulated by the state wildlife agencies.[1] The reader is referred to other publications for additional details about regulations involving non-avian wildlife species.[1,4]

State regulations pertaining to wildlife are typically available on the state wildlife agency's website. It is imperative that veterinarians treating wildlife familiarize themselves with the regulations of the state in which they practice. Individual states may impose restrictions on what species are allowed to be rehabilitated (e.g., rabies vectors, deer, invasives, big game/dangerous species), and may impose restrictions on release sites (i.e., animals must be released within a certain radius of their origin).[1,4] Some states issue a Memorandum of Understanding to rehabilitators, with a set of standards by which they, and by extension their veterinarians, must abide. Depending on the state, some of these MOUs may include specific guidelines for animal care and requirements for euthanasia.

U.S. Laws Pertaining to Wild Birds

The laws and treaties that pertain to wildlife The United States Code (USC) contains all general and permanent U.S. laws passed by Congress. The Code of Federal Regulations (CFR) is the permanent rules defining implementation of the laws set forth by the federal government agencies. The CFR is organized into 50 titles representing different subject areas, with each title further divided into chapters, parts, and sections. Most of the wildlife regulations are in Title 16 of the USC and Title 50 of the CFR.[5]

Endangered Species Act, 16 USC §1531–1544, Title 50 CFR Part 17

The Endangered Species Act was passed by Congress in 1973 with the aim to conserve threatened and endangered plant and animal species in the United States and preserve ecosystems to support these species.[1] It prohibits the "take" of listed plants or animals, or their parts or products. "Take" is defined as "to harass, harm, pursue, hunt, shoot, wound, kill, trap, capture, or collect, or to attempt to engage in any such conduct."[5] Permitted wildlife rehabilitators and veterinarians can work with and euthanize endangered species when necessary, but they must notify their Migratory Bird Permit Office upon presentation of an endangered species.[5] Ideally, the permit office is notified prior to euthanasia of an endangered species; however, euthanasia can be performed prior to approval if done for welfare concerns.[1,5]

Convention on International Trade in Endangered Species of Wild Fauna and Flora, CITES, Title 50 CFR Part 23

This agreement was originally established in 1975 among 175 nations to prevent the endangerment or extinction of species due to international trade. Listed plant and animal species, live or dead, or their parts or products (such as blood samples), cannot be imported or exported without a permit.[1,5] Species are listed in either Appendices I, II, or III, with Appendix I species being in most danger of extinction. The website cites.org links to a searchable list of species.[6]

Lacey Act, 16 USC 3371–3378, 18 USC §42–43, Title 50 CFR Part 16

Passed in 1900, this is the first U.S. law to protect wildlife and is named after the bill's sponsor.[1] The Lacey Act prohibits the sale, acquisition, and interstate or international transport of any fish, wildlife, or plants, or their parts or products, that are indigenous to the United States or covered by CITES and that were obtained in violation of any treaty or regulation of the U.S. or any Native American tribal law.[1,5] This law also stipulates that live wildlife being transported to the United States be kept under humane conditions.[1,5] A 1981 amendment called the Injurious Wildlife Act prevents the import or interstate travel of species that could harm human beings, or the interests of agriculture, horticulture, forestry, wildlife, or wildlife resources of the United States. This act allows for transport of "injurious wildlife" by permit for zoo collections, or scientific, medicinal, or educational purposes.[1,5]

Wild Bird Conservation Act, Title 50 CFR Part 15

While not relevant to U.S.-native birds, the Wild Bird Conservation Act is an important law that is designed to ensure that all U.S. trade in exotic birds is sustainable. This act was signed into law in 1992 and limits or prohibits the importation of exotic bird species into the United States, assists in wild bird conservation programs in countries native to those birds, and establishes the Exotic Bird Conservation Fund.[1,5]

Bald and Golden Eagle Protection Act, Title 50 CFR Part 22

Although the bald eagle (*Haliaeetus leucocephalus*) did not become the official national bird of the United States until December 2024, it became a national symbol in 1782. The species was on the verge of extinction in most states due to a combination of hunting for sport and as pests, display, as well as habitat loss and pollutants such as DDT.[1] The Bald Eagle Protection Act was passed in 1940 to protect the species by prohibiting the take of eagles and their parts and products. The golden eagle (*Aquila chrysaetos*) was added in 1962.[1] Provisions are made for Native American tribes to have live eagles, eagle parts, and eagle feathers as part of their religious ceremonies.[5] All deceased eagles must be shipped to the National Eagle Repository, who then distribute eagle feathers and parts to Native Americans and Alaskan Natives. Permission from the Migratory Bird Permit Office can be granted to perform necropsies on deceased eagles. Wildlife rehabilitators with a permit to rehabilitate migratory birds and veterinarians must also notify their Migratory Bird Permit Office upon receiving an eagle for care. Similarly, as with endangered species, the permit office should be notified prior to euthanasia of a bald or golden eagle; however, euthanasia can be performed prior to approval if done for welfare concerns.[1,5]

Migratory Bird Treaty Act, 16 USC 703-712, Title 50 CFR Part 21

Written into law in 1918, the Migratory Bird Treaty Act (MBTA) is the most important law protecting native North American birds and is the most crucial for veterinarians of U.S.-native bird species to understand. The MBTA protects all bird species that live, reproduce, or migrate within or across the United States or its territories, making it illegal to "pursue, hunt, take, capture, kill, attempt to take, capture, or kill, possess, offer for sale, sell, offer to barter, barter, offer to purchase, purchase, deliver for shipment, ship, export, import, cause to be shipped, exported, or imported, deliver for transportation, transport or cause to be transported, carry or cause to be carried, or receive for shipment, transportation, carriage, or export, any migratory bird, any part, nest, or egg of any such bird, or any product".[5] At this point, the term "migratory" in the name of the MBTA is a bit of a misnomer, as it now protects over 1,000 species of birds, some of which do not actually migrate.

The MBTA does allow for "take"' of some wild birds in North America. The MBTA allows for regulations regarding hunting of game birds and crows, and does not cover non-native birds, even if they are migratory or are now endemic to the United States, such as Eurasian starlings, house sparrows, or rock doves.[1,5] The MBTA also does not fund conservation or prevent habitat loss.

It is extremely important to note that because some native birds may be legally hunted in the United States, these species need to be considered food animals.[7] In general, if a species has a legal hunting season in the state, it should be treated as if it were a food animal and should not be administered drugs prohibited in food animals (e.g., chloramphenicol, clenbuterol, diesthylstilbestrerol, fluoroquinolones, glycopeptides, nitroimidazoles, nitrofurans).[7,8] See farad.org for additional information or to request extra-label advice.[8] The use of human antiviral drugs is also highly discouraged in wildlife species due to the concern for antiviral resistance in potentially zoonotic pathogens.[9]

History of the Migratory Bird Treaty Act

The MBTA was created largely in response to the decimation of many U.S. bird populations by hunters.[10] "High fashion" in the 1800's and early 1900's included women's hats embellished with ornamental feathers.[10] Overhunting led to the extinction of many native bird species, including the Labrador duck (*Camptorhynchus labradorius*), great auk (*Pinguinus impennis*), heath hen (*Tympanuchus cupido cupido*), Carolina parakeet (*Conuropsis carolinensis*), and of course, the passenger pigeon (*Ectopistes migratorius*).[10] In 1916, the United States signed a treaty with Canada (at the time controlled by Great Britain), and then a similar treaty with Mexico in the 1930's, and Japan and the Soviet Union in the 1970's.[1,10] The 1970's is also when U.S. prosecutors began charging oil, gas, timber, mining, chemical, and electrical companies for incidental violations of the MBTA.[10] The first time a wind farm operator was charged for violations of the MBTA occurred in 2013, when a wind farm operator in Wyoming was charged $1 million.[10] The MBTA also allowed $100 million in penalties against the companies responsible for the Deepwater Horizon oil well blowout that killed approximately 1 million birds.[11]

Unfortunately, deregulation of the MBTA is possible, as we first saw in 2017 when the Trump Administration changed the interpretation of the law to no longer enforce incidental takes.[10,11] These changes protect industries from accountability, for example, as in the event of an oil spill. The Biden Administration reinstated incidental take in October 2021,[11] but that has since been rescinded for the second time by the Trump Administration with Executive Order 14154 "Unleashing American Energy".[12] Some states, however, maintain their protections for native birds and allow them to be enforced.

Falconry regulations, Title 50 CFR Part 21.29: Falconers require federal and state permits to possess native birds for falconry.[5] An in-depth review of requirements for falconers is beyond the scope of this article, but veterinarians should ensure that falconer clients have all the necessary permitting. Falconers can trap wild raptors (generally hatch-year birds) for use in falconry, or they can use captive-bred birds.[5] There are three levels of falconry licenses: Apprentice, General, and Master. Each one has certain requirements for level of experience and restrictions on the number and species of birds they are allowed to keep.[1,5] Apprentices must have a letter of commitment to mentor from a General or Master falconer, and can only possess one raptor at a time, either a red-tailed hawk (*Buteo jamaicensis*) or an American kestrel (*Falco sparverius*).[5] There are many specifications regarding the care and banding that falconers must follow. Wildlife rehabilitators can have a falconer assist in reconditioning a patient without it counting towards the falconer's limit to the number of birds they may keep if it is kept on the rehabilitator's permit.[1,5]

Rehabilitation, Title 50 CFR Part 21.31

To obtain a permit to rehabilitate native birds in the United States, you must have your state permits first, >100 hours of experience for over 1 year, access to appropriate facilities, and a relationship with a veterinarian.[1,5] There are various requirements for housing of native birds. Rehabilitators must maintain their records for at least 5 years, and patients should not be able to see members of the public at any time (i.e., rehabilitators can not put their rehab patients on "display").[5] Permission is needed from the regional Migratory Bird Permit Office if birds need to be held for greater than 180 days.[1,5]

Rehabilitators, as well as veterinarians, are required to notify the Migratory Bird Permit Office within 24 hours if they receive an endangered or threatened species, or an eagle.[1,5] Ideally, permission is obtained prior to euthanasia of these species, but it is allowed to euthanize endangered or threatened species without prior approval if necessary for humane reasons.[1,5] Rehabilitators and veterinarians are allowed to necropsy and collect samples from native birds for research purposes but must obtain additional permits to do so with eagles or threatened/endangered species.[5] Rehabilitators should also notify prior to the release of a threatened or endangered bird to ensure they comply with regulations and state requirements, which may need to be individualized.[1,5]

Other situations in which rehabilitators must notify their Migratory Bird Permit Office include intake of an animal that has been electrocuted, gunshot, intentionally poisoned, or showing evidence of any other criminal activity.[5] Depending on the state's regulations and current interpretation and enforcement of incidental take in the MBTA, companies may be held financially responsible for the injury and death of birds. They may be required to correct their equipment responsible for the injury or death of a bird, fined, and in some cases, they may be required to pay the rehabilitator for the cost of the bird's care. The author has been employed at rehabilitation centers that have received payments from electric companies and solar farms for the care of birds that their equipment injured.

Notification of state or local authorities is also required if wildlife rehabilitators identify or suspect a contagious disease or public health hazard, particularly in case of reportable diseases or if an agency is currently collecting data on particular diseases.[1,5] There is likely significant underreporting of many of these situations, as most wildlife rehabilitators lack the resources for laboratory testing and may have limited access to radiography equipment.

The MBTA states that wildlife rehabilitators must euthanize (and NOT place into a zoo, education program, or sanctuary) any bird that can not self-feed, perch upright, or ambulate without injury that medical or rehabilitative care will not correct.[1,5] Rehabilitators must also euthanize birds that are blind or require amputation of a leg, foot, or wing at the elbow or more proximally.[1,5] No differentiation is made for different taxa. An exception can be made to the above if the rehabilitator has written testament from a licensed veterinarian stating that they do not think the bird will have complications in captivity due to its disabilities and that the veterinarian commits to providing life-long veterinary care, including annual examinations, for that animal.[5] If the animal is being transferred, the veterinarian who will be providing veterinary care must be listed. The rehabilitator must also show that they have appropriate placement for the bird with someone authorized to possess it AND have approval from the USFWS. If any one of these conditions are not met, the bird must be euthanized.[5]

Education, Title 50 CFR Part 21.27

A Special Purpose Possession (Educational Use-Live Specimens) permit must be obtained to possess native birds for educational purposes.[5] Prior written approval from the USFWS is required to obtain, transfer, or dispose of any migratory bird for educational purposes. The birds remain the property of the USFWS.[5] An additional permit is required for the possession of eagles for educational purposes. Housing and care of educational raptors should abide by standards put forth by *Raptors in Captivity: Guidelines of Care and Management* written by Lori Arent.[13]

All education birds must be captive-bred or deemed non-releasable. Permit holders are not allowed to keep birds that they imprinted themselves.[1,5] Education permit holders are required to hold a minimum of 12 public education programs per year.[1,5] These programs must educate on biology, conservation, and/or ecology of the species and cannot imply personal ownership of the birds.[5] The public must not have any physical contact with the birds, and death or escape of an education bird must be reported in 48 hours.[5]

Veterinarians working with rehabilitators and nature centers are likely to be asked to provide a "Veterinary Certification of Condition that Renders Migratory Bird/Eagle as Non-Releasable."[14] This form is required for all rehabilitated birds that are placed in zoos, sanctuaries, nature centers, or education programs. This form must fully describe why the bird is non-releasable and why it is suitable for life in captivity. If the bird is imprinted on humans, the form must explain how and under what circumstances this occurred, and who made that determination.[14] A veterinary examination must occur within 30 days of transfer. This form is still required even if the rehabilitation center is keeping the animal in their own education program. The form and instructions can be found online.[14]

The ethics surrounding permanent placement of non-releasable wild-born birds have been shifting over the past several years. Many groups are advocating against permanent captive placement of birds with chronically painful conditions, permanent neurologic abnormalities, and partial wing amputations, as well as parent-reared owls.[15,16] Careful consideration should be given to the individual animal's temperament and disabilities to make a decision that prioritizes animal welfare. The International Association of Avian Trainers and Educators provides several position statements on their website regarding training, husbandry, and welfare of birds in public education programs, and there are other publications that discuss selection of ambassador birds.[13,15,16]

Scientific Collection and Salvage Permits

Separate USFWS permits are required for individuals engaging in collection of biological samples (e.g., blood, feathers, GI contents, tissues, oral swabs, cloacal swabs, eggs) from migratory birds and eagles for scientific or educational purposes.[5,17] Several exceptions exist regarding when these permits are required. For example, licensed migratory bird rehabilitators are allowed to bank feathers for the purpose of imping, and collect samples for scientific purposes, but must apply for additional permitting to do so on eagles.[5,17] Licensed bird banders may also not require a permit to engage in scientific collection or salvage.[5,17]

USFWS Permits

Additional information on permits under the USFWS can be found online.[17] Permits are submitted and managed on the ePermits platform.[18]

USDA Animal and Plant Health Inspection Service

Prior to 2023, birds were not covered at all under the Animal Welfare Act (AWA), and birds bred for use in research or teaching are still not covered by the AWA.[19] Breeders, brokers, transporters, and exhibitors of birds, may be required to be licensed under the USDA Animal and Plant Health Inspection Service (APHIS).[19] Those with a USFWS Migratory Bird Special Purpose Possession- Education (Live) permit are required to obtain a Class C (exhibitor) license if they have more than four raptors.[4,19] They would need to abide by the AWA Standards for Birds, in addition to any state and USFWS requirements. Wildlife rehabilitators and veterinarians who do not keep permanently captive birds are not required to be licensed with USDA-APHIS. No license is required for birds used for food/fiber, pigeons, falconry birds, pets, or exhibits of domestic birds. Birds and mammals covered under the AWA that are being shipped across state lines or internationally may also need a Health Certificate completed by a USDA Accredited Veterinarian. To view the AWA Standards for Birds and to learn additional information about USDA license and registration requirements, see the USDA-APHIS website and other resources available from the Association of Avian Veterinarians.[19,20]

What Can Veterinarians Do with Wild Birds Without Permits?

There are many ways veterinarians can work with native wild birds in the United States without any required permits. Licensed veterinarians, including private practitioners, can see wildlife brought to their clinic by a permitted rehabilitator, treat wildlife at a permitted rehabilitation facility, or triage wildlife brought to them by the public, animal control, or other agency.[5,21] This encompasses all aspects of veterinary care, including examination, diagnostics, treatment, surgery, and euthanasia.[5,21] For their permits, wildlife rehabilitators require a veterinarian of record, but other veterinarians can still work with animals under that rehabilitator's permit, even if they are not the veterinarian of record.

Depending on the preferences of the veterinarian, rehabilitator, and clinic, veterinarians may help wildlife rehabilitators on volunteer basis (sometimes only charging for the cost of drugs or other non-reusable supplies), or some clinics charge rehabilitators as though they were a typical client. Larger rehabilitation centers may employ a staff veterinarian, or a veterinarian may work with a rehab facility as an independent contractor.

Veterinarians can still triage and treat native wildlife that are brought to them from the public, even without a relationship with a licensed rehabilitator.[5,21] They can provide medical, surgical, and husbandry care, or euthanize wildlife if indicated.[5,21] This includes all native wildlife, even eagles and endangered species, but the Migratory Bird Permit Office must be notified if the latter are seen, and other notification requirements for wildlife rehabilitators under the MBTA must be followed. Veterinarians can find the contact information for the Migratory Bird Permit Office of their region online.[22] Wildlife also must be transferred to a permitted rehabilitator once they are no longer in need of veterinary care, or within 24 hours of stabilization.[1,4,5,21]

Veterinarians also must maintain medical records for wildlife seen in their care, even if they die or are euthanized.[1,5] Those records must include, at minimum, the species, injury, date of acquisition, date of death, and cause of death.[1,5] Veterinarians triaging wildlife should also always obtain the location the animal was found, as well as the name and contact information for those who found that animal. This information is particularly important if the animal must be released back to its origin (as is typically the case), is a healthy orphan and can be reunited with its family or is diagnosed with an infectious disease that is reportable and/or of public health, agricultural, or ecological importance.

If a veterinary clinic triages wildlife from the public, it is advisable to have an established relationship with a licensed rehabilitator. Under the MBTA, a veterinarian cannot perform any procedure (e.g., amputation) that renders the bird non-releasable, unless directed by a rehabilitator.[1,5] Additionally, unless the animal is in care for under 24 hours, wildlife must be transferred to a licensed rehabilitator prior to release.[5,21] Plans for continued care and reconditioning should drafted prior to any potentially painful or stressful medical or surgical treatments in a wild animal. Wildlife rehabilitators are also likely to have a more thorough understanding of an animal's natural history, requirements for release, and husbandry. There is a wealth of information that veterinarians can gain from experienced wildlife rehabilitators, and an integrated approach that incorporates the knowledge of rehabilitators, veterinarians, biologists, and other wildlife professionals to the management of wildlife patients is most likely to provide the best welfare and outcome.

References

1. Duerr R, Whittington JK. Legal responsibilities and restrictions on veterinarians working with wildlife. *Wildl Rehabil Bull.* 2017;35(1):25-37.

2. Miller EA, Schlieps J, eds. *Standards for Wildlife Rehabilitation.* Bloomington, MN: National Wildlife Rehabilitators Assocation; 2021.

3. Hagen CA, Goodell JM, Millsap BA, et al. 'Dead birds flying': can North American rehabilitated raptors released into the wild mitigate anthropogenic mortality? *Wildl Biol.* 2024;2024(5):1-22.

4. Willette M, Rosenhagen N, Buhl G, et al. Interrupted lives: welfare considerations in wildlife rehabilitation. *Animals.* 2023;13(11):1836.

5. Code of Federal Regulations, Title 50. Available at: https://www.ecfr.gov/current/title-50. Accessed March 18, 2025.

6. Convention on International Trade in Endangered Species of Wild Fauna and Flora. Available at: https://cites.org/eng. Accessed March 18, 2025.

7. FARAD. Wildlife. Available at: http://www.usfarad.org/wildlife.html. Accessed March 19, 2025.

8. FARAD Prohibited and Restricted Drugs in Food Animals. Available at: http://www.farad.org/prohibited-and-restricted-drugs.html. Accessed March 19, 2025.

9. National Wildlife Rehabilitators Association. Position statement regarding the use of human antiviral drugs in wildlife patients. Available at: https:// www.nwrawildlife.org/resource/resmgr/position_statements/NWRA_Antiviral_statement_Vet.pdf. Accessed March 19, 2025.

10. Audubon. The history and evolution of the Migratory Bird Treaty Act. Available at: https://www.audubon.org/news/the-history-and-evolution-migratory-bird-treaty-act. Accessed March 19, 2025.

11. All About Birds. Migratory Bird Treaty Act Sees Incidental Take Protections Reinstated. Available at: https://www.audubon.org/news/the-history-and-evolution-migratory-bird-treaty-act. Accessed March 19, 2025.

12. Presidential Actions, Unleashing American Energy. Available at: https://www.whitehouse.gov/presidential-actions/2025/01/unleashing-american-energy/. Accessed March 19, 2025.

13. Arent LR. *Raptors in Captivity: Guidelines for Care and Management.* Blaine, WA: Hancock House Publishers; 2007.

14. USFWS Template Veterinarian Certification of Condition the Renders Migratory Bird/Eagle as Non-Releasable. Available at: https://www.fws.gov/sites/default/files/documents/Template_Vet_Letter.pdf. Accessed March 19, 2025.

15. International Association of Avian Trainers and Educators. Position Statement: Selection Considerations for Non-Releasable Birds. March 2023. Available at: https://iaate.org/wp-content/uploads/2023/03/Position-Statement_Non-releasable-Birds-2023.pdf. Accessed March 19, 2025.

16. International Association of Avian Trainers and Educators Position Statements. Available at: https://iaate.org/position-statements/. Accessed March 19, 2025.

17. USFWS Permits. Available at: https://www.fws.gov/service/permits. Accessed March 19, 2025.

18. USFWS ePermits. Available at: https://permits.fws.gov/fws. Accessed March 19, 2025.

19. USDA-APHIS AWA Standards for Birds. Available at: https://www.aphis.usda.gov/awa/bird-standards. Accessed March 19, 2025.

20. Association of Avian Veterinarians. USDA Resources. Available at: https://www.aav.org/page/usda. Accessed March 19, 2025.

21. USFWS Migratory Bird Permits 2024 updates. Available at: https://www.fws.gov/program/migratory-bird-permits/authorized-activities-no-permit-required?utm_source=nwra&utm_medium=newsletter&utm_campaign=February25. Accessed March 19, 2025.

22. USFWS Regional Migratory Bird Permit Office Contacts. Available at: https://www.fws.gov/program/migratory-bird-permits/contact-us. Accessed March 19, 2025.

Clinical Lead Toxicosis in Raptors

Dana Franzen-Klein, DVM, MS, and
Annette Ahlmann-Garcia, DVM, MS

Session #140

Affiliation: From The Raptor Center, University of Minnesota College of Veterinary Medicine, St. Paul, MN 55108, USA.

Abstract: Lead toxicosis is commonly diagnosed in bald eagles admitted to The Raptor Center. While lead toxicosis can be treated with intensive chelation therapy, in many cases of clinical lead poisoning, treatment is not in the bird's best interests. Lead toxicosis often causes irreparable damage to the central nervous and cardiovascular systems, resulting in either incomplete resolution of clinical signs or a patient that cannot regain the necessary physical fitness for release to the wild. This 1-hour presentation discusses a clinical approach to lead poisoning including diagnosis, treatment options, prognosis, and triage criteria for when to consider humane euthanasia. Video examples of clinically affected bald eagles is provided, and the prognosis associated with different clinical presentations is discussed. Euthanasia recommendations are correlated with advanced imaging and post-mortem findings from previous studies.

Introduction

Lead toxicosis is common in scavenging raptor species in North America, including the bald eagle (*Haliaeetus leucocephalus*), turkey vulture (*Cathartes aura*), black vulture (*Coragyps atratus*), and California condor (*Gymnogyps californianus*).[1] Lead toxicosis is a relatively common reason for admission of these species to wildlife rehabilitation facilities. These birds can present with a range of clinical signs and carry a variable prognosis. This presentation discusses diagnosis, treatment, prognosis, and triage criteria of raptors with lead toxicosis.

While lead toxicosis may impact any raptor, the most common species admitted to The Raptor Center (TRC) at the University of Minnesota with clinical lead toxicosis is the bald eagle. The majority of bald eagles suffering from clinical lead toxicosis as their primary reason for admission are admitted from fall through early spring. The fall time frame is associated with hunting seasons in the upper Midwest, as ingestion of spent lead ammunition from gut piles left in the field by hunters is the most common route of exposure.[2] In our clinical practice we often see a small increase in cases during the spring thaw, which we presume to be due to gut piles that were previously covered with snow becoming accessible to scavengers again. TRC does admit bald eagles with detectable levels of lead in their blood year-round.

Clinical Presentation and Diagnosis of Lead Toxicosis

Clinical signs

While lead has numerous deleterious effects on the avian body, the central nervous system and cardiovascular system are often severely affected by this toxin.[3] Bald eagles that are suffering from lead poisoning may be displaying a variety of clinical signs, including but not limited to: weakness, quiet or dull mentation, head tremors, nystagmus, visual impairment, abnormal respiratory pattern (possibly with vocalization), green urates and fecal output, anemia, thin body condition, gastrointestinal stasis, regurgitation, inappetence, seizures, and death.[3-5]

Differential diagnoses

There are many potential causes of central nervous system signs in bald eagles. The causes we see most commonly in wild raptor patients at TRC include lead toxicosis, highly pathogenic avian influenza (HPAI), West Nile virus, head trauma, or other toxins such as pentobarbital ingestion.

Diagnostics

Below is a summary of the recommended diagnostics when lead poisoning is suspected, as well as the expected results when lead poisoning is the primary cause of clinical signs.

Complete blood count

Anemia, hemolysis, and abnormal red blood cell structure may be seen.

Blood lead levels

This analysis requires whole blood, not plasma or serum. The LeadCare II™ machine (Meridian Bioscience), although intended for use in humans, can be used to measure blood lead levels in wildlife species, and provides results in under 5 minutes. Staff at TRC run blood lead levels on all bald eagles and turkey vultures (*Cathartes aura*) at the time of admission, because these are our most affected species. Blood lead levels are run on any other bird that shows clinical signs consistent with lead toxicosis or has a metal opacity in the gastrointestinal tract as seen on radiographs.

Radiographs

Most lead toxicosis cases no longer have the ingested lead in the gastrointestinal tract at the time of admission,[3] but if metal opacities are present, further intervention such as gastric lavage may be required pending patient status.

Prognosis and Treatment

Prognosis

In the experience of the clinicians at TRC, the severity of clinical neurologic signs at the time of presentation is the most reliable indicator of prognosis associated with lead poisoning in bald eagles, and the primary determining factor when making treatment versus euthanasia decisions. If the patient presents with severe neurologic signs, that generally indicates irreversible, severe brain damage and humane euthanasia should be elected.[5] Severe neurologic signs include ataxia, muscle or head tremors, a characteristic abnormal respiratory pattern caused by damage to the brainstem (shown via video in the lecture), nystagmus, central blindness, significantly altered mentation, or seizures. Blood lead levels can also help assess prognosis, though they are not always definitive. Birds with blood lead levels of > 120 µg/dl (1.2 ppm) at the time of admission often have a poor prognosis for recovery to release.[4] If using a LeadCare II™ analyzer (an in-house option used by many wildlife facilities), the analyzer will report "HIGH" for any sample with a blood lead level greater than 65 µg/dl, which is outside the machine's reportable range of 3.3 to 65 µg/dl (0.033 to 0.65 ppm). The clinical signs are always the primary determination of prognosis, and when the blood lead level reads as "HIGH", that confirms that lead poisoning is the most likely cause of the neurologic signs. Whole blood samples may be serially diluted and re-run to try to determine the exact blood lead level or may be sent to a reference laboratory for analysis.

Treatment

Based on the experiences of the clinicians at TRC, we have developed the following criteria and guidelines for treating bald eagles that are intoxicated by lead. If treatment is elected, blood lead levels should be checked 2 days after discontinuing chelation medications. If at any point the patient's neurologic status declines during treatment, euthanasia should be considered. If the blood lead levels are still elevated after the first round of chelation therapy, treatment is repeated according to the guidelines outlined below. Lead chelation medications can be harsh on the body and do have significant side effects, including stomach upset and the potential for kidney damage. The treatment is quite stressful for the patient and costly to perform. Clinical signs, stress level, welfare, and prognosis for recovery should be assessed daily while treating the lead poisoning patient.

Blood lead levels < 10 µg/dl

- No chelation treatment is indicated.

Blood lead levels 10-20 µg/dl

- Chelation with oral DMSA (dimercaptosuccinic acid) 25 mg/kg twice per day for 5 days (10 doses)[6] is recommended.

- Provide other supportive care as needed based on clinical signs and concurrent medical conditions.

Blood lead levels > 20 µg/dl with no or mild neurologic symptoms

- Chelation with both calcium EDTA (edetate calcium disodium) and DMSA (dimercaptosuccinic acid) is recommended.
  - Oral DMSA 25 mg/kg twice per day for 5 days (10 doses).[6]
  - Injectable calcium EDTA 50 mg/kg (diluted in sterile 0.9% saline) twice per day for 5 days (10 doses).[6]
    - Access to this medication for veterinary use has become challenging due to changes in regulations for its use in human medicine.
    - This medication can be given via subcutaneous (SC) or intravenous (IV) routes.
    - Intramuscular (IM) administration should be avoided due to pain and risk for muscle irritation.
    - In the experiences of the clinicians at TRC, SC administration is just as effective as IM, with fewer injection site complications.
      - Administer additional subcutaneous fluids for diuresis at a dose of ~90-120 ml per bald eagle twice daily pending hydration status, urine and urate output, weight, and fluid retention.
      - Provide other supportive care as needed based on clinical signs and concurrent medical conditions.

Blood lead levels > 20 µg/dl with severe neurologic symptoms

- It is uncommon to see severe neurologic signs secondary to lead toxicosis unless the blood lead level is > 65 µg/dl ("HIGH" on a LeadCare II™ analyzer), but this can vary by individual.

- Humane euthanasia is recommended due to irreversible neurologic damage.
  - A study evaluating MRI scans and histopathology has shown that severely clinically affected bald eagles with blood lead levels > 100 µg/dl (1.0 ppm) had severe neurologic damage secondary to lead toxicosis.[5]

- If metallic opacities (suspected lead) are present in the gastrointestinal tract (i.e., esophagus, crop, proventriculus, ventriculus) on radiographs, removal is recommended. Metal opacities that have already passed into the lower gastrointestinal tract (i.e., intestines) cannot be removed using either of the methods listed below:
  - Casting material
    - Force-feed rat skins or other casting material to encourage bird to cast metallic foreign bodies.
    - This is the method of choice for birds that are not systemically stable, or staffing is not available, to perform general anesthesia and gastric lavage.
    - Lead toxicosis does cause ileus, therefore this method is not always successful.
    - Consider administering metoclopramide (Metoclopramide injection, USP 5 mg/ml, Hospira, Inc; Metoclopramide tablets, Teva USA) at a dose of 1 mg/kg[6] SC or PO twice daily and ensure the patient is well hydrated to promote GI motility.
  - Gastric lavage
    - Lavage is not recommended if the patient is too unstable for general anesthesia or adequate staff are not available to perform the procedure.
    - Induce general anesthesia and intubate with a snugly fitting, uncuffed endotracheal tube.
    - Secure the patient to a table and tilt the table so the head end of the body is pointed down towards the floor at approximately a 45-degree angle.
    - Pass a 6- to 10-mm diameter stomach tube (18-French red rubber catheter, Kendall™ Covidien™; medium canine stomach tube, 3/8 inch outer diameter, JorVet™) into the ventriculus using lubrication and manually directing the tube into the ventriculus; the tube will often ball up in the crop and can be manually directed externally into the thoracic inlet.
    - Flush large volumes of warm water (often 3-5 L or more) using either a veterinary stomach pump and a bucket of warm water (e.g., JorVet™ stomach pump or equine stomach pump by AAProTools) or multiple 60-ml catheter tip syringes into the ventriculus.

- If using catheter tip syringes, have one staff member re-filling syringes in a rotating fashion so that a full syringe is readily available for the person performing the lavage.

- The goal is to create a steady flow of water into the stomach and back out of the mouth.

- Have a staff member keep a hand on the ventral coelomic wall to palpate the ventriculus and ensure it is not becoming overly distended without fluid passing back out through the mouth.

- If fluid is not passing back out of the mouth, try passing a smaller stomach tube to prevent the tube from blocking the flow of water back out of the mouth.
  - Allow fluid and stomach contents to passively flow from the stomach out the mouth during the lavage procedure.
    - Collect the fluid and material from the lavage into a bucket covered with a towel; this allows for manual inspection of the effluent.
    - Take repeat radiographs after the procedure to ensure that all metal objects have been successfully removed.

- Include the head and neck in the radiographs. Metallic fragments can become lodged in the mouth, allowing the bird to re-swallow them after the procedure if they are not manually removed.

References

1. Katzner TE, et al. Lead poisoning of raptors: state of the science and cross-discipline mitigation options for a global problem. *Biol Rev.* 2024;99(5):1672-1699.

2. Cruz-Martinez L, Redig PT, Deen J. Lead from spent ammunition: a source of exposure and poisoning in bald eagles. *Hum Wildl Interact.* 2012;6(1):94-104.

3. Manning LK, Wünschmann A, Armién AG, et al. Lead intoxication in free-ranging bald eagles (*Haliaeetus leucocephalus*). *Vet Pathol.* 2019;56(2):289-299.

4. Redig PT, Arent LR. Raptor Toxicology. *Vet Clin North Am Exot Anim Pract.* 2008; 11:261-282.

5. de Francisco ON, Feeney D, Armién AG, et al. Correlation of brain magnetic resonance imaging of spontaneously lead poisoned bald eagles (*Haliaeetus leucocephalus*) with histological lesions: a pilot study. *Res Vet Sci.* 2016;105:236-242.

6. Carpenter JW, Marion C. *Exotic Animal Formulary.* 5th Ed. St. Louis, MO: Elsevier Saunders; 2017.

Goothanasia: Evaluating the Use of Cloacal Transmucosal Pentobarbital Administration for Euthanasia of Canada Geese

Crystal Matt, DVM

Session #171

Affiliation: From the Avian and Exotic Animal Clinic of Indianapolis, Indianapolis, IN 46268, USA.

Abstract: Due to their frequent presence in human-populated areas, Canada geese are commonly presented to wildlife rehabilitation centers and veterinary hospitals with injuries sustained through human interactions (e.g., vehicle collisions, fishing line entanglement, lead toxicity).[1-3] In recent months, a rise in highly pathogenic avian influenza cases has further increased their admission rates.[4,5] These medical conditions often necessitate euthanasia, which can be challenging due to the birds' large size and strength. Intravenous (IV) administration of pentobarbital is the most widely used euthanasia method, but it requires skilled personnel, precise restraint, and often results in perimortem behaviors—such as flapping and vocalization—that can be distressing to observers if additional sedation is not provided beforehand. This study evaluates intracloacal administration of pentobarbital as an alternative to IV administration in Canada geese. Geese admitted through wildlife rehabilitation and deemed in need of euthanasia for humane reasons were assigned to receive pentobarbital either intravenously or per cloaca (5 ml via teat cannula) for transmucosal absorption. Variables recorded included ease of administration, observed stress during handling, time to cardiopulmonary arrest, physical responses (e.g., flapping, vocalizations), and whether a rescue dose was required. Although data collection is ongoing, preliminary results suggest the intracloacal method is associated with reduced movement, vocalization, and handling stress. The primary limitation observed to date is a longer time to cardiopulmonary arrest. Final statistical comparisons of euthanasia characteristics are presented.

References

1. Miller TK, Pierce K, Clark Jr EE, et al. Wildlife rehabilitation records reveal impacts of anthropogenic activities on wildlife health. *Biol Conserv.* 2023;286:110295.

2. Dunne R, Miller E. Post-release survival of oiled, rehabilitated waterfowl. *Proc Ninth Int Effects Oil Wildl.* 2007;27-37.

3. Titchenell MA, Lynch Jr WE. Coping with Canada geese: conflict management and damage prevention strategies. Ohioline. The Ohio State University Extension. 2010. Available at: https://ohioline.osu.edu/factsheet/W-3. Accessed June 22, 2025.

4. Avery-Gomm S, Barychka T, English M, et al. Wild bird mass mortalities in eastern Canada associated with the highly pathogenic avian influenza A (H5N1) virus, 2022. *Ecosphere.* 2024;15:e4980.

5. Giacinti JA, Signore AV, Jones ME, et al. Avian influenza viruses in wild birds in Canada following incursions of highly pathogenic H5N1 virus from Eurasia in 2021-2022. *Mbio.* 2024;15:e03203-03223.

Long-Term Medical Care of the Raptor Ambassador

Dana Franzen-Klein, DVM, MS

Session #141

Affiliation: From The Raptor Center at The University of Minnesota College of Veterinary Medicine, St. Paul, MN 55108, USA.

Abstract: Raptors are long-lived, charismatic birds that can relay powerful environmental education messages to the public. The majority of raptors become educational ambassadors due to an injury or disability that prohibits their release and requires periodic medical assessment. Routine medical care helps educational ambassador raptors maintain a long, healthy, and comfortable life in captivity. This lecture will discuss recommended preventive medical care plans for captive raptors, will review common medical conditions in captive raptors and how these conditions may present, and will discuss quality of life assessment tools that facilitate end of life decisions.

Preventive Medical Care Plan

Native North American raptors frequently enter education programs due to a permanent medical injury that prevents them from being released to the wild. These permanent injuries require regular medical care to ensure the health and welfare of these animals in captivity. The Raptor Center (TRC) recommends annual veterinary examinations for all education raptors regardless of their historical injury and reason for placement. These examinations allow the veterinarian to obtain baseline normal information for the individual animal, as well as closely monitor the birds' existing medical problems. Annual examinations also present an opportunity to catch any subtle changes or new problems early on, as raptors are so adept at hiding clinical signs associated with pain or illness.

TRC's recommended preventive medicine plan includes obtaining a baseline data set for each new bird that enters a collection. This includes a physical examination, fecal, complete blood count (CBC), blood chemistry panel (chem) and whole-body radiographs. This allows the veterinarian to establish baseline normals for that individual. There are limited published reference intervals for raptor species; therefore, obtaining baseline blood work while the bird is clinically healthy gives the veterinarian normal values to compare to in the future. Additionally, the baseline examination and diagnostics will allow the veterinarian to more accurately monitor existing medical conditions throughout the bird's life. In the experience of the veterinarians at TRC, having baseline normal diagnostics for individual raptors has proven highly beneficial on multiple occasions, and has allowed for more rapid and accurate diagnoses when medical concerns arise.

The other component of the preventive medical care plan is an annual wellness examination. TRC recommends a thorough physical examination, fecal examination to check for gastrointestinal parasites, and a complete blood count be performed on an annual basis. If baseline results were normal, we recommend performing recheck radiographs and a blood chemistry panel every 3-5 years depending on the species, age and health status of the individual bird. Recheck radiographs and chemistry panels may be performed more frequently if the bird is on chronic medications, any previous abnormalities were noted, or there is a change in behavior or health status.

West Nile Virus Vaccination

The Raptor Center recommends vaccinating captive raptors against West Nile virus (WNV). We are currently using the West Nile-Innovator equine vaccine from Zoetis. It is important to note that the product contains only the killed West Nile virus. Do not use products that contain vaccines for other viruses (e.g., 3- or 4-way vaccines).

Vaccine administration protocol

For new birds (not previously vaccinated). Single dose (1.0 ml) intramuscularly (IM) or subcutaneously (SC); repeat every 3 weeks for a total of 3 doses.

For established birds with a previous vaccination history (regardless of type of West Nile vaccine). Single dose (1.0 ml) IM or SC annually.

For single-packaged doses use a 22- to 25-gauge needle instead of the 20-gauge needle that comes inside the package.

Pre-warm the loaded syringe before administering the product by gently holding it in your hand until it reaches room temperature.

Ideally, birds should be fully vaccinated prior to the beginning of mosquito season in the region. *Culex* species are the main species of mosquito that spread WNV, and these species peak at different times of the year in different regions. Information regarding when different mosquito species peak in different regions is commonly available online from state public health departments. If the mosquito season in your area begins early in the spring and extends late into the fall, you should consider a booster vaccination in late summer. For Minnesota, we aim to have the initial vaccine series complete, or administer the annual booster vaccine, between April and mid-May.

The efficacy of WNV vaccination has not been studied in all raptor species. Vaccines for WNV may not completely prevent the disease, but there is some scientific evidence in a few representative species that shows vaccination may decrease the severity of the disease.

Allergic reactions to the vaccine are possible and are more likely to occur after the first dose. The veterinarians at TRC have not seen an allergic reaction to a WNV vaccination in our practice to date.

IM versus SC vaccine administration

The studies that looked at the efficacy of WNV vaccine administration in preventing clinical disease in raptors gave the killed vaccine IM.[1-3] The efficacy of the vaccine when given SC has not been evaluated in an experimental model.

Vaccines in dogs and cats are given SC instead of IM, as it is more comfortable for the animal and the immune system still finds the vaccine and has an appropriate response to it. Horses and humans do not have a good subcutaneous space, so vaccines are given IM. Birds do have a good SC space, so based on the information available in other species, giving the vaccine SC in birds should still be efficacious while being more comfortable for the bird, as the immune system will still see the vaccine because it has been administered into the body and mount an appropriate immune response.

In one of the studies that looked at WNV vaccine in falcons, when muscle tissue was evaluated after the birds were euthanized, they found severe inflammation and muscle necrosis in response to the vaccine.[1] Based on anecdotal evidence collected via Dr. Dana Franzen-Klein's clinical experience at TRC, it seems that IM administration of WNV vaccines may be associated with focal pectoral muscle wasting (also called thin keel syndrome) in small to medium sized raptors.

TRC has decided to give the vaccine subcutaneously since 2019 and have not had any side effects nor clinical WNV infections since we transitioned to this method.

Ethograms and Behavior Logs

Ethograms or behavior logs can be used as an objective way to monitor and assess the quality of life of captive education raptors.[4] Baseline ethograms are helpful to monitor long-term quality of life in geriatric birds, and when performed on a regular basis, they can help detect subtle signs of pain that may be related to arthritis or other age-related conditions. When these abnormalities are noted, birds can be started on appropriate pain medications via consultation with the veterinarian. Ethograms can be used to monitor the efficacy of oral analgesics and titrate the medications to the most effective dose for that species and individual bird. Ethograms can also be helpful when birds reach a point in their life when end-of-life decisions need to be made surrounding deteriorating quality of life.

Below is a stepwise process that the veterinarians at TRC have found beneficial for performing systematic behavior logs to assess the efficacy of oral pain medications in raptor ambassadors.

- First perform baseline ethograms once daily for ~10 days to determine the bird's normal behavior pattern. If the bird is hesitant or nervous with people in front of the enclosure, you may need to perform monitoring sessions remotely either with binoculars or using a video camera outside the enclosure.

- Behaviors to monitor vary depending on the individual and the underlying medical condition. Options include frequency and severity of wing droop, which perch the bird is using, how often the bird moves around, does the bird fly to move around the enclosure or does it use a ladder system to reach high perches, leaning to one side and consistently bearing more weight on one leg or foot than the other, comfort behaviors (e.g., foot tucking, preening, bathing), willingness to engage with trainer, and willingness to be viewed by the public, among others.

- Once the baseline ethograms are established, perform daily ethograms after the bird is started on pain medications for 10-14 days.

- Compare the ethograms before and after the additional pain medications to determine if there are any changes in behavior. Discuss the results with the veterinarian to determine if the owner or caretaker would like to try additional or increased pain medications, or decrease or stop pain medications.

- Repeat ethograms after medication changes are made to monitor the response, including discontinuing a medication. In some instances, stopping pain medication is more revealing than the changes that occurred after starting it.

- Once baseline ethograms are established and the bird is doing well, monitoring frequency can be decreased.

References

1. Angenvoort J, Fischer D, Fast C, et al. Limited efficacy of West Nile virus vaccines in large falcons (*Falco* spp.). *Vet Res*. 2014;45:1-12.

2. Fischer D, Angenvoort J, Ziegler U, et al. DNA vaccines encoding the envelope protein of West Nile virus lineages 1 or 2 administered intramuscularly, via electroporation and with recombinant virus protein induce partial protection in large falcons (*Falco* spp.). *Vet Res*. 2015;46. doi: 10.1186/s13567-015-0220-1 .

3. Redig PT, Tully TN, Ritchie BW, et al. Effect of West Nile virus DNA-plasmid vaccination on response to live virus challenge in red-tailed hawks (*Buteo jamaicensis*). *Am J Vet Res*. 2011;72:1065-1070.

4. Rose P, O'Brien M. Welfare assessment for captive Anseriformes: a guide for practitioners and animal keepers. *Animals*. 2020;10:1132. doi: 10.3390/ani10071132.

Reflecting on the Pathology of Bird-Window Collisions

Jason D. Struthers, DVM, MVetSc, DACVP,
Alanna P Tonelli-Raylove, DVM,
Morgan Hirsh, DVM,
Melissa Kasai, DVM,
Christopher Olson, PhD

Session #199

Affiliation: From Department of Pathology, College of Veterinary Medicine (Struthers, Tonelli-Raylove, Hirsh, Kasai), College of Graduate Studies (Olson), Midwestern University, 19555 N 59th Ave, Glendale, AZ 85308, USA.

Abstract: Building collisions kill 1.28 to 5.19 billion birds per year in the United States.[1] Some collision victims are admitted to wildlife rehabilitation centers with sublethal or permanent injury. Due to limited knowledge of window-caused pathology that may benefit clinical treatment and inform better building design, we surveyed the blunt force trauma of building window collisions in birds collected from downtown Phoenix and our university campus in Arizona. Between 2020-25, building window-killed birds (n = 323) were collected and were subjected to postmortem examination. These included 39 species, 40.2% were white-winged doves, 11.8% mourning doves, and the remainder mostly songbirds. Specimens were confirmed as building window-strike death based on their proximity to a window, some leaving visible imprints. Birds frequently died of hemocoelom (35.6%) often secondary to liver fractures (25.1%). Cardiovascular pathology (29.7%) also precipitated death and it implicated myocardial hemorrhage, hemopericardium, chamber rupture, and in 10.5% of cases it was due to fractured rib, coracoid, or sternum. Fatal lower respiratory pathology (17.5%) involved lung contusions and tracheal luminal clots. Cranial injury was the cause of death in 15.2% of collisions. There were no incidents of cervical vertebral fractures. Postmortem examinations often documented multiple concomitant lesions. In our analysis death was attributed to the most significant pathology, but simultaneous non-fatal lesions attributable to collision were common. Live birds admitted to wildlife rehabilitation centers that may have suffered a window collision should be evaluated for such lesions, including feather loss of the cranioventral neck, crop ruptures, pectoral girdle and sternum or keel fractures, soft tissue hemorrhage, and coelomic wall intestinal herniation.

References

1. Klem Jr D, Saenger PG, Brogel BP. Evidence, consequences, and angle of strike of bird–window collisions. *Wilson J Ornithol.* 2024;136(1):113-119.

Section 4

Cardiology and Coelomic Cavity Session

Nicole Sands, DVM
Session Chair

Update on Avian Atherosclerosis

Hugues Beaufrère, DVM, PhD, Dipl ACZM, DABVP (Avian Practice), DipECZM (Avian)

Session #120

Affiliation: From Department of Medicine and Epidemiology, School of Veterinary Medicine, University of California, Davis, CA 95616, USA.

Abstract: Avian atherosclerotic lesions consist of the progressive accumulation of inflammatory cells, cholesterol, fat, and cellular debris in the intima and luminal side of the media. They are classified into seven lesion types. What we know about the pathogenesis of atherosclerosis is mainly derived from human research and is consistent with the progression of lesions in birds. A comprehensive review of comparative atherosclerosis pathogenesis with a focus on birds is available and a summary is provided here, including advances in blood lipid testing, vascular imaging, and pharmacotherapy. Recent research in clinical lipidology diagnostics and therapeutics allows for the derivation of some basic recommendations for avian practitioners.

Background Information

Avian atherosclerotic lesions consist of the progressive accumulation of inflammatory cells, cholesterol, fat, and cellular debris in the intima and luminal side of the media and are classified into seven lesion types. Early lesions (types I-III) are mainly composed of subendothelial aggregates of foam cells and mild disruption of the arterial wall. In advanced lesions, responsible for clinical signs, there is formation of a lipid core (atheroma; type IV lesion) covered by a fibrous cap (fibroatheroma; type V). Complications such as fissures, hematomas, and thrombosis (type VI) may occur. Type VII lesions are calcific lesions with minimal fat.

What we know about the pathogenesis of atherosclerosis is mainly derived from human research,[1] but is largely consistent with the progression of lesions in birds.[2,3] A comprehensive review of comparative atherosclerosis pathogenesis with a focus on birds is available and only a brief summary is given here.[4] Lesions are initially caused by the migration of macrophages to the subendothelial space, which start to uptake fat and become foam cells. These are triggered by some levels of endothelial dysfunction and activation by proinflammatory mediators or stimuli and low-density-lipoproteins entrapment and aggregation within the arteries. As such, atherosclerotic lesions are inflammatory lesions. Eventually, with continuous exposure to risk factors, the lesion progresses to large amounts of extracellular fat, necrosis, and calcification. Lesions will also form fibrous tissue, especially over the necrotic core of an atheromatous lesion to encapsulate it, this is the fibrous cap. The proliferative nature of the lesions leads to flow-limiting stenosis of the lumen. In humans, unstable lesions may rupture this fibrous cap and expose the content of the necrotic core to the blood stream, which may trigger thrombosis and subsequent thromboembolic events. There is evidence to suggest that thromboembolic events are less common in spontaneous diseases of birds, probably due to different properties of avian thrombocytes when compared to mammalian platelets.[5] This progression of lesions has been confirmed in parrots in studies looking at the different lesion types and experimental induction of lesions through cholesterol feeding.[2,3]

In parrots, atherosclerotic lesions are central and commonly found in the great arteries at the base of the heart, brachiocephalic arteries, ascending aorta, and pulmonary arteries.[3,6] Lesions in the abdominal aorta, coronary arteries, carotid arteries and many peripheral arteries also occur.[7-10] Lesions also seem more likely to develop at areas of arterial bifurcation, likely due to mechanical factors.[8] Complications and clinical signs are usually due to either severe stenosis from the continuously growing fibroatheromatous plaque or thrombosis and haemorrhage caused by plaque disruption that can decrease or interrupt blood flow or provoke emboli. Atherosclerotic lesions are silent and asymptomatic until such complications arise. Stenosis secondary to atherosclerotic lesions is common in birds but atherothrombosis and emboli are rare and, in a post-mortem study, were found in only 1.9% of clinically relevant atherosclerotic lesions.[6] Clinical signs are only related to advanced lesion types (type IV-VI).

The occurrence of histologic lesions in other organ systems in association with atherosclerosis has been reported.[6,11] A study identified an association between atherosclerosis of the ascending aorta and organ alterations such as myocardial hypertrophy, myocardial fibrosis, pulmonary congestion and fibrosis.[11]

Another report documented a significant association between advanced atherosclerosis and reproductive diseases and myocardial fibrosis.[6] The association with myocardial lesions found in two studies suggest that atherosclerotic lesions may induce cardiac pathologic changes that could lead to myocardial dysfunction, cardiac arrhythmia, and further degenerative changes. In addition, other ischemic disorders have been reported with atherosclerosis such as intermittent claudication and strokes.[7,12,13] Xanthomatous masses and systemic xanthogranulomatosis are also frequently associated with atherosclerotic lesions.[14,15]

Associations between atherosclerosis and other disorders such as hepatic lipidosis and adipocytic tumors have not been found despite studies looking specifically at these associations.[9,14,16]

The raw prevalence of advanced atherosclerosis in parrots is around 7% but depends on many factors.[6] Females of most species are more susceptible to atherosclerosis than males,[6] except in Quaker parrots where the opposite was found.[16] It is interesting to note that psittacine birds exhibit a reversed sex effect from mammals in which males experience higher prevalence of atherosclerosis. In female birds, estrogens have tremendous physiologic effects on lipid, protein, and calcium metabolism related to reproduction and egg formation. In reproductively active females, estrogens induce an increase in plasma total calcium, proteins, cholesterol and triglycerides and hepatic synthesis of two specific lipoproteins, vitellogenin and VLDLy (yolk labeled), which target the developing oocyte and are protected against the normal action of plasma lipoprotein lipase.[17] Increased plasma cholesterol, VLDL, VLDL-remnants, and non-HDL cholesterol promote atherogenesis, providing a plausible explanation for the enhanced predisposition found in female psittacine birds.[6] Reproductive diseases were also found to be significantly associated with increased prevalence of atherosclerotic lesions and may be a target for prevention.[6] Increasing age is also a strong risk factor for advanced atherosclerotic lesions and geriatric birds of some species may have prevalence of up to 30-50%. Grey parrots (*Psittacus erithacus*), Amazon parrots (*Amazona* spp.), and cockatiels (*Nymphicus hollandicus*) appear to be atherosclerosis-prone, while cockatoos (*Cacatua* spp.) and macaws (*Ara* spp.) species appear to be somewhat atherosclerosis-resistant.[6] In a large retrospective study on plasma samples, genus-specific differences in plasma total cholesterol values were found to mirror these identified trends in the prevalence of clinically important atherosclerotic lesions.[18] This suggests that the differences observed in prevalence among psittacine genera may partly be explained by differences in plasma total cholesterol levels.

Within birds of prey, atherosclerosis is seen most commonly in falcons and kites.[9,10,19,20] The practice of feeding day-old chicks to susceptible raptor species (falcons, kites) or insectivorous or piscivorous raptor species may be a risk factor for atherosclerosis.[9,20] The disease seems less prevalent in owls, particularly in barn owls where lesions are seldom found. As in parrots, females tend to have a higher prevalence than males.[10]

Atherosclerosis is also found routinely in many other species, particularly in wild turkeys, hornbills, flamingos, ibises, and species previously used as models such as the pigeon.

Most of the knowledge on avian atherosclerosis pertains to pathology and epidemiology as the disease is difficult to study ante-mortem. For this reason, clinical management of White Carneau atherosclerotic diseases is still extremely challenging because of significant knowledge gaps in diagnostic methods and treatment options. We will focus on notable updates in this knowledge over the past 5 years in blood lipid testing, vascular imaging, and lipid-lowering therapeutics.

Recent Advances in Blood Lipid Testing

Dyslipidemic changes with notably hypercholesterolemia and high LDL are thought to predispose to the development of atherosclerotic lesions as in other animal species. While robust confirmatory evidence linking specific types of dyslipidemia to atherosclerosis in birds is lacking, cholesterol still plays a central role in lesion initiation and progression. To better study the associations between dyslipidemic risk factors and atherosclerosis, having reliable blood analytical tools is essential. For this reason, several studies have been published to further our understanding of lipid diagnostic tests in birds and explore new lipidologic assays in parrots.

While total cholesterol and total triglycerides assays are reliable using standard reference analyzers in birds, this is not the case for lipoprotein diagnostics. Avian lipoproteins have many differences from mammals and lipoprotein diagnostic test designed for humans cannot always be used on birds.[21] For these reasons, several studies have focused on lipoprotein diagnostics using different tools and attempting to characterize the different fractions and subfractions in birds.[22,23]

The gold standard for lipoprotein analysis is separation by density gradient ultracentrifugation, wherein lipoproteins are separated along a density gradient in a vertical tube. This technique is extremely time consuming, requiring 6-30 hours of very high-speed centrifugation and specialized equipment. It is not offered commercially for animals, but has been used for research in pigeons, chickens, quails, geese, Quaker parrots, and cockatiels.[17,24-27]

Recently, a high-resolution technique using gel-permeation HPLC (Liposearch®) has provided full lipoprotein profiling of parrots with a panel of 20 subclasses and particle sizes reported from just one drop of plasma.[22] In healthy Quaker parrots, it was shown that most VLDLs are of the small types, LDLs are large to medium in size with 37% being small, dense LDL, and HDL are mostly of the medium and small categories.

Another technique that has been used by the author to separate lipoprotein particles and subfractions is high-definition polyacrylamide gel electrophoresis (Lipoprint®). While subfractions can be difficult to decipher using this technique, VLDL, intermediate-dense lipoproteins (IDL or VLDL remnants), LDL, and HDL can be easily separated and analyzed.

The Liposearch and Lipoprint techniques are very reliable to characterize dyslipidemia and changes in lipoprotein distribution patterns. Both techniques can also provide lipoprotein particle size. Reference values were provided in Quaker parrots[22] and cockatiels (in preparation) using these techniques.

Although these reference techniques are applicable to any animal species, they are mainly used in research and not offered by veterinary clinical reference laboratories, which prefer rapid and more automated tests. Direct LDL and HDL wet-chemistry tests were developed for human plasma and typically only measure cholesterol content. These tests focus on HDL and use selective precipitation or complexation methods. As direct LDL measurement is challenging, indirect formulas have been proposed (such as the Friedewald formula). Unfortunately, direct LDL measurement using standard biochemistry analyzers and indirect LDL calculations using various formula have been shown to be unreliable in parrots and likely in other birds;[22,23] however, HDL can be measured reliably by standard analyzers and atherogenic lipoproteins can be estimated using non-HDL cholesterol which encompasses VLDL, IDL, and LDL. Hypertriglyceridemia may interfere with lipoprotein measurements using standard laboratory analyzers. Total cholesterol:HDL ratio is also a useful biomarker that allows assessment of the relative importance of atherogenic lipoproteins. Non-HDL cholesterol and total cholesterol:HDL ratios are effective biomarkers in people and better predict risks of cardiovascular disease than LDL alone.[28] As HDLs are the main plasma lipoproteins in parrots versus LDLs in humans, human guidelines are not useful in psittacine species.

Plasma lipids are incredibly diverse and standard lipid testing is restricted in terms of assessing broad lipid changes and lipid metabolic pathway disruption in relation to disease. For this reason, comprehensive lipid characterization with lipidomics has been investigated in parrots. In the context of dyslipidemia, specific lipid species of complex lipids may be better targets or biomarkers than crude measurements of a single lipid class or subclass such as with cholesterol and glycerol (for triglycerides) analysis. Thousands of species of triglycerides, phospholipids, and cholesteryl esters with a variety of saturated and unsaturated fatty acyl chains can be measured in plasma. Approximately 1000-1500 distinct lipid species were detected in Quaker parrot plasma using high performance liquid chromatography-tandem mass spectrometry.[29] Some of these lipids could serve as novel biomarkers for several lipid-related and metabolic diseases. Lipidomic analysis provides an important alternative approach that bypasses many limitations of conventional tests while providing considerably more information. The lipidome of male and female parrots also tend to be different. While it is not yet used on clinical patients, lipidomic research is underway in parrots and may lead to improved lipidologic diagnostic tests in the future. Specific biomarkers or lipidomic signature for specific diseases may be discovered with these techniques. For instance, plasma concentrations of several lipid metabolites including many cholesteryl esters, ceramides, phosphatidylcholines, and sphingomyelins increased in Quaker parrots following 0.3% dietary cholesterol supplementation.[30] Further studies investigating lipidomic profiling in association with various diseases are needed to explore and propose new biomarkers or combination of biomarkers as diagnostic for disease in avian species.

Recent Advances in Vascular Imaging

As blood lipid diagnostics only assess risk factors, the diagnosis of atherosclerosis mainly relies on imaging to detect either arterial luminal stenosis through angiography protocols or arterial calcification, strongly associated with the lesions, with radiographs, CT scan or PET scan.

Since an initial study developing and validating a CT angiography protocol in Hispaniolan Amazon parrots,[31] several studies have been published in grey parrots optimizing CT angiography protocols through different injection speeds and iodinated contrast volumes and doses.[32-34] These studies have shown that arterial diameter measurements under CTA differs depending on the technique used, specifically the dose of contrast agent, the speed of administration, and the measured arteries.[34]

While fluoroscopic angiography has been used before in parrots and other birds,[35-37] it was mainly reported in case reports and no standardized fluoroscopic angiography protocol was published. This was remedied by a study on fluoroscopic angiography performed in Hispaniolan Amazon parrots using iohexol at 3 ml/kg manually injected at a speed of 1-2 ml/s with images acquired using a C-arm fluoroscopic machine.[38] Intraobserver agreement was excellent, but interobserver agreement was moderate with 10 to 20% variability.

Angiographic protocols show that measuring arterial luminal diameter in birds may lead to high measurement variability, especially when motions artefacts are taken into considerations. Arterial luminal diameters of medium-sized parrots are only 2-3 mm for the largest arteries. Therefore, angiographic protocols are likely insensitive in detecting all but the most severely stenotic lesions, which is probably why only scant information is available on using angiography on clinical atherosclerotic cases in live birds.

The better strategy to identify atherosclerotic lesions antemortem is related to detecting and quantifying arterial wall calcification. Arterial calcification is a hallmark of atherosclerosis in mammals and birds and ranges from microcalcification to complete bone tissue formation.[3,39,40] In parrots, arterial calcification is also strongly associated with advanced and clinically significant lesions (types IV-VII lesions) and is correlated to lesion area, but microcalcification can be seen in earlier lesions.[3] Arterial calcification is also present in diet-induced atherosclerotic lesions in experimental parrot models and can be detected as early as 4 months.[2] Arterial calcification can be detected on radiographs, but CT is more sensitive. On CT-scan, calcification can also be quantified, which is then used to estimate overall atherosclerosis burden. CT calcium score systems used in parrots are extrapolated from humans and have been applied to Amazon parrots and grey parrots.[8,41] These scores are relatively easy to obtain using HOROS, an open-source diagnostic imaging software, that can include a free downloadable calcium score plugin.[42] Areas with an attenuation higher than 130 Hounsfield units are included in calculation. CT calcium scoring cannot be reliably done using cone-beam CT scans (available in many private practices for use in birds), even if spatial resolution is higher than multirow detector CT, as attenuation assessment using Hounsfield units is not available (replaced with grey values or HU approximation instead) and also streak artefacts caused by adjacent bones are common. Semi-quantitative assessment needs to be performed instead on CBCT images. On the other hand, the higher resolution of CBCT may detect decreased fat attenuation of thick and fatty atheromatous plaques. Published information is lacking on this approach and it is likely to lack sensitivity except for severe protruding or stenotic lesions. Further research is needed on the use of various CT scanners for plaque composition assessment in psittacine birds.

Because only macrocalcification can be detected on CT scans, making sensitivity still relatively low for tiny animals such as birds, other methods are currently being investigated to better detect arterial calcification at an earlier stage (i.e., microcalcifications). A positron-emission tomography protocol using ^{18}F sodium fluoride was recently published in two species of Amazon parrots showing high sensitivity of this imaging technique when assessed against histopathologic diagnosis.[8] Another interesting finding of this study was the fact that lesions were more common at arterial bifurcation sites of large arteries and that tracer uptake was harder to quantify in arteries directly adjacent to bones (i.e., coracoid-sternal junction). This technique also allowed to detect lesions in smaller arteries such as the coronary arteries. Current research is ongoing to optimize ^{18}F-NaF PET scan imaging of atherosclerosis for easier use in clinics on companion parrots.

Recent Advances in Pharmacotherapy

For most lipid-accumulation disorders, dyslipidemia is a strong contributing factor, thus treating dyslipidemia still remains a treatment focus. While improving the diet, incorporating omega-3 fatty acid supplements, limiting animal products and high-fat diet, and increasing physical activity can be recommended. Pharmacologic interventions may be necessary in some cases.

Statin drugs are considered a mainstay of treatment for human hyperlipidemia and atherosclerosis. These medications are 3-hydroxy-3-methylglutaryl coenzyme A (HMG-CoA) reductase inhibitors. By inhibiting this enzyme, statins inhibit a key step in the mevalonate pathway of cholesterol synthesis in the liver. Statins also have direct anti-atherosclerotic effects independent from their lipid-lowering effects.[43] The use of statins has recently been explored in different avian species including atorvastatin in Amazon parrots, Quaker parrots, and cockatiels and rosuvastatin in Hispaniolan Amazon parrots and Quaker parrots.[44-48] Administration of atorvastatin at 20 mg/kg orally reach high plasma concentrations in orange-winged Amazon parrots and cockatiels with approximately 70% of the predicted activity being attributable the two active metabolites (para- and hydroxy-atorvastatin). In cockatiels, plasma concentrations were much higher than in Amazon parrots, but with a shorter half-life (4-7 hours). Atorvastatin is difficult to compound and keep in a liquid formulation due to its hydrophobic properties. For this reason, pravastatin, a highly hydrosoluble statin, has also been investigated in cockatiels. A dose of 40 mg/kg orally reached high plasma concentrations with a short half-life of half an hour (Beaufrère, in preparation). Some preliminary results also suggest that mixing pravastatin powder with pellets is effective in reaching adequate plasma concentrations in cockatiels whereas adding pravastatin to the drinking water results in subtherapeutic plasma levels (Beaufrère, in preparation).

While statin pharmacokinetic profiles seem adequate in psittacine species, their lipid-lowering efficacy is still questionable. Several pharmacodynamic studies using atorvastatin and rosuvastatin failed to show significant decrease in blood lipids in Quaker parrots and Amazon parrots.[44,48] An ongoing experiment in cockatiels using atorvastatin and rosuvastatin doses of up to 80 mg/kg q24h for 2 weeks is also not showing appreciable differences in plasma lipids and lipoproteins when compared to controls, but analysis of results is still ongoing (Beaufrère, in preparation). Most of these studies were performed on psittacine birds with normal or subnormal plasma lipids and it is possible that statins work best on birds with more severe hyperlipidemia or when used for longer periods. Finally, direct effects of statins on atherosclerosis lesion development have not been studied yet in birds.

In female birds, vitellogenesis and reproductive disorders are linked to plasma lipid changes that could potentiate atherogenesis. Therefore, GnRh agonists, such as deslorelin acetate implants, could be used to decrease reproductive-associated hyperlipidemia. An ongoing study using deslorelin implants in female cockatiels showed overall reduction in plasma lipids including cholesterol, triglycerides, and HDL for 1 month and IDL for 2 months. As deslorelin implants have been reported to suppress egg laying in cockatiels for at least 6 months,[49] it is surprising to only see a significant effect on plasma lipids for 1-2 months. There is tremendous variability in plasma lipids in cockatiels and this could have masked a longer duration of action. In addition, deslorelin resulted in a lower incidence of hypertriglyceridemia during the study period when compared to controls. The main caveat is that deslorelin is an indexed drug in the United States so a product other than the ferret product should be imported from abroad to be able to legally use this treatment in companion birds.

Conclusions and Recommendations

Research over the past 5 years in clinical lipidology diagnostics and therapeutics allows for some basic recommendations for avian practitioners to be derived.

1. A complete lipid panel that is both reliable and practical to obtain on clinical patients should include the following: cholesterol, triglycerides, HDL, and calculated non-HDL in non-lipemic samples. Direct and indirect LDL measurements using standard analyzers should not be performed in birds. If samples are lipemic, they should be (ultra)centrifuged to clear the lipemia.

2. The best method to diagnose atherosclerosis is to detect or quantify arterial calcification either using radiographs or CT scans. While a CT scan is more sensitive, it is still considered to have low sensitivity in birds including micro-CBCT. Fat attenuation of advanced thick plaques may be detectable on CBCT. Molecular imaging using ^{18}F-NaF PET scan is a promising new technique that is highly sensitive.

3. Treatment recommendations are difficult to make at this stage owing to the lack of evidence for lipid-lowering efficacy of studied statin protocols. A starting dose of 20 mg/kg of atorvastatin q12-24h is safe and blood cholesterol can be monitored for efficacy monthly. Deslorelin seems to have short-lived hypolipidemic effects but can be incorporated into the treatment regimen of female birds providing it can be used legally.

References

1. Libby P, Buring JE, Badimon L, et al. Atherosclerosis. *Nat Rev Dis Primers.* 2019;5:56.

2. Beaufrere H, Nevarez J, Clubb S, et al. Diet-induced experimental atherosclerosis in Quaker parrots (*Myiopsitta monachus*). *Vet Pathol.* 2013;50:1116-1126.

3. Beaufrere H, Nevarez JG, Holder K, et al. Characterization and classification of psittacine atherosclerotic lesions by histopathology, digital image analysis, transmission and scanning electron microscopy. *Avian Pathol.* 2011;40:531-544.

4. Beaufrere H. Atherosclerosis: comparative pathogenesis, lipoprotein metabolism and avian and exotic companion mammal models. *J Exot Pet Med.* 2013;22:320-335.

5. Schmaier AA, Stalker TJ, Runge JJ, et al. Occlusive thrombi arise in mammals but not birds in response to arterial injury: evolutionary insight into human cardiovascular disease. *Blood.* 2011;118:3661-3669.

6. Beaufrere H, Ammersbach M, Reavill D, et al. Prevalence and risk factors in psittacine athersoclerosis: a multicenter case-control study. *J Am Vet Med Assoc.* 2013;242:1696-1704.

7. Beaufrere H, Holder KA, Bauer R, et al. Intermittent claudication-like syndrome secondary to atherosclerosis in a yellow-naped Amazon parrot (*Amazona ochrocephala auropalliata*). *J Avian Med Surg.* 2011;25: 266-276.

8. Megan L, Brust K, Spriet M, et al. ^{18}F-sodium fluoride positron emission tomography is a sensitive imaging technique to detect atherosclerosis in Amazon parrots (*Amazona* spp). *Am J Vet Res.* 2024;1:1-10.

9. Beaufrere H. Avian atherosclerosis: parrots and beyond. *J Exot Pet Med.* 2013;22:336-347.

10. Lujan-Vega C, Keel MK, Barker CM, et al. Evaluation of atherosclerotic lesions and risk factors of atherosclerosis in raptors in northern California. *J Avian Med Surg.* 2021;35:295-304.

11. Fricke C, Schmidt V, Cramer K, et al. Characterization of atherosclerosis by histochemical and immunohistochemical methods in African grey parrots (*Psittacus erithacus*) and Amazon parrots (*Amazona* spp.). *Avian Dis.* 2009;53:466-472.

12. Beaufrère H, Nevarez J, Gaschen L, et al. Diagnosis of presumed acute ischemic stroke and associated seizure management in a Congo African grey parrot. *J Am Vet Med Assoc.* 2011;239:122-128.

13. Grosset C, Guzman DSM, Keating MK, et al. Central vestibular disease in a blue and gold macaw (*Ara ararauna*) with vasculopathy and cerebral infarction and hemorrhage. *Proc Annu Conf Assoc Avian Vet.* 2012;257.

14. Megan L, Guzman DS-M, Keel K, et al. Prevalence, anatomical distribution, and risk factors of adipocytic tumors and xanthomas in Psittaciformes: 1096 cases (1998-2018). *J Avian Med Surg.* 2024;38:108-115.

15. Donovan TA, Garner MM, Phalen D, et al. Disseminated coelomic xanthogranulomatosis in eclectus parrots (*Eclectus roratus*) and budgerigars (*Melopsittacus undulatus*). *Vet Pathol.* 2022;59:143-151.

16. Beaufrère H, Reavill D, Heatley J, et al. Lipid-related lesions in Quaker parrots (*Myiopsitta monachus*). *Vet Pathol.* 2019;56:282-288.

17. Alvarenga RR, Zangeronimo MG, Pereira LJ, et al. Lipoprotein metabolism in poultry. *World Poult Sci J.* 2011;67:431-440.

18. Beaufrère H, Vet DM, Cray C, et al. Association of plasma lipid levels with atherosclerosis prevalence in Psittaciformes. *J Avian Med Surg.* 2014;28:225-231.

19. Facon C, Beaufrere H, Gaborit C, et al. Cluster of atherosclerosis in a captive population of black kites (*Milvus migrans* subsp.) in France and effect of nutrition on the plasma lipidogram. *Avian Dis.* 2014;58: 176-182.

20. Potier R. Lipid blood profile in captive brahminy kite (*Haliastur indus*) as a possible indication of increased susceptibility to atherosclerosis. *J Zoo Wildl Med.* 2013;44:549-54.

21. Beaufrère H. Blood lipid diagnostics in psittacine birds. *Vet Clin North Am Exot Anim Pract.* 2022;25: 697-712.

22. Beaufrere H, Gardhouse S, Ammersbach M. Lipoprotein characterization in Quaker parrots (*Myiopsitta monachus*) using gel-permeation high-performance liquid chromatography. *Vet Clin Pathol.* 2020;49: 417-427.

23. Beaufrère H, Wood D. Comparison of lipoprotein analysis using gel-permeation high-performance liquid chromatography and a biochemistry analyzer in normolipidemic and dyslipidemic Quaker parrots (*Myiopsitta monachus*). *J Avian Med Surg.* 2023;36:345-355.

24. Belcher C, Heatley J, Petzinger C, et al. Evaluation of plasma cholesterol, triglyceride, and lipid density profiles in captive monk parakeets (*Myiopsitta monachus*). *J Exot Pet Med.* 2014;23:71-78.

25. Oku H, Ishikawa M, Nagata J, et al. Lipoprotein and apoprotein profile of Japanese quail. *Biochemica et Biophysica Acta.* 1993;1167:22-28.

26. Barakat HA, Clair RWS. Characterization of plasma lipoproteins of grain- and cholesterol-fed white Carneau and show racer pigeons. *J Lipid Res.* 1985;26:1252-1268.

27. Hermier D, Saadoun A, Salichon MR, et al. Plasma lipoproteins and liver lipids in two breeds of geese with different susceptibility to hepatic steatosis: changes induced by development and force-feeding. *Lipids.* 1991;26:331-339.

28. Pokharel Y, Negi S, Virani SS. Cholesterol: concentration, ratio, and particle number. In: Ballantyne CM, ed. *Clinical Lipidology*. 2nd Ed. St. Louis, MO: Elsevier; 2015:91-98.

29. Beaufrère H, Gardhouse SM, Wood RD, et al. The plasma lipidome of the Quaker parrot (*Myiopsitta monachus*). *PLOS One* 2020;15(12):e0240449.

30. Beaufrere H, Stark KD, Wood RD. Effects of a 0.3% cholesterol diet and a 20% fat diet on plasma lipids and lipoproteins in Quaker parrots (*Myiopsitta monachus*). *Vet Clin Pathol.* 2022;51:376-384.

31. Beaufrère H, Rodriguez D, Pariaut R, et al. Estimation of intrathoracic arterial diameter by means of computed tomographic angiography in Hispaniolan Amazon parrots. *Am J Vet Res.* 2011;72:210-218.

32. Yu PH, Lee YL, Chen CL, et al. Comparison of three computed tomographic angiography protocols to assess diameters of major arteries in African grey parrots (*Psittacus erithacus*). *Am J Vet Res.* 2018;79:42-53.

33. Lee Y-L, Yu P-H, Chen C-L, et al. Determination of the enhancement effect and diameters of the major arteries of African grey parrots using a dual-head power injector for computed tomography angiography. *Taiwan Vet J.* 2015;41:165-175.

34. Wang W-L, Chang F-L, Yu P-H. Effect of different injection rates and doses of contrast medium on the image quality of computed tomographic angiography in African grey parrots (*Psittacus erithacus*). *BMC Vet Res.* 2022;18:428.

35. Phalen DN, Hays HB, Filippich LJ, et al. Heart failure in a macaw with atherosclerosis of the aorta and brachiocephalic arteries. *J Am Vet Med Assoc.* 1996;209:1435-1440.

36. Vink-Nooteboom M, Schoemaker NJ, Kik M, et al. Clinical diagnosis of aneurysm of the right coronary artery in a white cockatoo (*Cacatua alba*). *J Small Anim Pract.* 1998;39:533-537.

37. Beaufrere H, Pariaut R, Rodriguez D, et al. Avian vascular imaging: a review. *J Avian Med Surg.* 2010;24: 174-184.

38. Aller TL, Guzman DS-M, Stern JA, et al. Evaluation of a fluoroscopic angiography protocol in Hispaniolan Amazon parrots (*Amazona ventralis*). *J Avian Med Surg.* 2022;36(2):178-186.

39. Shioi A, Ikari Y. Plaque calcification during atherosclerosis progression and regression. *J Atheroscler Thromb.* 2018;25(4):294-303.

40. Alexopoulos N, Raggi P. Calcification in atherosclerosis. *Nat Rev Cardiol.* 2009;6:681-688.

41. Louste E, Huynh M, Beaufrère H. Use of calcium score to quantify atherosclerotic lesion in African grey parrots (*Psittacus erithacus*) using CT-scan: preliminary findings. *Proceedings ICARE.* 2024.

42. Nappi C, Megna R, Volpe F, et al. Quantification of coronary artery atherosclerotic burden and muscle mass: exploratory comparison of two freely available software programs. *Appl. Sci.* 2022;12:5468.

43. McKenney JM, Roth EM. Statins. In: Ballantyne CM, ed. *Clinical Lipidology*. 2nd Ed. St. Louis, MO: Elsevier; 2015:227-254.

44. Robertson JA, Guzman DSM, Graham JL, et al. Evaluation of orally administered atorvastatin on plasma lipid and biochemistry profiles in hypercholesterolemic Hispaniolan Amazon parrots (*Amazona ventralis*). *J Avian Med Surg.* 2020;34:32-40.

45. Mikoni N, Guzman DS-M, Knych H, et al. Pharmacokinetics of single-dose oral atorvastatin and its metabolites support therapeutic use in orange-winged amazon parrots (*Amazona amazonica*). *Am J Vet Res.* 2024;85(1):ajvr.23.10.0227.

46. Mitchell M, Guzman DS-M, Knych H, et al. The pharmacokinetics of single-dose oral atorvastatin and its metabolites support therapeutic use in cockatiels (*Nymphicus hollandicus*). *Am J Vet Res.* 2024;1:1-7.

47. Beaufrère H, Papich MG, Brandão J, et al. Plasma drug concentrations of orally administered rosuvastatin in Hispaniolan Amazon parrots (*Amazona ventralis*). *J Avian Med Surg.* 2015;29:18-24.

48. Beaufrère H, Barboza T, Burnett A, et al. Effects of atorvastatin and rosuvastatin on blood lipids in Quaker parrots (*Myiopsitta monachus*). *J Avian Med Surg.* 2023;37:199-208.

49. Summa NM, Guzman DS-M, Wils-Plotz EL, et al. Evaluation of the effects of a 4.7-mg deslorelin acetate implant on egg laying in cockatiels (*Nymphicus hollandicus*). *Am J Vet Res.* 2017;78:745-751.

Serum Lipid Reference Intervals of High-Density, Low-Density and Non-High-Density Lipoprotein Cholesterols and Their Association with Atherosclerosis and Other Factors in Psittaciformes

Matthias Janeczek, med vet,
Rüdiger Korbel, DipECZM (Avian), Prof, Dr med vet habil,
Friedrich Janeczek, Dr med vet,
Helen Alber, MSc,
Helmut Küchenhoff, Prof, Dr,
Monika Rinder, Prof, Dr med vet habil

Session #177

Affiliation: From Centre for Clinical Veterinary Medicine, Clinic for Birds, Small Mammals, Reptiles and Ornamental Fish, Ludwig-Maximilians-Universität München, Sonnenstr. 18, 85764 Oberschleißheim, Germany (Korbel, Rinder), Veterinary Practice for Parrots, Hans-Cornelius-Str. 2c, 82166 Gräfelfing, Germany (Janeczek M, Janeczek F), Department of Statistics, LMU Munich, Ludwigstraße 33, Room 344, D-80539 München (Alber) Department of Statistics, Statistical Consulting Unit StaBLab, Ludwig-Maximilians-Universität München, Ludwigstr. 33, 80539 München, Germany (Küchenhoff).

Abstract: The prevalence of atherosclerosis is high in captive psittacine populations[1,2] and the lesions and subsequent heart problems have become a common reason for consultations in avian veterinary practices. To this day, ante-mortem diagnosis in birds remains challengingdie in Gefangenschaft mit unzureichenden Haltungsbedingungen leben. Die Diagnose beim lebenden Vogel bleibt jedoch schwierig und die Krankheit wird oft erst post mortem erkannt. In diesem Zusammenhang sollte der diagnostische Wert beschriebener Methoden bei lebenden Graupapageien (Psittacus erithacus;[3,4] however, the serum lipoprotein-panel has been suggested to potentially aid in the diagnosis of the disease and provide better understanding of the pathogenesis.[4] In parrots, unlike in humans, a prospective association between serum lipoproteins and spontaneous atherosclerosis has not been proven. The present retrospective cohort study aimed to establish multi-genera serum reference intervals for high-density lipoprotein-cholesterol (HDL-C), low-density lipoprotein-cholesterol (LDL-C) and non-high-density lipoprotein-cholesterol (non-HDL-C) in various parrot species according to ASVCP guidelines. HDL-C and LDL-C were obtained via photometric measurement based on an enzymatic, colorimetric assay. An analysis of 1199 blood samples originating from 692 birds of 14 different parrot genera and 46 species revealed genus-specific variations in lipoprotein levels. In addition, lipoproteins were studied for their association with independent variables of age, sex, genus, diet, reproductive and breeding status, atherosclerosis and body conditioning score. Multivariable statistical methods in the form of linear mixed models and general linear models revealed a significant association between LDL-C levels and endoscopically diagnosed prevalence of atherosclerosis, similar to findings in humans. Diet was statistically associated with both serum lipoprotein levels and severity of atherosclerosis. Results showed that LDL-C appears to be a potential predictor of atherosclerosis, with non-HDL-C showing a similar, but less pronounced trend, while the role of HDL-C remained less conclusively defined. The results of the study provide a foundational framework for the potential future use of lipoprotein analysis in parrot medicine; however, the results of this study should be interpreted with caution, because LDL-C measurements have been shown to be unreliable[5] and the use of endoscopy for the diagnosis of atherosclerosis has not been validated in parrots.

References

1. Fitzgerald BC, Beaufrère H. Cardiology. In: Speer BL, ed. *Current Therapy in Avian Medicine and Surgery.* St. Louis, MO: Elsevier; 2016:252-328.

2. Beaufrère H. Blood lipid diagnostics in psittacine birds. *Vet Clin North Am Exot Anim Pract.* 2022;25(3): 697-712.

3. Megan L, Brust K, Spriet M, et al. 18F-sodium fluoride positron emission tomography is a sensitive imaging technique to detect atherosclerosis in Amazon parrots (*Amazona* spp). *Am J Vet Res.* 2024;82(2):1-10.

4. Krautwald-Junghanns ME, Schulz U, Konicek C, Pees M. Evaluation of diagnostic criteria in grey parrots (*Psittacus erithacus*) with suspected atherosclerosis. *Tierarztl Prax Ausg K Kleintiere Heimtiere.* 2022;50(6):411-422.

5. Beaufrère H, Wood D. Comparison of lipoprotein analysis using gel-permeation high-performance liquid chromatography and a biochemistry analyzer in normolipidemic and dyslipidemic Quaker parrots (*Myiopsitta monachus*). *J Avian Med Surg.* 2023;36(4):345-355

Characterization and Reference Intervals for Plasma Lipids and Lipoproteins in Cockatiels (*Nymphicus hollandicus*)

Rachel M. Hirota, BS,
Marcel Gómez Ponce, DVM,
Hugues Beaufrère, DVM, PhD, Dipl ACZM, DABVP (Avian Practice), DipECZM (Avian)

Session #212

Affiliation: From the Department of Medicine and Epidemiology (Beaufrère), School of Veterinary Medicine (Hirota, Gómez Ponce), University of California Davis, Davis, CA 95616, USA.

Abstract: Lipid disorders such as obesity, atherosclerosis, and hepatic lipidosis are common in psittacine birds under human care. Dyslipidemias may be associated with the prevalence of these lipid accumulation disorders in birds, akin to mammals.[1] Species-specific reference intervals for reliable lipidologic assays are lacking even in common companion species. Furthermore, female lipoprotein metabolism involves periods of vitellogenesis, which may contribute to sex differences in avian lipid and lipoprotein values.[2] Among companion psittacine birds, cockatiels are particularly susceptible to dyslipidemias and atherosclerosis.[3-4] A study population of 70 clinically healthy adult cockatiels (37 females and 33 males) was used to establish reference intervals for plasma lipids and lipoproteins. Total cholesterol and triglycerides were obtained from a Roche Cobas Integra 400+ biochemistry analyzer. Lipoprotein profiles were determined using high-resolution polyacrylamide gel electrophoresis (HR-PAGE) and gel-permeation high-performance liquid chromatography (GP-HPLC). Sex differences in lipid and lipoprotein values were assessed via pairwise permutation tests at 2.5 and 97.5 percentiles. Besides total triglycerides and low-density lipoproteins, no analytes varied by sex when using a biochemistry analyzer or HR-PAGE. Most lipids and lipoprotein fractions significantly differed across sexes using GP-HPLC. Thus, lipidologic assay granularity may obscure or elucidate sex differences within birds. Sex-stratified reference intervals across techniques will contribute to our understanding of avian lipidology and diagnosing, treating, and monitoring psittacine dyslipemidias and lipid-accumulation disorders, particularly those associated with female reproduction.

References

1. Arvanitis M, Lowenstein CJ. Dyslipidemia. *Ann Intern Med.* 2023;176(6):ITC81-ITC96. doi: 10.7326/aitc202306200.

2. Ravich M, Cray C, Hess L, Arheart KL. Lipid panel reference intervals for Amazon parrots (*Amazona species*). *J Avian Med Surg.* 2014;28(3):209-215. doi: 10.1647/2013-055.

3. Beaufrère H, Ammersbach M, Reavill DR, et al. Prevalence of and risk factors associated with atherosclerosis in psittacine birds. *J Am Vet Med Assoc.* 2013;242(12):1696-1704. doi: 10.2460/javma.242.12.1696.

4. Beaufrère H. Avian atherosclerosis: parrots and beyond. *J Exot Pet Med.* 2013;22(4):336-347. doi: 10.1053/j.jepm.2013.10.015.

Effects of Atorvastatin on Blood Lipids in Cockatiels (*Nymphicus hollandicus*)

Marcel Gómez-Ponce, DVM,
Rachel M. Hirota BS,
Krista A. Keller, DVM, Dipl ACZM,
Minh Huynh, DVM, MRCVS, DipECZM (Avian), Dipl ACZM,
David Sanchez-Migallon Guzman, LV, MS, DipECZM (Avian), DipECZM (Small Mammal), Dipl ACZM,
Hugues Beaufrère, DVM, PhD, Dipl ACZM, DABVP (Avian Practice), DipECZM (Avian)

Session #162

Affiliation: From the Graduate Group in Integrative Pathobiology (Gómez-Ponce, Beaufrère), School of Veterinary Medicine (Hirota), Department of Medicine and Epidemiology (Guzman, Keller, Beaufrère), Veterinary Medical Teaching Hospital (Huynh), University of California Davis, Davis, CA 95616, USA.

Abstract: Dyslipidemias and lipid-accumulation disorders such as atherosclerosis and hepatic lipidosis are prevalent in psittacine birds. Statins, particularly atorvastatin, have shown promise as potential therapeutic agents, but their efficacy in avian species remains poorly characterized.[1-3] Twenty-four healthy adult cockatiels were randomly assigned to two treatment groups (atorvastatin 40 or 80 mg/kg PO q24h) and a control group (ORA-Plus® PO q24h at a volume equal to the mean of the two treatment volumes) in a balanced crossover study design for 14 days. A 5-week washout period separated treatments. Atorvastatin (10 mg/ml; prepared daily as a suspension of crushed tablets in ORA-Plus®/water) was given by oral gavage for 14 days per phase. A 1 ml blood sample was collected at the end of each trial and plasma cholesterol, triglycerides, and a lipoprotein panel were obtained using polyacrylamide gel electrophoresis (Lipoprint®). Plasma lipids decreased over time across all groups; however, no statistically significant hypolipidemic effects were observed in the treatment groups compared to the control. The use of normolipidemic healthy birds without increased non-HDL cholesterol may have limited the detectable effects of atorvastatin, as statins are typically more effective in dysregulated lipid metabolism. In addition, stress associated with the experiment may have masked potential hypolipidemic responses. In conclusion, this study did not show significant hypolipidemic effects of a 2-week high-dose course of atorvastatin in normolipidemic cockatiels. More pharmacodynamic studies are needed to assess the hypolipidemic and non-lipid effects of statins and other drugs to provide evidence-based therapeutic options for the management of lipid-accumulation disorders in psittacine birds.

References

1. Mitchell M, Guzman DS, Knych H, Beaufrère H. The pharmacokinetics of single-dose oral atorvastatin and its metabolites support therapeutic use in cockatiels (*Nymphicus hollandicus*). *Am J Vet Res.* 2024;85(7):ajvr.24.03.0068.

2. Mikoni N, Guzman DS, Knych H, Beaufrère H. Pharmacokinetics of single-dose oral atorvastatin and its metabolites support therapeutic use in orange-winged amazon parrots (*Amazona amazonica*). *Am J Vet Res.* 2024;85(1):ajvr.23.10.0227.

3. Beaufrère H, Barboza T, Burnett A, Stark KD, Wood RD. Effects of atorvastatin and rosuvastatin on blood lipids in Quaker parrots (*Myiopsitta monachus*). *J Avian Med Surg.* 2023;37(3):199-208.

Atherosclerosis, Aortic Stenosis, and Myxomatous Valvular Degeneration of the Aortic Valve in a Male Eclectus Parrot (*Eclectus roratus*)

Davia Kot, DVM,
Katherine Quesenberry, DVM, MPH, DABVP (Avian Practice),
Andrea Winkel, DVM, MPH,
Taryn A. Donovan, DVM, DACVP

Session #160

Affiliation: From the Schwarzman Animal Medical Center, New York, NY 10065, USA.

Abstract: Cardiovascular diseases are reported in many avian species and common diseases include atherosclerosis, inflammatory disease, and infectious disease.[1] Myxomatous valvular degeneration (MVD) has been less commonly reported in several avian species, with several reports on left atrioventricular valve MVD and one report describing aortic MVD in bald eagles (*Haliaeetus leucocephalus*).[2-4] In this case, chronic aortic stenosis was documented in an eclectus parrot antemortem with transcoelomic echocardiography. Serial examinations showed progressive changes over a period of 3 years. At necropsy, the aortic valve showed myxomatous degeneration and expansion of the aortic valve leaflets in combination with valvular stenosis and arterial mineralization. These findings are considered clinically significant, and the probable cause of this patient's death from presumed congestive heart failure.

References

1. Jäger K, Cerezo-Echevarria A, Pohl A, et al. Histopathological findings in the cardiovascular system of Psittacidae in routine diagnostics. *Vet Clin North Am Exot Anim Pract*. 2022;25(2):453-467. doi: 10.1016/j.cvex.2022.01.009.

2. Oglesbee BL, Lehmkuhl L. Congestive heart failure associated with myxomatous degeneration of the left atrioventricular valve in a parakeet. *J Am Vet Med Assoc*. 2001;218(3):376-380.

3. Vuong KS, Jones M, Craig LE. Postmortem evaluation of cardiac valvular disease in bald eagles (*Haliaeetus leucocephalus*) and a golden eagle (*Aquila chrysaetos*). *J Avian Med Surg*. 2021;37(1):22-31. doi: 10.1647/1082-6742-37.1.22.

4. Kagohara A, dos Santos Filho M, Marinho JPM, et al. Myxomatous degeneration of the left atrioventricular valve in a true parrot (*Amazona aestiva*): a case report. *Rev Bras Med Vet*. 2020;42(1). doi: 10.29374/2527-2179.BJVM104720.

Radiographic Morphometric Correlation of the Heart and Adjacent Structures in Potoo (*Nyctibius griseus*)

Nelson Dias Lucas,
Luana Célia Stunitz da Silva,
Rafael Rostirolla Debiage,
Cristiane Maria Montilho,
Juliana Ascenço Lopes,
Anaisa de Lima Grapiglia,
Anderson Luiz de Carvalho

Session #135

Affiliation: From Medicine and Wildlife Conservation, Universidade Federal do Paraná, Brazil (Dias Lucas), Federal University of Paraná, Palotina Campus, Brazil (Stunitz da Silva, Debiage, Lopes, Grapiglia, de Carvalho).

Abstract: The common potoo (*Nyctibius griseus*) is a nocturnal bird with a geographic distribution restricted to tropical regions and limited representation in the scientific literature. This study aimed to establish reference intervals for cardiac size by analyzing radiographs of nine adult free-living specimens, previously evaluated through clinical examination, using ventrodorsal and mediolateral projections. In the ventrodorsal radiographs, the following structures were measured: cardiac width, thoracic width, heart length, distance between the third and fourth left ribs, width of the left coracoid bone, and the distance between the right and left clavicles. In the mediolateral projection, measurements included sternum width, cardiac width, and heart length. Ratios between these measurements were calculated, and the results showed positive and statistically significant correlations between cardiac width and thoracic width, as well as between cardiac width and heart length in the ventrodorsal projection. The values and proportions obtained in this study may serve as radiographic reference parameters for the assessment of cardiac size in *Nyctibius griseus*, aiding in the diagnosis of potential cardiac disorders in this species.

Anatomy, Form, and Function of the Coelomic Cavities of Birds and Their Interactions with the Respiratory System

Brian Speer, DVM, DABVP (Avian Practice), DipECZM (Avian),
Emma Schachner, PhD,
M. Scott Echols, DVM, DABVP (Avian Practice),
Casey Holliday, PhD,
Wes Baumgartner, DVM, PhD, DACVP

Session #168

Affiliation: From Medical Center for Birds, 3807 Main Street, Oakley, CA 94561, USA (Speer, Echols), Department of Physiological Sciences, College of Veterinary Medicine, University of Florida, 2015 SW 16th Ave, Gainesville, FL 32608, USA (Schachner), Parrish Creek Veterinary Hospital, Centerville, UT 84014, USA (Echols), Pathology and Anatomical Sciences, University of Missouri, Columbia, 1 Hospital Dr., Columbia, MO 65201 USA (Holliday), Illinois Veterinary Diagnostic Laboratory, University of Illinois Urbana-Champaign, Urbana, IL 61801, USA (Baumgartner).

Abstract: The clinical significance of the unique coelomic cavities of avian species and their interrelationship with the respiratory system is great, and their anatomy has been well described for years. Unfortunately, there remains a widespread lack of understanding of these structures, even as represented in the peer-reviewed literature, particularly over the past approximately 20 years. Coelomic cavities are, by definition, lined with mesothelium and produce lubricating serosal fluids. The coelomic cavities of birds are considerably more complex than is seen in other vertebrates. Problems stemming from deficits in a clear understanding of their anatomy and interrelationships with the respiratory tract can lead to several clinical challenges. These may include problems in the formulation of differential diagnostic thought processes pertinent to observed clinical signs, inaccuracies in interpretation of diagnostic imaging findings, inaccurate descriptions of clinical and diagnostic procedures including surgery, endoscopy, and even necropsy findings. The purpose of this masterclass is to review the most current knowledge regarding the eight coelomic cavities of birds and, the detailed interrelationships that these cavities have with the respiratory system, and the clinical significance of these structures.

The Coelomic Cavities of Birds

Birds have a total of eight coelomic cavities, as compared to two currently described in squamates (lizards and snakes) and three or more described in Archelosauria (turtles, crocodilians and birds) and mammals.[1] These distinct cavities include the pericardial cavity, two pleural cavities, and five peritoneal cavities (four hepatic, and one intestinal peritoneal). Nile crocodiles have ten serous cavities and bursae, which are likely at least largely homologous to these coelomic cavities.[2] The embryonic coelom of birds is rapidly divided into its component eight cavities by day 5 of incubation.[3] From this point in time, the cavities remain distinct. Although the term "coelom" refers to the embryonic or primordial body cavity, more specific terminology can and should be applied whenever possible. From a clinical perspective, when the specific cavity or cavities of concern are not yet identified, regional anatomic terminology may be valid. Localizing the area of interest or concern to the thoracic region, or abdominal region are very fair general terms to use, and in and of themselves are more accurate than the use of the term coelom.[1] Inflammation of the peritoneal mesothelium of mammals has, by convention, been called peritonitis. There is no indication why this should not be so in birds, even though they have more complex partitioning of their peritoneum into four hepatic peritoneal cavities and a single intestinal peritoneal cavity.

Inflammation of the non-peritoneal lined cavity linings (i.e., pericardial, right and left pleural) would be termed pericarditis, or pleuritis. The use of the term coelomitis has become more commonly used in the literature over the past 20 years, which is inaccurate because it is too generalized and non-specific (i.e., vague and ambiguous), and it is not the cavity (i.e., coelom) that is inflamed, it is the tissues that compose the lining that are inflamed. A non-tissue (i.e., cavity) cannot truly be inflamed, although it may contain inflammatory exudates or transudates originating from their inflamed or abnormal linings.

Pericardial cavity

The pericardial cavity is situated ventrally on the midline, just dorsal to the sternum, and is restrained by several fibrous attachments to surrounding structures, most notably the sternum. The apex of the sac is intimately associated with, and positioned between, the two lobes of the liver. The base of the pericardial sac is attached dorsolaterally to the horizontal septum near the hilus of each lung. The ventral pericardium is attached to the inner surface of the sternum by numerous fibrous strands. The pericardium also lies in contact dorsally with the esophagus and primary bronchi. The outer layer of the sac is lined by parietal pericardium, but at the heart base where the great vessels issue, the pericardium reflects on itself to produce the visceral pericardium, most referred to as the epicardium, which is adherent to the outer surface of the myocardium.[1]

Pleural cavities

The parietal pleura that surrounds each lung and the mesothelial coating of the visceral pleura applied to the lung itself form a small, potential pleural cavity due to the fixed dimensions of the lung of birds. The horizontal septum in birds, which develops from a splitting of the post-pulmonary septum helps to maintain the immobility of the gas-exchanging lung during ventilation.[4]

Peritoneal cavities

Birds have five distinct peritoneal cavities, which line specific regions in the abdominopelvic region.[1] There are four hepatic peritoneal cavities: right dorsal and right ventral hepatic peritoneal cavities, left dorsal and left ventral hepatic peritoneal cavities. The intestinal peritoneal cavity contains most of the gastrointestinal tract, the gall bladder (if present), the reproductive organs, and the spleen.

Hepatic peritoneal cavities

The left and right ventral hepatic peritoneal cavities are larger than their left and right dorsal cavity counterparts. In the chicken, the left dorsal hepatic peritoneal cavity communicates with the intestinal peritoneal cavity through a small opening that is covered on the intestinal peritoneal cavity only by the abdominal air sac.[5-7] This connection is small and not easy to visualize, but it is likely present in most birds.[1] The ventral wall of the ventral hepatic peritoneal cavity is the parietal peritoneum lining the sternum. The lateral walls of the ventral hepatic peritoneal cavity are formed by the oblique septa, and the dorsal wall is formed by the post hepatic septum. These two hepatic peritoneal cavities are divided on the midline into right and left by the ventral mesentery. In taxonomic orders with a gall bladder (Falconiformes, Anseriformes), the gall bladder is located within the intestinal peritoneal cavity but is attached to the post hepatic septum by the terminal aspect of the ventral mesentery. The post hepatic septum separates the hepatic peritoneal cavities from the intestinal peritoneal cavity and runs from a dorsal attachment at the vertebral column just caudal to the hilus of the lung and continues gradually in a caudoventral direction toward the caudal boarder of the sternum or the ventral body wall, depending on species.[1]

Intestinal peritoneal cavity

The intestinal peritoneal cavity is centrally located in the mid-to-caudal abdomen, and contains the proventriculus, ventriculus, gall bladder (if present), spleen, intestines, and male or female reproductive tracts.[8,5] The intestinal peritoneal cavity is the only coelomic cavity which contains air sacs (left and right abdominal air sacs). These abdominal air sacs enter the intestinal peritoneal cavity through an ostium in the small, unsplit portion of the horizontal and oblique septa and run dorsally and caudally toward the pelvic region of the bird. The volumes of these air sacs vary considerably in volume amongst species, largely being affected by the volume and relative location of the viscera. [9,10] The intestinal peritoneal cavity is lined dorsally by the parietal peritoneum of the dorsal body wall. The kidneys lie retroperitoneally, dorsally.[5] The entire gastrointestinal system from the proventriculus to the rectum is suspended by the dorsal mesentery. The ovary is suspended dorsally by the mesovarium, the oviduct is suspended by the mesosalpinx, and the testicles are suspended dorsally by mesorchia. The ventral mesentery ends at or before the caudal wall of the ventriculus. This means that the intestinal peritoneal cavity is a large single chamber for the bulk of its central to caudal abdominal locations.

Septa

A septum is a dividing wall or membrane, which divides cavities or structures into smaller ones. There are three key septa which function in this manner in birds to help separate, divide and stabilize the coelomic cavities and segments of the respiratory system.

Horizontal septum

The right and left sides of the horizontal septum form the ventral walls of the pleural cavities. This structure supports and strengthens the ventral surface of the lung. The visceral and parietal pleura on this surface of the lung also fuse to the horizontal septum, providing additional strength. The horizontal septum also includes the attached walls of portions of the caudal thoracic and cranial thoracic air sacs. The base of the pericardial sac is attached dorsolaterally to the horizontal septum near the hilus of each lung. The abdominal air sac pushes through the horizontal septum to expand within the intestinal peritoneal cavity of all known birds except the Kiwi (*Apteryx* sp.).[5] The horizontal septum is composed of a thick layer of connective tissue attached to the dorsal parietal pleura. Fan shaped, striated muscle bundles (*m. costoseptalis*) run from the ribs of the lateral body wall and attach to the surface of the horizontal septum.[4,11] These costoseptal muscles contract during expiration to help maintain the volume constancy of the lung during ventilation. The horizontal septum is penetrated by several structures, most importantly the ostia of the cervical, clavicular, cranial thoracic and caudal thoracic air sacs.

Oblique septum

The oblique septum is composed of connective tissue and in most species is supported at its caudal attachment to the vertebral column by smooth muscle. It is composed of the parietal peritoneum on its dorsal and medial surfaces and the air sac epithelium on the ventral and lateral surfaces.[5] The walls of the cranial thoracic air sac and caudal thoracic air sacs are fixed to the horizontal and oblique septa so that their shapes are maintained.[1] The oblique septum forms the lateral walls of the ventral hepatic peritoneal cavity. In passerine birds, the left and right oblique septa meet over the liver and form the ventral surface of the dorsal hepatic peritoneal cavities.[5]

Post hepatic septum

The post hepatic septum separates the hepatic peritoneal cavities from the intestinal peritoneal cavity and is composed of the paired linings of the hepatic and intestinal peritoneal cavities. It originates dorsally, caudal to the hilus of the lung between the origins of the oblique septa and forms the dorsal wall of the hepatic peritoneal cavity. From its dorsal origin, the post hepatic septum descends gradually in a caudoventral direction to reach the ventral body wall a variable distance caudal to the sternum, depending on the species.[1,9]

The Respiratory System

The lung-air sac system

Birds are an extraordinarily diverse clade, with over 11,000 known species.[12] While the anatomy of the respiratory system has been superficially described in many species from latex casts (e.g., Duncker, 1971), a detailed three-dimensional (3D) comparative analysis of the respiratory anatomy of avian pulmonary anatomic diversity not been conducted.[9] Accurate description of the avian coelomic cavities is challenging, particularly due to the complex nature of the formation of lung and air sac system during embryonic development and the complete functional decoupling of the gas-exchanging lung and the ventilatory air sacs. Sauropsids (reptiles and birds), all have structurally heterogeneous lower respiratory systems, and vast structural and functional diversity of their pulmonary tissues exists across the clade;[13] however, it is only birds that have evolved a complete functional decoupling of the gas-exchanging lung from the ventilatory air sacs (For a review comparing the lungs of birds to their non-avian sauropsid relatives, see Schachner and Moore.[11]). This heterogenous morphology found in sauropsids is in contrast to the predominantly homogenous, tidally ventilated mammalian bronchoalveolar lung, in which the bronchial branches following a tree-like pattern terminating in alveoli where gas-exchange occurs across a flexible membrane.[14,15]

The avian respiratory system is composed of a volume constant, immobilized gas-exchanging lung (here termed just the lung), unidirectionally ventilated by a series of compliant air sacs that expand as dilations from the primary and secondary airways during development.[11,16-18]

The lung is bounded ventrally by a horizontal septum anchored to the vertebral hypapophyses medially and the proximal heads of the sternal ribs laterally, which keeps the exchange parenchyma completely immobilized in the right and left pulmonary cavities and isolates the lungs from the ventilatory air sacs.[4,11,19,20]

In most birds, there are nine air sacs: a single unpaired sac that sits around the syrinx and heart, and four paired sacs.[5] The unpaired interclavicular sac (also called the clavicular sac) branches off of ventrobronchus three (a secondary bronchus) early in development and fuses along the ventral midline with its contralateral counterpart.[16] The remaining four air sacs are paired: (1) cervical air sacs emerge from the first ventrobronchus and extend cranially towards the thoracic inlet; (2) cranial thoracic air sacs branch from ventrobronchus three and extend laterally towards the inner wall of the thoracic ribs and sternum; (3) caudal thoracic air sacs expand from a large ventrally directed laterobronchus and project caudally; and (4) the abdominal air sacs balloon off of the caudal margin of the primary bronchus and/or adjacent neopulmonic parabronchi in some taxa.[5,21-23]The ventilatory sacs pass through ostia in the horizontal septum during embryonic development and all of them, with the exception of the abdominal air sacs, pass into the subpulmonary cavity, which is located ventral to the lungs with the horizontal septum acting as the dorsal ceiling, and the oblique septum serving as the medial and caudal boundaries.[8,20,23]

The oblique septa are composed of a continuation of the horizontal septum and retain the sacs, with the exception of the abdominal sacs, into their own separate cavity unique to birds. The abdominal air sacs are the only sacs that, after passing through the horizontal septum, expand into the intestinal peritoneal cavity (often referred to as the abdominal cavity).[8,11] Kiwis are the only known birds in which the abdominal air sacs are retained within the subpulmonary cavity, and do not extend into the peritoneal cavity.[5] While this is the standard avian *bauplan*, many birds have more than the typical nine air sacs, or fewer, and sacs frequently fuse together.[9,21] In all other avian taxa, the abdominal air sacs interweave freely with the abdominal viscera; however, depending on the species, their size is highly variable with large abdominal air sacs being recorded in grey parrots (*Psittacus erithacus*)[21] and very small ones in zebra finches (*Taeniopygia castanotis*) even when inflated to maximum inspiratory capacity.[22]

In addition to the ventilatory air sacs, both these sacs and the lung give rise to a highly variable series of non-ventilatory diverticula that can leave the body (i.e., the primordial coelomic) cavity entirely, pneumatize the postcranial skeleton, wrap around joints, dive between muscles, and extend around the body and limbs.[11,24-27]

These diverticula are poorly understood, are not well mapped across Aves, yet are clinically relevant, particularly with respect to planning surgical approaches. In ostriches these diverticula extend out to at least the knee joint, and in many large plunge diving birds like pelicans, the diverticula wrap around the body and limbs between the muscles and the skin.[13,28] Thus, any insult or trauma that penetrates air sacs and diverticulae can drastically impact the ventilatory biology and the health and welfare of the animal.

Diagnostic Imaging Techniques

There are a variety of imaging techniques available for use in shaping a diagnosis in avian practice. Projection radiographic imaging, fluoroscopic imaging, ultrasound imaging, CT imaging are the most common options reflected in the literature to-date.[29] Somewhat contradictory to some thought processes, these imaging modalities are not mutually exclusive; a patient may need to be imaged in multiple methods to best shape up a functional working diagnosis and regionalize a particular area of anatomic pathology.

Projection radiographic imaging

Traditional projection radiographic techniques remain one of the most common and cost-effective diagnostic procedures in avian medicine. Projection radiographic imaging may be either screen-film or digital radiographic methods.[29] Of the digital methods, there are computed radiography (CR) and direct radiography (DR), with direct radiography seeming to be the more commonly utilized in practice at present time. Patient positioning is important for optimizing the diagnostic value of projection radiographic imaging studies.[29] Under normal circumstances, the coelomic cavities of birds have minimal serosal fluid content. Air sacs are visible, with this imaging modality, particularly when there is pathology present. The accumulation of effusion and debris, an increase in visceral organ size, or an increase in other intraperitoneal tissues may increase the interpretive sizes of the five peritoneal cavities radiographically. Air sac pathology may make interpretation of the coelomic cavities more challenging, and in some cases, can extend into those cavities (e.g., aspergillosis). Particularly with fluid production and accumulation, projection radiographic imaging can result in a decrease in detail, rendering the study more challenging to interpret. These challenges in interpretive detail can make effective localization of which organ system or which coelomic cavities are involved with a suspected disease process difficult. Without additional supportive imaging studies, some of these interpretations are vulnerable to both over assessment and under assessment.

Fluoroscopic imaging

Fluoroscopic imaging, often using barium or iohexol contrast can provide substantial information through the demonstration of real-time gastrointestinal (GI) contractions. Unlike a series of still barium contrast images using projection radiographic methods, fluoroscopy offers continual motion assessment, helping to pinpoint issues with contractility and to localize functional deficits or specific lesions of interest. In addition, fluoroscopy can be used to define, by subtraction of the GI elements, the shape of the hepatic peritoneal cavities and space-occupying structures within the intestinal peritoneal cavity. Unlike with projection radiography, there is no anesthesia, sedation or patient positioning required for most fluoroscopic imaging studies. Patients are imaged either in a standing or perching position, without the need for restraint, offering the diagnostician a minimally stressful imaging option and modality. Fluoroscopy also allows for the effects of gravity to be evident, as compared with standard contrast projection radiographic imaging.[29] Although fluoroscopy is still not often utilized in avian private practice, it is an almost daily diagnostic tool that this author uses in practice. Used C-arm fluoroscopes can be relatively inexpensive to obtain, and for those colleagues who have CT imaging capabilities, the fluoroscopic imaging function of most machines can be used.

Ultrasound

Ultrasound is an increasingly common imaging modality in avian practice. Linear array transducers are typically used for B-mode (brightness) imaging, which is the typical grey scale mode used for anatomic/structural evaluation of organs, and phased array for M-mode (motion) imaging, which is the mode used during echocardiography. Transducers can have varying footprints/shapes and sizes. Microconvex transducers that have a small, curved footprint are most amenable to avian patients.[29] Ultrasound imaging offers the potential to evaluate the coelomic cavities of birds as well as their individual organs within. Air filled structures, such as the lung and air sacs, are not conducive to ultrasound evaluation. The findings of ultrasonographic examinations can help localize exactly which specific coelomic cavity or cavities may be involved in a specific disease process and can help guide a targeted centesis, fine needle aspirate, or a specific surgical approach if indicated. The doppler function can help identify the presence or absence of vascularity in the area, and direction of blood flow. Ultrasound is the only imaging modality currently that can provide functional cardiac data (echocardiography) in birds.

Surgical Procedures

Paracentesis

Techniques for withdrawing a sample of fluids from the abdomen have been described and are relatively simple to perform. A detailed understanding of the regional anatomy of relevant regional body cavities is important to minimize inadvertent organ trauma, and blind centesis is becoming increasingly discouraged for this reason. Fluid distension that is recognized can be in the pericardial,[31] hepatic peritoneal or intestinal peritoneal cavities, or combinations of these locations.[1] Transillumination through the abdominal wall in small bird species prior to and during paracentesis can greatly facilitate locating a site for puncture and can aid in helping the operator(s) avoid causing inadvertent trauma to underlying structures. Ultrasound imaging can often help localize the specific cavity to be targeted, and to help minimize risk of iatrogenic trauma. Without guidance with additional imaging (e.g., transillumination, ultrasound), there is an increased risk of perforation or laceration of the intestine, perforation or laceration of the liver, or other internal lesions and vessels within the abdomen. Pericardiocentesis is best performed with ultrasound guidance.

Endoscopy

The air sac system is often used in birds to allow movement of the endoscope within the body so that structures can be examined through their normally thin, clear walls. Endoscopic entry directly into the left caudal thoracic air sac from the body wall can allow that the associated cranial thoracic air sac and abdominal air sac walls may be punctured and all three air sacs can be entered from a single body wall incision, which would allow a minimally traumatic, yet efficient unilateral examination.[1] From this left caudal thoracic air sac endoscopic approach, the left abdominal air sac is entered by puncturing the saccoperitoneal membrane and the gonad, oviduct or ductus deferens, rectum, coprodeum, urodeum, spleen, the left portions of the proventriculus and ventriculus, and some loops of the small intestine are visible suspended within the intestinal peritoneal cavity. Located in a dorsal retroperitoneal position, the adrenal gland, the three divisions of the left kidney, and the left ureter are also visible. The intestinal peritoneal cavity can be entered directly from the body wall without penetrating an air sac by using one of two approaches. The intestinal peritoneal cavity is a potential space and may need to be insufflated or the abdominal sac otherwise manipulated to reduce its size so that the intestinal peritoneal organs can be visualized and accessed. A midline, paramedian, or transverse incision in the ventral body wall will enable one to enter the intestinal peritoneal cavity when the incision is made caudal to the end of the post hepatic septum and medial to the distal limits of the cranial thoracic air sac and abdominal air sac. This location varies somewhat between species. The other site for direct entry from the body wall to the intestinal peritoneal cavity is the left or right post ischial approach.[1] A skin incision dorsal to the pubis and just caudal to the ischium on either the right or left side allows for a blunt puncture through the thin body wall dorsolateral to the coprodeum. The tip of the endoscope first enters the lateral and caudal intestinal peritoneal cavity, and the caudal wall of the respective abdominal air sac can be seen directly ahead of the telescope.[1]

Diseases and Pathophysiology

Pericardial effusion

Pericardial effusion and active pericarditis are relatively common pathologic findings in birds and are characterized by inappropriate accumulation of fluid within the pericardial sac.[32] They can develop in birds because of inflammatory changes of the pericardium (pericarditis), trauma or aneurismal rupture, cardiac or pericardial neoplasia or as a manifestation of metabolic derangements, toxic insult, or, commonly, right-sided congestive heart failure. In some cases, an underlying cause cannot be identified, and the condition is ruled idiopathic.[32] Depending on etiology, effusion may be a pure transudate, modified transudate, or an exudate. An acute onset of clinical signs would be expected with rapid development of pericardial effusion, as this would cause sudden, severe circulatory compromise because of cardiac tamponade. These patients may die suddenly or be presented in a profound state of debilitation and in respiratory distress. A more chronic onset would accompany slow accumulation of fluid over time, because this would permit stretching and increased compliance of the pericardium.[32] If the rate of development of pericardial effusion is slow and is not associated with much fibrosis, the pericardial sac may greatly increase in size before functional cardiac deficits occur.[31,33] Many systemic bacterial infections cause pericardial inflammation as part of their pathogenesis, with *Chlamydia psittaci* being the most well-known. Hydropericardium occurs frequently with pulmonary hypertension, often with ascites of the hepatic peritoneal cavities. Systemic viral infections, including polyomavirus infection in juvenile psittacine birds, will also often cause serous pericardial effusion.[1]

Ascites

Ascites is defined as the accumulation of an effusion of excess fluid within the peritoneal cavities. In birds, these peritoneal cavities include the intestinal peritoneal cavity, and the four hepatic peritoneal cavities. The fluid may contain variable amounts of protein and various inflammatory or neoplastic cells, and/or biochemical constituents, depending on the cause of the effusion.

Hepatic peritoneal cavities

Ascites of the hepatic peritoneal cavities occur in cases of slowly decompensating cardiomyopathy and pulmonary hypertension, leading to right sided heart failure which often progresses slowly, leading to portal hypertension.[1,32] Chronic interstitial fibrosis of the lung and/or chronic obstructive pulmonary disease, resulting in hypoxemia, which is followed by secondary polycythemia, and gradual development of pulmonary hypertension, also can result in cor pulmonale and right heart failure.

As the right heart begins to fail, increasing oncotic pressure within the hepatic portal system leads to the leakage of plasma from the liver sinusoids into the hepatic peritoneal cavities. The four hepatic peritoneal cavities can become very enlarged if the fluid accumulation occurs slowly enough. Once the fluid pressure in the hepatic peritoneal cavities increases to a certain point, overflow to the intestinal peritoneal cavity will occur through the known connection between the left dorsal hepatic peritoneal cavity to the intestinal peritoneal cavity.[1]

Chronic hepatic inflammation from various causes leads ultimately to fibrosis of the hepatic parenchyma as part of the normal attempted healing response. This fibrosis can in turn lead to restriction of portal blood flow, creating portal hypertension.[1] Since most hepatic diseases are diffuse conditions, and the ventral hepatic peritoneal cavities are larger than the dorsal hepatic peritoneal cavities, bilateral ventral hepatic peritoneal cavity ascites is more commonly encountered. The amount of ascitic fluid present in these ventral hepatic peritoneal cavities will vary considerably due to the degree of hepatic damage. Unilateral, or focal hepatic peritoneal effusion into one specific cavity is more likely to be associated with a focal hepatic lesion; hence an accurate description and understanding of which cavities contain effusion is important to guide more focused diagnostic maneuvers such as hepatic biopsy, when indicated.

Primary or secondary metastatic hepatic neoplasia can lead to ascites due to disruption of hepatic portal blood flow. Hepatic carcinoma, advanced biliary duct carcinoma, hemangiosarcoma of the liver, and lymphosarcoma have all been described, associated with the development of ascites in the hepatic peritoneal cavities.[1]

Intestinal peritoneal cavity

Neoplastic diseases of the spleen, ovary, testes, and pancreas have been reported to be associated with the development of ascites in the intestinal peritoneal cavity. It is believed that this develops most commonly by a disruption of drainage of the lymphatic system.[1] Cystic tumors, particularly those of the ovary, testes, and pancreas may also be present which can complicate a diagnostic picture, sometimes mimicking the presence of ascites, as well as sometimes causing the development of ascites.

Peritonitis

Inflammation of the peritoneal lining is, by convention and definition, termed peritonitis. Regardless of the complex partitions of the coelomic cavity of birds, when a disease process associated with inflamed mesothelial linings of the peritoneal cavities is identified, optimal terminology would be peritonitis, as opposed to coelomitis.

Regarding the latter term, there are in addition to the five peritoneal cavities, the pericardial and pleural cavities. All of these eight cavities are true coelomic cavities, and a broad, general term to which could include any or all of these cavities is anatomically, diagnostically and often interventionally inaccurate and inadequate. In addition, the term coelomitis is inaccurate as cavity cannot be inflamed. Inflammation is of the lining(s) of these cavities.

Yolk peritonitis

Inflammation of the parietal or visceral peritoneum of the intestinal cavity may occur with the spillage of yolk at ovulation, the retro propulsion of an unfinished egg back into the intestinal peritoneal cavity, or the rupture of egg and its contents from the oviduct into the cavity.[1] Not all free egg yolk in the intestinal peritoneal cavity causes medically actionable yolk peritonitis. Small amounts of internally ovulated yolk which are absorbed are described in several taxonomic orders of birds. Yolk peritonitis, once the process is initiated and progressing, can result in the production of considerable amounts of fluid, resulting in the distention of the intestinal peritoneal cavity and abdominal walls. Many clinically significant cases of yolk peritonitis will be most often the result of more significant anatomic pathology, including infectious oophoritis, oviductal impaction, neoplastic conditions of the ovary and oviduct, and in some cases, marked obesity. Although a clinical diagnosis of yolk peritonitis may be accurate in an individual patient, this diagnosis may be the secondary result of a primary disease problem which still requires diagnosis and management.

Bacterial peritonitis

Bacterial peritonitis is comparatively uncommon companion psittacine species as compared to Galliformes, Anseriformes and other taxonomic orders, but can occur following foreign body puncture of the GI tract, proventricular ulceration, ascending infection of the oviduct (usually with retropropulsion of egg material back into the IPC), or penetrating wounds.[1] Perforations of the gastrointestinal tract tend to lead to ongoing contamination of the intestinal peritoneal cavity, in many cases, with higher rates of morbidity, sepsis, and mortality.[34]

Hernias

A hernia is defined as the abnormal exit of tissue or an organ, such as the bowel, through the wall of the cavity in which it normally resides. True hernias occur when tissue bulges through a torn or ruptures dividing area of the body or the body wall. These can be internal hernias (i.e., within the body cavity and between two compartments), or externally visible hernias, typically involving the body wall of the abdomen. Herniated organs or tissues may be reducible or may become entrapped. An incarcerated hernia cannot be replaced. A strangulated hernia is a medical emergency because the area bulging through loses blood supply. Pseudo hernias, on the other hand, are an abnormal bulging of the external body wall or an internal dividing wall, but without an associated underlying tissue or muscle defect. Both true hernias and pseudo hernias occur in birds.

Diagnosis of some of these true or pseudo hernias in birds can be challenging, viewing the complexity of their coelomic cavities, their associated septa and the interactions of these structures with the respiratory system.

True hernias involving the brain and abdominal body wall have been described in a number of cases. These have involved the brain and meninges,[35] of the cerebrum in crested Polish chickens,[36] of the ventral body wall in pigeons,[37] in a yellow naped Amazon parrot with cloacal entrapment,[38] in a yellow crowned Amazon parrot,[39] in a nanday conure and eclectus parrot with oviductal entrapment,[40] of the ventral abdominal wall in association with hepatic lipidosis in a red lory,[41] the ventral abdominal wall in a rhea with intestinal entrapment,[42] of the umbilicus and abdominal wall in turkeys,[43,44] and of the caudal ventral abdominal wall and associated with development of a urate concretion.[45] This author has seen a handful of true hernias in birds where the abdominal air sac is being exteriorized through the rent in the body wall. Post surgical cases of abdominal wall herniation are described, including one case in an ostrich following castration[46] In a number of the peer-reviewed case reports, there are incomplete, sometimes inaccurate and at times misleading descriptions of which specific body cavities and their dividing walls were involved. Often, the coelom is used to describe a particular area of herniation or anatomic pathology, rather than the use of more anatomic acuity in word choice. Some of these cases were taken to surgery successfully, others which were taken to surgery were unsuccessful or met with fatal outcomes.

Internal hernias are much more infrequently described in birds, but reported cases include herniation of the post hepatic septum[47] and herniation of the duodenum through the post hepatic septum into the right ventral hepatic peritoneal cavity in a chicken.[48] In uncommon cases, internal herniation of the post hepatic septum with intestinal entrapment in the hepatic peritoneal cavities has been seen by this author in companion psittacine birds. Pseudo hernias also have been described. In a pituitary-dependent case of hyperadrenocorticism in a Moluccan cockatoo, a pseudo hernia was noted.[49]

References

1. Taylor M. Pleura, pericardium, and peritoneum: the coelomic cavities of birds and their relationship to the lung-air sac system. In: Speer B., ed. *Current Veterinary Therapy in Avian Medicine and Surgery.* St Louis, MO: Elsevier; 2016:345-362.

2. Mushonga B, Horowitz A. Serous cavities of the Nile crocodile (*Crocodylus niloticus*). *J Zoo Wildl Med.* 1996;27(2):170-179.

3. Bellairs R, Osmond R. *The Atlas of Chick Development.* San Diego, CA: Elsevier; 2005:77.

4. Fedde MR, Burger RE, Kitchell RL. Anatomic and electromyographic studies of the costo-pulmonary muscles in the cock. *Poult Sci.* 1964;43(5):1177-1184.

5. Duncker HR. Coelomic cavities. In: King AS, McLelland J, eds. *Form and Function in Birds.* London, UK: Academic Press; 1979;1:39-67.

6. Taylor WM. Endoscopic examination and biopsy techniques. In: Ritchie BW, Harrison GJ, Harrison LR, eds. *Avian Medicine: Principles and Application.* Lake Worth, FL: Wingers Publishing; 1994:327-354.

7. Kajigaya H, Kamemura M, Tanahara N, et al. The influence of celomic membranes and a tunnel between celomic cavities on cancer metastasis in poultry. *Avian Dis.* 1987;31(1):176-186.

8. McLelland J, King AS. Aves coelomic cavities and mesenteries. In: Getty R, ed. *Sisson and Grossman's the Anatomy of the Domestic Animals.* 5th Ed. Philadelphia, PA: WB Saunders; 1975:1849-1856.

9. Duncker HR. Coelomic cavities. In: King AS, McLelland J, eds. *Form and Function in Birds.* London, UK: Academic Press; 1979:39-67.

10. Klein W, Riberio VP, DeSouza RBP. Avian air sacs and neopulmo: their evolution, form and function. *Philos Trans R Soc Lond B Biol Sci.* 2025;380(1920):20230421.

11. Schachner ER, Moore AJ. Unidirectional airflow, air sacs or the horizontal septum: what does it take to make a bird lung? *Philos Trans R Soc Lond B Biol Sci.* 2025;380(1920):20230418.

12. Clements JF, Schuelenberg TS, Iliff JJ, et al. The eBird. *The Clements Checklist of Birds of the World.* v2022. Available at: https://www.birds.cornell.edu/clementschecklist/download/.

13. Schachner ER, Hedrick BP, Richbourg HA, et al. Anatomy, ontogeny, and evolution of the archosaurian respiratory system: a case study on *Alligator mississippiensis* and *Struthio camelus*. *J Anat.* 2021;238(4): 845-873.

14. Maina P, vanGils P. Morphometric characterization of the airway and vascular systems of the lung of the domestic pig, Sus scrofa: comparison of the airway, arterial and venous systems. *Comp Biochem Physiol A Mol Integr Physiol.* 2001;130(4):781-798.

15. West B. Comparative physiology of the pulmonary blood-gas barrier: the unique avian solution. *Am J Physiol Regul Integr Comp Physiol.* 2009;297(6):R1625-634.

16. Locy WA, Larsell O. The embryology of the bird's lung. Part 1. *Am J Anat.* 1916;19:447-504.

17. Locy WA, Larsell O. The embryology of the bird's lung. Part 2. *Am J Anat.* 1916;19:1-44.

18. Maina J. Development, structure, and function of a novel respiratory organ, the lung-air sac system of birds: to go where no other vertebrate has gone. *Biol Rev Camb Philos Soc.* 2006;81(4):545-579.

19. Maina JN. Spectacularly robust! Tensegrity principle explains the mechanical strength of the avian lung. *Respir Physiol Neurobiol.* 2007;155(1):1-10.

20. Poole M. Development of the subdivisions of the pleuroperitoneal cavity in birds. 1909;79(2):210-235. doi: 10.1111/j.1096-3642.1909.tb01866.x.

21. Lawson AB, Hedrick BP, Echols S, et al. Anatomy, variation, and asymmetry of the bronchial tree in the African grey parrot (*Psittacus erithacus*). *J Morphol.* 2021;282(5):701-719.

22. Martinez A, Diaz RE, Grand Pre CA, et al. The lungs of the finch: three-dimensional pulmonary anatomy of the zebra finch (*Taeniopygia castanotis*). *Philos Trans R Soc Lond B Biol Sci.* 2025;380(1920):20230420.

23. Klein W, Ribeiero P, Bueno deSouza RB. Avian air sacs and neopulmo: their evolution, form and function. *Philos Trans R Soc Lond B Biol Sci.* 2025;380(1920):20230421.

24. Casler CL. The air-sac systems and buoyancy of the anhinga and double-crested cormorant. *The Auk.* 1973;90(2):324-340.

25. Daoust PY, Dobbin G, Ridlington Abbott RC, et al. Descriptive anatomy of the subcutaneous air diverticula in the northern gannet *Morus bassanus*. *Seabird.* 2008;21:64-76.

26. Moore AJ, Schachner ER. When the lung invades: a review of avian postcranial skeletal pneumaticity. *Philos Trans R Soc Lond B Biol Sci.* 2025;380(1920):rstb20230427.

27. Schachner ER, Moore AJ, Martinez A, et al. The respiratory system influences flight mechanics in soaring birds. Nature. 2024 630(8017):671-676.

28. Richardson F. Functional aspects of the pneumatic system of the California brown pelican. *The Condor.* 1939;41(1):13-17.

29. Dennison S. Common imaging modalities: selecting the right modality and getting the most out of your imaging. In: Speer B, VanZeeland YRA, eds. *Current Therapy in Avian Medicine and Surgery.* Vol. 2. St. Louis, MO: Elsevier; 2025:29-51.

30. Campbell TW. Cytology. In Ritchie BW, Harrison GJ, Harrison LR, eds. *Avian Medicine: Principles and Application.* Lake Worth, FL: Wingers Publishing; 1994:199-222.

31. Straub J, Pees M, Enders F, et al. Pericardiocentesis and the use of enalapril in a Fischer's lovebird (*Agapornis fi scheri*). *Vet Rec.* 2003;152:24-26.

32. Fitzgerald BC, Beaufrere H. Cardiology. In: Speer B, ed. *Current Veterinary Therapy in Avian Medicine and Surgery.* St Louis, MO: Elsevier; 2016:252-328.

33. Straub J, Pees M, Krautwald-Junghanns ME. Diagnosis of pericardial effusion in birds by ultrasound. *Vet Rec.* 2001;149:86-88.

34. Hoefer H, Levitan D. Perforating foreign body in the ventriculus of an umbrella cockatoo (*Cacatua alba*). *J Avian Med Surg.* 2013;27(2):128-135.

35. Gartrell, BD, Argilla LS, Chatterton J, et al. Surgical repair of a meningoencephalocoele in a kākāpō (*Strigops habroptilus*). *N Z Vet J.* 2021;69(4):247-254.

36. Yoshimura K, Kinoshita K, Mizutani, et al. Inheritance and developmental pattern of cerebral hernia in the crested Polish chicken. *J Exp Zool B Mol Dev Evol.* 2012 318(8):613-620.

37. Amer MS, Hassan E, Torad FA. Radiographic and ultrasonographic characteristics of ventral abdominal hernia in pigeons (*Columba livia*). *J Vet Med Sci.* 2018;80(2):292-296.

38. Murphy B, Fitzgerald B, Olsen G, et al. Diagnosis and surgical repair of an acute abdominal wall hernia and partial cloacal strangulation in a yellow-naped Amazon parrot (*Amazona auropalliata*). *J Avian Med Surg.* 2018;32(4):336-341.

39. Barboza TK, Beaufrere H, Chalmers H. True coelomic hernia and herniorrhaphy in a yellow-crowned Amazon parrot (*Amazona ochrocephala*). *J Avian Med Surg.* 2018 32(3):221-225.

40. Anderson K, Brandao J, Mans C. Lateral body wall herniation involving the oviduct in two psittacine birds. *J Avian Med Surg.* 2018;32(4):328-335.

41. Langlois I, Jones MP. Ventral abdominal hernia associated with hepatic lipidosis in a red lory (*Eos bornea*). *J Avian Med Surg.* 2001;15(3):216-222.

42. St Jean G, Anderson DE. Ventral abdominal hernia with intestinal incarceration in a rhea. *Agri-Pract.* 1995;16(10):20-24.

43. Ranck FM. Umbilical hernias in turkeys. *Poult Sci.* 1973;52(5):1991-1992.

44. Carlson HC. Abdominal hernia in turkey poults. *Can Vet J.* 1962 3(8):263-264.

45. Martin HD. Abdominal hernia with formation of a urate concretion in a cockatiel. *J Am Vet Med Assoc.* 1986;189(10):1332-1333.

46. Pye GW. Intestinal entrapment in the right pulmonary ostium after castration in a juvenile ostrich (*Struthio camelus*). *J Avian Med Surg.* 2007;21(4):290-293.

47. Goodchild WM. Hernia of the posthepatic septum of *Gallus domesticus*. *Avian Pathol.* 1977;6(4):307-311.

48. Haridy M, Sasaki J, Goryo M. Herniation of duodenum into the right ventral hepatic peritoneal cavity with groove formation at the ventral hepatic surface in a 2-week-old chicken. *J Vet Med Sci.* 2013;75(10): 1405-1407.

49. Starkey SR, Morrisey JK, Stewart JE, et al. Pituitary-dependent hyperadrenocorticism in a cockatoo. *J Am Vet Med Assoc.* 2008 232(3):394-398.

Section 5

Technician Session

Bri Bryan, BS, RVT
Session Chair

Overcoming Fear and Building Confidence in Avian Handling: Training Strategies for Veterinary Teams

Daniel Scrivener, MEd, CVT, LVT

Session #170

Affiliation: From Arizona Exotic Animal Hospital ,744 N Center St #101, Mesa, AZ 85201, USA, Veterinary Emergency Group 44 S. Broadway, LL3, White Plains, NY 10601, USA.

Abstract: In specialty exotic animal clinics, one of the greatest challenges is finding and developing veterinary professionals who are confident in handling and treating avian patients. Many individuals in the field experience hesitation or fear when working with birds, often due to a lack of familiarity, concerns about causing harm, or even underlying ornithophobia. This presentation explores the psychological and technical barriers that contribute to these fears and provides a structured approach to training veterinary teams in avian handling and care. Through a combination of progressive exposure, hands-on skill-building, and a strong foundation in avian anatomy and behavior, veterinary professionals can develop the confidence necessary to provide high-quality medical care. By addressing these challenges head-on, veterinary teams can enhance their skills, reduce patient stress, and improve overall outcomes in avian medicine.

Introduction

Handling birds in a veterinary setting can be challenging, even for seasoned professionals. Birds, whether parrots, raptors, or poultry, have unique anatomies and behaviors that require gentle, skillful handling. Thirty-minute training sessions can significantly boost your team's comfort and competence with avian patients, but keep in mind, repetition will ultimately increase the capabilities and comfort levels over time.

Veterinary professionals need to create a supportive learning environment, continuously update handling techniques when appropriate, and constantly adjust training strategies to build staff confidence through practice and positive reinforcement. Our goal is to help veterinary managers, technicians, and doctors train their teams to handle birds safely and empathetically, ensuring both staff and avian patients have a low-stress experience. By the end, you'll have a roadmap for fostering patience, empathy, and continuous learning in your practice.

Creating a Supportive and Inclusive Learning Environment and Overcoming Fear

A successful training program begins with a supportive, inclusive, patient atmosphere. Team members must feel safe to admit their uncertainties and learn at their own pace. Research in low-stress handling emphasizes that calm, leadership-led demonstrations during training are key: leadership should model gentle animal handling in a safe, constructive space and constantly encourage questions from their team.[1] In practice, this means managers and senior clinicians should openly share their own learning experiences and reassure staff that hesitation or mistakes during training are normal, also, including previous inhibitions regarding handling and noting that a graded approach to exposure is the better evidence-based fix.[2] An inclusive environment values every team member's input, whether they are veterinarians, technicians, or assistants, and encourages mutual respect. Empathy and support go a long way. When trainees feel respected and understood, they are more likely to stay engaged and gradually build confidence. Remember, everyone learns and copes differently, so incorporating various teaching styles (discussion, demonstration, hands-on practice), along with stair-stepping exposure to uncomfortable situations, ensures no one is left behind. Above all, patience is critical: allow each person the time they need to become comfortable without pressure or judgment.

Understanding Avian Behavior and Stress

Before diving into techniques, your team needs to understand why birds require special handling considerations. Most avian patients are prey species by nature, and a visit to the clinic can be extremely stressful for them. Unfamiliar environments, being handled by strangers, and loud noises can trigger fear responses. Birds are also highly intelligent and often neophobic; they may react fearfully to new objects or people. For example, a parrot that has never seen a stethoscope may lean away or try to escape when one approaches. Signs of avian stress can be subtle: birds might vocalize in alarm, pant, visibly tense their body, or even play dead by freezing . In extreme cases, acute stress can lead to serious health consequences like cardiac arrest in already ill birds.[3] This highlights why our handling must be gradual and mindful. As you train your team, emphasize empathy, encourage them to put themselves in the bird's shoes and recognize the bird's perspective. A calm, quiet setting can alleviate fear: for instance, use a dedicated exam room away from barking dogs and sudden disruptions. The more your staff appreciate these behavioral insights, the more patient and careful they will be when approaching and restraining avian patients.

Humane Handling Techniques Across Avian Species

Up-to-date handling and restraint techniques that prioritize safety and welfare for both birds and handlers. Modern avian handling is grounded in the Fear-Free or low-stress approach: use the minimal amount of restraint necessary, applied gently and confidently.[3] Every team member should learn to recognize when a bird is becoming excessively stressed and know how to pause or adjust their technique.

- General principles
- When not to handle
- Towel restraint for parrots and small birds

In training, demonstrate this technique step-by-step. Let team members first practice the towel motion on a stuffed animal or model, then on a calm bird under supervision. Highlight how the towel is held and how to quickly wrap and unwrap the bird, always watching the bird's response.

Building Confidence Through Hands-On Training

Technical knowledge alone isn't enough; building your team's confidence in handling birds is crucial. Confidence comes from practice and positive experiences. A proven approach is to give team members gradual, hands-on training opportunities without the pressure of immediate performance (4). Start your training program with a demonstration by an experienced handler so staff can observe proper technique. Then move to low-stakes practice: for instance, using a stuffed bird or model to rehearse hand positions, or practicing towel wraps on a dummy. This may sound silly, but it allows people to focus on technique without a live animal reacting. Next, introduce a real bird that is calm or well-socialized (perhaps a staff member's gentle pet parrot or a reliably tame clinic bird). Let each trainee practice basic holds and maneuvers under supervision, with no expectation of perfection on the first tries. As one veterinary leadership coach puts it, "One of the best ways to help team members build confidence is to give them opportunities to practice without the pressure of being evaluated or judged." [4] Repetition is key. The more times someone successfully handles a bird, even a compliant one, the more their fear will diminish.

In these practice sessions, encourage stepwise progression for each skill. For example, when learning towel restraint, one might first practice approaching the bird with the towel, then pausing. Next, practice draping the towel without fully capturing the bird, and finally move to a full restraint. This incremental approach lets the handler and bird acclimate to each step.[4] Similarly, for raptor handling, one might first get comfortable wearing the glove and approaching the bird, then practice the grip on the legs with an instructor guiding, before doing it solo. Breaking tasks into smaller steps helps reduce anxiety by focusing on one challenge at a time.

Another confidence-building method is to use role-playing and group exercises. After hands-on attempts, bring the team together to discuss what happened. Create an open forum where they can share feelings like "I was nervous the bird would bite me when I reached for its head," or "I wasn't sure how tightly to hold." Facilitating a group discussion helps team members process these experiences and learn from each other's perspectives. A trainer or manager can moderate, addressing misconceptions and highlighting what went right. It's also beneficial to role-play scenarios. For example, act out how to catch an escaped bird in the exam room, or how to transfer a bird from one handler to another safely. These simulations allow staff to think through their responses in a controlled setting. According to veterinary education experts, active learning techniques, like small group discussions, role play, and case-based scenarios, improve retention and skill acquisition. By engaging in these interactive exercises, staff not only learn techniques but also build mental preparedness to handle unexpected real-life situations calmly.

Positive Reinforcement and Encouragement

Just as animals respond to positive reinforcement, people learn best in an environment of encouragement and support. Make it a point to celebrate progress, not just perfection. If a typically timid technician successfully holds a cockatiel for an exam, applaud that achievement and provide specific praise like "Great job keeping your hold gentle while still secure, the bird stayed pretty calm because of that." Recognizing these small wins builds confidence. Research on veterinary team performance shows that confidence builds when someone sees themselves improving, even if they haven't mastered the skill yet.[4] So, acknowledge every improvement, no matter how minor: a steadier grip, a calmer approach, a better read of the bird's body language. This kind of positive feedback reinforces the desired behavior (gentle, confident handling) similar to how we would reward an animal learning a new trick.

It is also important to provide constructive feedback in a supportive way when mistakes occur. If someone restrains a bird too tightly or hesitates and allows an escape, frame the discussion as a learning opportunity rather than a failure. For instance, "I noticed the parrot started panting, a sign we were holding too tight. Next time, we'll adjust your hand position to avoid that." This approach maintains psychological safety and encourages team members to keep trying. You can further boost morale by pairing less experienced handlers with mentors who coach them through early attempts. Over time, as skills improve, each person's responsibilities gradually increase, like having them be the primary handler for a procedure with a bird and afterward debrief to highlight what they did well. By consistently reinforcing progress and providing patient coaching, you create a positive cycle: competence breeds confidence, which in turn breeds more willingness to take on challenges.

Goal Setting and Individualized Training Plans

Every staff member comes into avian handling training with a different baseline comfort level and skill set. Some may have prior experience with pet birds; others may be completely new and quite nervous. It's essential to identify individual training needs and tailor the program accordingly. One effective approach is to use goal setting as a tool for personalized development. Work with each team member to set SMART goals: specific, measurable, attainable, relevant, and time-bound, related to avian handling. For example, a specific goal might be, "Within the next month, Jane will confidently restrain a medium-sized parrot for a routine exam without assistance." Goals give learners a clear target and a sense of purpose, and achieving them builds confidence. Establishing SMART goals for veterinary staff can help them "develop their interests, acquire new skills, become more confident, and advance their careers."[5] Invite each person to voice their own interests and concerns when setting these goals. Perhaps one technician is eager to learn raptor handling (interest), while another mostly wants to overcome her fear of being bitten by parrots (concern). Collaborative goal setting allows you to address both enthusiasm and anxiety: let them pursue skills they're excited about, but also include goals that strengthen any weak spots.[5] For instance, you might pair an interest-based goal (e.g., learn to apply falconry gloves and hold a hawk on the fist) with a challenge goal (e.g., perform a nail trim on a parrot using towel restraint with supervision).

Once goals are set, outline a custom training plan for each person. Someone very anxious may need extra one-on-one sessions in a quiet setting, while a more confident person might progress through tasks faster or benefit from advanced workshops. Check in periodically on these goals to measure progress and adjust as needed. If a goal is met early, set a new one to keep momentum; if a goal is proving too difficult, break it into smaller steps or extend the timeline. The point is to ensure continuous, tailored growth. And remember to celebrate when goals are achieved, not only does this reward the individual, but it also shows the whole team that the training process yields real, positive results.[5]

In group training sessions, you can also incorporate individualized attention by having each person demonstrate a skill they've been working on and receive gentle feedback. This recognizes that different people may excel in different areas. Perhaps one technician has become great at holding small budgies, while another has a knack for calming larger parrots. Acknowledge these strengths and consider assigning go-to roles (e.g., Emilie is our small-bird expert), which can further build confidence and peer recognition. Be careful to rotate responsibilities so everyone continues to improve in all areas; no one should be pigeonholed (pun intended) into only one type of handling. The ultimate aim is a well-rounded team where each member feels competent and comfortable handling any avian patient that comes through the door.

Fostering Patience, Empathy, and Continuous Learning

As your team grows more skilled, reinforce the core values of patience, empathy, and continuous learning. Handling birds, and indeed all animals, requires a great deal of empathy, not just for the animals, but for each other. Make it a mantra that empathy and compassion are as important as technical skill in veterinary care. By tuning into a bird's subtle cues of discomfort, as well as a colleague's anxieties or frustrations, the team can respond with understanding and adjust their actions accordingly. Encourage staff to remain patient with difficult cases; for example, a fractious macaw that screams and resists might require slow desensitization over multiple visits. Rushing or losing temper will only set back progress. Instead, an empathetic approach, like pausing the exam to give the bird a break or having the same technician handle the bird each time to build trust, can yield better outcomes in the long run. Reinforce that it's ok to take a step back and breathe if a situation is getting tense. This patience will prevent mistakes and injuries and model to newer staff that staying calm is part of the protocol.

Empathy within the team is also critical. If a colleague is struggling with a skill, teammates should offer help or at least moral support. Fostering a non-judgmental culture, where asking for help is seen as a strength, not a weakness, will make the learning process continuous and sustainable. For instance, a veterinarian might openly empathize with a technician by saying, "I used to get nervous restraining birds, too; it gets easier with time, and I'm here to help you." Such statements build trust and encourage open communication.

Finally, instill a mindset of continuous learning and improvement. Veterinary medicine is always evolving, especially in exotic animal care. What we consider best practice today (e.g., fear-free handling techniques) might be refined further tomorrow. Urge your team to stay curious, attend continuing education courses or webinars on avian care, read articles (e.g., new research on bird anesthesia or handling), and even pursue advanced certifications if interested (e.g., Fear Free^{SM} Avian certification.[6] As a manager or training leader, you can set the example by sharing new things you've learned and perhaps hosting mini-refreshers during staff meetings. Emphasize that no one ever finishes learning animal handling, even experts encounter new scenarios and can learn new tricks. This humility in learning ensures your team will keep honing their skills rather than becoming complacent. We should never be finished with learning; it is part of being a veterinary professional. Encourage team members to reflect on cases involving birds. What went well and what could be improved next time? Perhaps start a quick debrief after each avian appointment, turning everyday work into learning moments. Over time, this continuous improvement attitude will significantly raise your team's proficiency and confidence with birds.

Conclusions

Training your veterinary team to handle and restrain avian species confidently is an investment that pays off in numerous ways. With a supportive, patient training environment, staff members feel safe to grow and ask questions. By learning humane, low-stress handling techniques, like gentle towel restraint and careful control of wings and talons, your team can ensure the safety and well-being of both birds and them. Through hands-on practice, guided repetition, and positive reinforcement, even the most apprehensive team members can gain the confidence to manage feisty parrots or powerful raptors. We've discussed how setting individualized goals and nurturing each person's development leads to steady improvement and empowerment. Perhaps most importantly, we've highlighted that cultivating patience, empathy, and a love of learning creates a team culture where everyone is motivated to continually improve their skills.

By implementing these strategies in a 30-minute training format (and follow-up refreshers), you create a ripple effect. Your staff become more comfortable and skilled, which means avian patients experience less stress and better care, and clients see a competent, compassionate team handling their cherished birds. Over time, your clinic builds a reputation for excellence with avian patients, opening doors to helping more feathered friends. The journey to competency is gradual, much like training a parrot to step onto a hand; it takes time and trust, but with the right approach, every team member can become confident in avian handling. Encourage your team to celebrate each success and learn from each challenge. With supportive leadership and continuous practice, handling birds will become as routine as handling cats and dogs, and your team will take pride in this hard-earned skill set. Together, by fostering an inclusive and encouraging training environment, you empower your veterinary team to spread their wings and truly shine in avian care.

References

1. Yin S. A visit to the Medical Center for Birds. CattleDog Publishing. Available at: https://cattledogpublishing.com/blog/a-visit-to-the-medical-center-for-birds. Accessed July 14, 2025.

2. Virtual CBT Psychotherapy. Overcoming ornithophobia: effective bird phobia treatment. Available at: https://www.virtualcbt.ca/therapy-blog/bird-phobia-treatment-ornithophobia. Accessed July 14, 2025.

3. Hollwarth A. Fear-free avian practice–minimising stress for birds in veterinary practice. Improve Veterinary Practice. Available at: https://www.veterinary-practice.com/article/minimising-stress-for-birds. Accessed July 14, 2025.

4. Hall R. How to help your veterinary team build confidence. VetLead. Available at: https://www.vetlead.com/help-veterinary-team-build-confidence. Accessed July 14, 2025.

5. Rumple S. SMART Goals for veterinary technicians: how they boost your practice performance. Practice Live. Available at: https://www.practicelife.com/en/latest/smart-goals-for-veterinary-technicians-how-they-boost-your-practice-performance/. Accessed July 14, 2025.

6. Fear Free LLC. Veterinary Certification Program-Avian. Fear Free. Available at: https://fearfreepets.com/fear-free-certification-overview-avian/. Accessed July 14, 2025.

Avian Trauma Triage: Comprehensive Assessment and Stabilization in Emergency Scenarios

Daniel Scrivener, MEd, CVT, LVT

Session #187

Affiliation: From the Arizona Exotic Animal Hospital,744 N Center St #101, Mesa, AZ 85201, USA, Veterinary Emergency Group 44 S. Broadway, LL3, White Plains, NY 10601, USA.

Abstract: Triage and stabilization of avian trauma cases present unique challenges due to the species' specialized anatomy and physiology, high stress response, and diverse injury presentations. From self-inflicted wounds such as broken nails and blood feathers to more severe cases like traumatic brain injuries, a structured approach to avian emergency care is essential. This presentation provides a step-by-step guide to effective avian triage, focusing on rapid visual assessment, hands-on physical evaluation, and prioritization of life-threatening conditions. Attendees learn key strategies for stabilizing critical patients, including oxygen therapy, fluid resuscitation, hemorrhage control, and pain management, while minimizing handling stress. In addition, case-based discussions highlight common avian trauma presentations and evidence-based treatment protocols.

Unique Challenges in Avian Trauma Triage

Emergency triage of avian patients presents distinct challenges due to birds' different anatomy, physiology, and behavior. Many avian species kept as pets are prey species, evolutionarily adapted to hide signs of injury or illness to avoid predation.[1] This means by the time a bird shows obvious distress (e.g., sitting on the cage bottom, open-beak breathing, severe weakness), it is often in critical condition. In addition, restraint stress can be life-threatening for birds; struggling or excessive handling can trigger dangerous catecholamine surges and oxygen demand that an injured bird cannot sustain.[2] Key anatomic and physiologic differences influencing triage include those noted below.

- Respiratory system
- High metabolic rate and low reserve
- Blood volume and hemorrhage
- Fragile, pneumatic skeleton
- Lack of easily visible mucous membranes
- Behavioral response to injury

Regarding behavioral response, a frightened or pained bird may thrash or freeze. They often enter a state of shock that can be exacerbated by human intervention. Thus, the triage approach must balance obtaining critical information with keeping the bird as calm as possible.

In summary, while fundamental emergency principles (i.e., airway, breathing, circulation) apply to birds as they do to cats and dogs,[3] veterinary teams must adapt techniques to accommodate avian patients' delicate physiology and intense stress responses. Early stabilization and a gentle, efficient approach are paramount.

Step-by-Step Triage Approach for the Avian Patient

A structured approach to avian trauma triage ensures that life-threatening problems are rapidly identified and addressed while minimizing additional stress. The following stepwise protocol is evidence-based and practical for veterinary nurses and technicians.

1. Initial rapid visual assessment (hands-off examination)

The first step is a swift, hands-off observation of the bird upon arrival.[4] Before touching the patient, note its respiratory pattern, posture, mentation, and any obvious injuries. Key observations to make are noted below.

- Respiratory distress
- Level of consciousness and posture
- Bleeding or obvious trauma
- Behavior and stance

During this visual assessment, do not disturb the bird unnecessarily. If the bird is in a carrier, observe through the carrier. You may even place the entire carrier into a warmed oxygen-rich incubator if the bird appears severely dyspneic or weak. This provides stabilization (heat and oxygen) while you continue the evaluation. The goal is to gather essential information on airway, breathing, and circulation (ABC) status and apparent injuries in just a few seconds.

2. Triage decision and prioritization

Based on the initial observation, decide how urgent the situation is and what needs to be done immediately. Triage in avian emergencies follows the same principles as small animals in determining urgency. Always prioritize life-threatening issues first. A useful memory aid is focusing on ABC: airway (is it open and clear?), breathing (is the bird ventilating effectively?), circulation (are there signs of shock or severe bleeding?). In birds, D for disability (i.e., neurologic status) is also crucial; if a bird is seizing or comatose, that is a critical problem requiring immediate intervention.

3. Gentle physical examination (minimizing handling stress)

Once immediate needs are identified and addressed as able, the veterinary team can perform a targeted physical examination. Prepare all necessary equipment before touching the bird (e.g., have your scale, light source, oxygen source, fluids, bandages, and other materials ready). The examination should be as thorough as the situation allows, but brief and gentle.

- Restraint
- Examination priorities
- Pain and trauma survey

Throughout the examination, keep the bird's stress minimal. Use a quiet room, dim the lights, and ensure no predators like cats or dogs are nearby. If the bird at any point has worsened breathing or becomes limp, stop and return it to oxygen and heat immediately. In some cases, it may be necessary to sedate the bird for examination if pain or anxiety is extreme. Midazolam (1-3 mg/kg IM or intranasal) with or without butorphanol (1-2 mg/kg IM/IN) is commonly used to provide a calming effect and analgesia.[5] Sedation can be lifesaving in an agitated bird, preventing them from thrashing themselves into cardiopulmonary collapse, and these medications are reversible (e.g., flumazenil can reverse midazolam). The need for sedation should be assessed during the hands-off examination and used judiciously if restraint itself is too risky.

4. Identification and prioritization of life-threatening conditions

As you complete the primary survey of the bird, identify any conditions that require immediate stabilization, even if you haven't finished a full examination. The following are common life-threatening findings in avian trauma.

- Airway obstruction
- Respiratory failure or distress
- Severe hemorrhage
- Profound shock
- Traumatic brain injury (TBI)/severe neurologic signs
- Other critical injuries

By the end of this triage and primary survey, you should have a clear idea of any immediate interventions needed and have begun them. The next phase is to actively stabilize the bird's vital functions so that it can survive long enough for secondary evaluation and definitive treatment. Remember, in avian emergencies, stabilization takes priority over diagnostic procedures; x-rays or extensive handling can wait until the bird's oxygenation, perfusion, and pain are under control.

Key Stabilization Techniques in Avian Emergencies

Stabilization in avian trauma focuses on supporting the critical needs of oxygenation, circulation (i.e., fluid volume), hemostasis, and analgesia. Each of these interventions must be tailored to birds' physiology.

Oxygen therapy and thermal support

Providing oxygen is often the first and most important stabilization step for an injured bird. Hypoxemia will rapidly worsen shock and metabolic acidosis. Techniques for oxygen supplementation are provided below.

- Oxygen cage or incubator
- Face mask or flow-by oxygen
- Intubation and ventilation
- Monitor oxygen needs

Alongside oxygen, thermal support is critical. Most injured birds present with hypothermia or are at risk of it.[6] Warm the bird gradually using incubators, warm towels, or heat lamps (at a safe distance). Aim for an ambient temperature in the bird's comfort range (e.g., around 80-85 °F for most pet birds, higher for very small or young birds). Avoid overheating, monitor if the bird starts panting, which could also be stress or pain. Never place an unstable bird in a cold examination room; always provide supplemental heat as part of stabilization.

Fluid resuscitation (routes, types, and volumes)

Fluid therapy is the cornerstone of treating shock in avian trauma. Due to birds' small size and potential dehydration or blood loss, judicious fluid administration can drastically improve perfusion. Key points for fluid resuscitation are noted below.

- Preferred routes
- Subcutaneous (SC) fluids
- Fluid types
- Volumes and rates
- Monitor response

By promptly restoring intravascular volume, we improve oxygen delivery and buy time for the bird's body to recover. Remember that birds in shock may not show dramatic vital sign changes as they improve, so also rely on trends (e.g., slowly improving activity or alertness, better capillary refill).

Hemorrhage control (including blood feather management)

Bleeding control is a top priority in any trauma. Even though birds can theoretically lose a lot of blood relative to their size and compensate, in practice, any substantial hemorrhage can rapidly weaken a small bird. Key techniques for controlling avian hemorrhage include the following actions.

- Direct pressure
- Styptics and cautery
- Broken blood feather management
- Pressure bandages
- Internal bleeding
- Blood volume support

Once bleeding is controlled, avoid unnecessary movement of the bird that might dislodge clots. Keep the patient calm and resting. Also, monitor the bleeding site continuously; birds might start bleeding again if blood pressure improves or if they peck at the bandage. Using an Elizabethan collar (e-collar) or similar might be necessary to prevent self-trauma to dressings or wounds once the bird is alert enough to preen. Care needs to be taken with using e-collars because of the potential for adding additional stress to the patient.

Pain management and pharmacologic support

Analgesia is a critical component of stabilization. Pain causes significant stress in birds, contributing to shock through increased catecholamines and metabolic demand. It also negatively impacts breathing and can even accelerate coagulopathies. Veterinary nurses play a key role in ensuring pain is recognized and treated. Best practices for avian pain management in emergencies are listed below.

- Analgesic medications
- NSAIDs
- Local anesthetics
- Sedatives and anxiolytics
- Other supports
- Monitoring and titration

Providing pain relief is not only ethical and humane, but it also physiologically helps the bird stabilize. As one expert put it, ask yourself, "Would I give analgesia to a dog or cat with these injuries?" If yes, then the bird should receive it.[5] Birds may be stoic, but they feel pain similarly, so always include analgesia in your triage plan.

Utilizing Flowcharts and Checklists for Avian Triage

During the stress of an emergency, having supportive materials like flowcharts and checklists can ensure no critical step is overlooked. Veterinary nurses and technicians should prepare and familiarize themselves with avian-specific emergency guides. Below are some useful tools and how to incorporate them.

- Avian triage flowchart
- Stabilization checklists
- Reference charts for doses and normals
- Triage forms
- Team drills and protocols

In essence, supportive materials serve as cognitive aids. Under the pressure of saving a tiny avian life, they can guide the team through the correct sequence, remind them about the unique needs of birds (for instance, a note "Do not delay warmth for a bird in shock, use an incubator!"), and ultimately improve patient outcomes. Developing these materials from evidence-based guidelines and updating them with the latest best practices (e.g., new drug dosages, new monitoring tools) keeps the clinic prepared for any avian emergency.

Conclusions

In avian emergency medicine, every minute and every detail counts. Veterinary nurses and technicians are often the first line in triaging and stabilizing an injured bird. By understanding the unique challenges of avian physiology and behavior, utilizing a systematic approach to triage, and employing effective stabilization techniques, the veterinary team can significantly improve a trauma patient's chance of survival. Always remember to keep the bird calm and supported (oxygen and warmth are your best friends), address the ABCs and pain control promptly, and use flowcharts or checklists if available to ensure no step is missed in the heat of the moment. With practice and adherence to these evidence-based guidelines, handling avian trauma becomes less daunting and more rewarding, ultimately leading to better outcomes for our feathered patients.

References

1. Sokoloff A. From parrots to peregrines: initial triage & stabilization of birds. *Annu Conf Auburn Univ Coll Vet Med.* 2023. https://www.vetmed.auburn.edu/wp-content/uploads/conference/2023/smallanimal/2/Thur-03-SA2-AUCVM%20Annual%20Conference%20Avian%20Triage%202023.pdf#:~:text=.

2. Stilwell N. Nursing management of rabbit and parrot emergencies. *DVM 360.* 2019.

3. Proença L. Triaging avian emergencies. *DVM 360.* 2023. https://www.dvm360.com/view/triaging-avian-emergencies.

4. Hollwarth A. Managing trauma and shock in exotic patients. *Improve Vet Pract.* 2023. https://www.veterinary-practice.com/article/managing-trauma-shock-exotic-patients#:~:text=cannula,be%20considered%20on%20a%20practical.

5. Gregory A. Pain management in the avian patient. *Improve Vet Pract.* 2022. https://www.veterinary-practice.com/article/pain-management-in-the-avian-patient#:~:text=Opioid%20analgesia%20is%20arguably%20one,important%20to%20note%20that%20buprenorphine.

6. Hoppes SM. Traumatic injury of pet birds. In: Hess L, Moses MA, eds. *Merck Veterinary Manual.* Rahway, NJ: Merck & Co; 2024. https://www.merckvetmanual.com/exotic-and-laboratory-animals/pet-birds/traumatic-injury-of-pet-birds.

Avian Behavior and Reproductive Health in Psittacine Birds: Clinical Insights for Veterinary Technicians

Avery Low, AAS, CVT

Session #224

Affiliation: From Arizona Exotic Animal Hospital, Tucson, AZ 85712, USA, Pima Community College, Tucson, AZ 85730, USA.

Abstract: Psittacine species, commonly kept as companion animals, remain undomesticated in their behavioral and reproductive tendencies. This presentation, led by Avery Low, certified veterinary technician at Arizona Exotic Animal Hospital and adjunct professor at Pima Community College, provides veterinary technicians with practical insights into parrot behavior and reproductive health. Understanding psittacine natural history is vital for interpreting common behaviors in captivity. From foraging and flock vocalizations to destructive tendencies and fear-based aggression, many behaviors deemed problematic by owners are rooted in instinct. We explore how unmet environmental and social needs, inappropriate human interactions, and hormonal fluctuations can manifest as excessive vocalization, feather destructive behavior, and sexual frustration. Technician notes highlight strategies for client communication, species-appropriate enrichment, and the importance of recognizing normal vs. maladaptive behaviors. The session also covers frequently encountered reproductive disorders in clinical settings, including excessive egg laying, dystocia, egg yolk coelomitis, cloacal prolapse, and neoplasia. Environmental triggers, nutritional factors, and inappropriate human interactions are discussed as contributors to disease. Attendees gain insight into diagnostic cues, clinical presentations, and the technician's role in supportive care, client education, and prevention strategies. Through behaviorally informed care and effective communication, technicians can help improve welfare outcomes for psittacine patients. This talk aims to empower technicians to be proactive advocates for these intelligent, high-maintenance birds and to strengthen the collaborative care between clients, technicians, and veterinarians.

Understanding Psittacine Natural History: The Key to Enhancing Parrot Care

The world of parrots is a vibrant one, full of intelligent, social, and highly curious creatures. As members of the veterinary community, understanding the natural history of psittacine species is not just a matter of academic interest, it's crucial for providing them with a healthy and fulfilling life, especially in a captive environment. The more we know about their natural history, the better equipped we are to create environments that meet their needs, promote their well-being, and encourage their natural behaviors.

In the wild, psittacine birds are incredibly active creatures that spend most of their time foraging, socializing, and grooming. These behaviors are deeply ingrained in their daily routines, and their survival in nature depends on them. Unlike domesticated dogs or cats, parrots are not truly domesticated. They retain many of the behaviors that have allowed them to thrive in the wild, and these instincts can have a significant impact on how they behave in a home setting.

When a parrot is kept as a companion animal, it exists in a completely different environment – one that often doesn't mirror the complex and dynamic systems they experience in their natural habitat. Parrots are highly social creatures that thrive on interaction. In their natural habitats, they live in large flocks and rely on constant communication to stay connected. This social drive means that companion parrots may vocalize frequently, seeking attention and companionship, sometimes in ways that can be disruptive to their human caregivers. Their natural curiosity can also lead them to investigate, and sometimes destroy, objects in their environment, as well as escape from enclosures that are not fully secure. These behaviors are deeply rooted in the bird's evolutionary need to explore and engage with its surroundings.

To ensure a companion parrot's well-being, it is essential to address these instincts and provide an environment that allows them to express their natural behaviors in safe and enriching ways. African grey parrots (*Psittacus erithacus*) have even been seen exhibiting contrafreeloading behaviors. A study conducted in 2021 found that if offered two food items, one that requires effort to obtain and one that does not, they will often choose the item that requires more extensive input to obtain, showing that they have positive associations with effortful foraging.[1] By appreciating and supporting their natural instincts, we can help parrots thrive and develop strong, positive relationships with their human caregivers.

Parrots In Captivity: An Overview of Behavioral Concerns and Caregiver Considerations

Knowledge for new caregivers

Potential parrot owners should be thoroughly educated on the complexities and challenges associated with caring for these highly intelligent and social birds. Due to parrots' highly charismatic nature, they are often highly sought after companion animals. Inexperienced prospective owners can be quick to purchase or adopt these birds without first doing the proper research. Their care requires a significant investment of time, attention, and resources to meet their physical, emotional, and environmental needs. Without the proper knowledge on the complexities of caring for parrot species, parrot caregivers can be setting themselves up for failure.

Parrots are known for their strong need for social interaction, mental stimulation, and physical exercise, all of which are essential to preventing behavioral issues such as excessive vocalization, aggression, and destructive tendencies. In addition, due to their fast metabolism and prey species nature, parrots can conceal illness until it becomes severe, making regular monitoring and preventive care essential. Veterinary technicians should emphasize to potential owners that adopting a parrot is a long-term commitment, often lasting several decades, and that they must be prepared to provide a suitable environment, enrichment, and care throughout the bird's lifetime. Educating prospective owners about these realities, along with providing accurate resources and guidance, can help ensure that they are fully prepared for the responsibility of parrot ownership and that the bird receives the appropriate care it needs to thrive.

Commonly encountered behavioral issues

Companion parrots often display a variety of behaviors that can be perceived as problematic by their owners. These behaviors, though natural and often rooted in the bird's instinctual needs, can create challenges in a domestic setting.

Excessive vocalization is one of the most reported concerns. Parrots, particularly those in species with strong flocking instincts, will vocalize to maintain contact with members of their flock. This is commonly referred to as a contact call, which is a vocalization associated with two individuals interacting.[2] In captivity, humans are often regarded as part of this flock, and birds will contact call when they cannot see or hear their caregivers, or when being reunited with them after they were away. While this is a natural behavior, it can become disruptive in the home environment. Another frequently encountered issue is excessively loud vocalizations, or alarm calls, which are more broadband and harsher in nature. These calls can stem from boredom, stress, discomfort, or hormonal changes and can be intended to warn nearby individuals of danger.[2] Depending on the species these vocalizations can reach over 100 decibels, making them a serious concern in apartment or close quarters living situations.

Destructive behaviors, including chewing and tearing apart toys or furniture, are also common. Parrots' high intelligence and natural curiosity drive them to use their powerful beaks and claws for exploration, often damaging objects. While climbing, parrots exhibit a unique tripedal motion, using both their legs and beak to navigate, which can cause harm to delicate items in their environment. Biting and aggression are issues linked to parrots' prey instincts. Fear-based aggression is common, and their sharp beaks can easily cause injury if they feel threatened.

A complex, multifactorial problem is feather destructive behavior (FDB), which can be caused by a combination of behavioral and medical factors, making it important to differentiate between the two. Feather destruction, when done behaviorally, is a type of stereotypy, an abnormal repetitive behavior, that can be seen in captive animals whose enrichment needs are not met.[3] This is thought to manifest from excessive and inappropriate grooming, as grooming is such a fundamental part of the natural history of parrots.[2]

Lastly, inappropriate sexual behaviors can pose many challenges. Parrots displaying inappropriate sexual behavior will often regurgitate toward a perceived mate, self-stimulate on objects in their environment, and can show excessive aggression and vocalization. These behaviors, often triggered by hormonal fluctuations, longer photoperiods, and inappropriate petting, may be misinterpreted by inexperienced owners as playfulness or acting out. Proper education on appropriate handling and understanding the bird's natural behaviors is essential for managing these issues and improving the bird's quality of life in a captive setting.

Corrections and prevention of behavioral issues

Behavioral issues in psittacine birds are often multifactorial, and their resolution requires a comprehensive approach involving environmental management, behavioral modification, and sometimes medical intervention. By providing consistent enrichment, promoting healthy socialization, and addressing the underlying causes of these behaviors, caregivers can help pet birds thrive in their home environments. Early detection and intervention are key to preventing these issues from becoming chronic, ensuring both the bird's well-being and a harmonious relationship with its human caregivers.

To prevent unwanted behavior, caregivers should strive to provide abundant enrichment opportunities that satisfy their parrots' instinctual behaviors. Food should be offered in a variety of ways and places. Puzzle feeders can help keep parrots cognitively engaged and moving food bowls around their environment encourages physical exercise and mimics the foraging behavior they would be performing in their natural habitat. Early socialization is key in reducing fear-based aggression and educating owners on how to recognize signs of aggression, such as puffed-up feathers, tail flicking, or dilated pupils, can help prevent escalation into biting or other aggressive behaviors. Reinforcing desirable behaviors through consistent boundaries and positive reinforcement should be emphasized. For instance, rewarding calm behavior with treats or praise can encourage non-aggressive interactions, while redirecting aggressive tendencies with alternative behaviors or toys can help mitigate aggression. Destructive tendencies, such as chewing on furniture, wires, or cage bars, are often driven by a parrot's need to explore and exercise its beak. These behaviors may also arise from boredom, anxiety, or lack of outlets for natural instincts. To manage this, providing a variety of safe chew toys and rotating them regularly will help satisfy the bird's desire to chew.

In addition, caregivers can reinforce desirable behavior by rewarding the bird when it engages with appropriate objects, while redirecting its attention from inappropriate items. Ensuring the environment is safe by removing dangerous items, like electrical cords, ingestible metal, or toxic plants, also minimizes the risk of injury while preventing unwanted destructive behavior. FDB is a more complex issue that can result from physical, environmental, or psychologic factors. In cases of FDB, a comprehensive veterinary evaluation is essential to rule out underlying medical causes, such as infections, parasites, or hormonal imbalances. Once medical causes are addressed, enrichment and mental stimulation are key components in managing FDB. Offering a variety of perches, toys, and foraging opportunities can help alleviate stress and boredom.

Inappropriate sexual behavior, such as excessive mating displays, regurgitation, or aggression tied to hormonal fluctuations, is a common challenge in captive parrots, particularly during breeding season. To address these behaviors, controlling light cycles by limiting exposure to no more than 12 hours of light per day can help simulate seasonal changes and manage hormonal surges.[3] Nesting materials or locations that could trigger territorial or excessive sexual behaviors should be removed. Redirecting attention away from regurgitation or other sexual behaviors through interactive playtime or training exercises can prevent reinforcing such behaviors. In some cases, discussing the option of hormone therapy may be beneficial, particularly for birds that are excessively hormonal or exhibiting disruptive sexual behaviors.

Equipping clients with reliable, credible resources is essential for fostering responsible pet ownership. Recommending educational materials from trusted sources, such as LafeberVet, can provide valuable insights into bird care, nutrition, and behavior. Directing clients to certified avian behaviorists can also be invaluable. Certified individuals from the International Avian Trainers Certification Board (IATCB) have undergone rigorous training and testing to ensure they are well-equipped to assist parrot owners with behavior management.

Managing Reproductive Health Issues in Parrots: Addressing the Impact of Inappropriate Sexual Behavior

Inappropriate sexual behavior in captive female parrots is a significant concern for both the well-being of the bird and its owners. Often, environmental cues and inappropriate human interaction or petting can inadvertently encourage reproductive behaviors, leading to secondary reproductive health issues. As veterinary technicians, understanding how these behaviors contribute to conditions like chronic egg laying, dystocia, egg yolk coelomitis, and cloacal prolapse is essential for providing comprehensive care and guidance to parrot owners.

Chronic egg laying is a frequent condition in parrots where a female lays more eggs than usual for a typical clutch, or continues to lay eggs and engage in reproductive activity even without a mate or outside the natural breeding season.[4] While not inherently pathogenic, chronic egg laying greatly predisposes parrots to secondary reproductive issues such as nutritional secondary hyperparathyroidism due to hypocalcemia,[4] pathologic fractures, ectopic eggs, egg binding, dystocia, coelomitis, and cloacal prolapse. Once it is noticed that a parrot is laying excessively, the importance of preventing the condition continuing should be stressed to owners. Parrots undergoing egg binding (delayed oviposition) or dystocia (the mechanical obstruction preventing normal oviposition),[4] can present to the clinic with non-specific signs and acute distress. In patients with egg binding, it is recommended to minimize handling, especially if severely debilitated and dyspneic.[5] If stable, medical management may be attempted but if unsuccessful manual removal may be necessary. Ectopic ovulation refers to the ovulation of an egg outside of the oviduct. This can result in egg yolk coelomitis if the egg is displaced into the coelum. Cloacal prolapse can occur in both male and female parrots and can be a result of excessive self-stimulatory sexual behavior.[6]

With any female reproductive issue, the importance of egg production ceasing should be emphasized to the caregiver, as once a parrot has shown reproductive disease, there is a high chance of recurrence. Although the detailed medical management and treatment of these conditions fall outside the scope of this presentation, understanding them can better equip veterinary technicians in providing comprehensive care for these patients and further exploration is encouraged.

Role of the Veterinary Technician: Bridging the Gap Between the Care Team and Parrot Caregivers

One of the key responsibilities of veterinary technicians is to ensure open communication between the caregiver and the veterinarian. If the technician has a concern about a particular case, whether it's regarding the bird's behavior, response to treatment, or new symptoms that have developed, it is critical to communicate with the rest of the veterinary care team. Open dialogue with the veterinarian helps ensure that all potential issues are addressed and that the treatment plan is tailored to the bird's specific needs. Technicians are often the ones spending the most face-to-face time with both the client and the patient, which gives them a unique opportunity to observe behaviors that the veterinarian may not see during the brief clinical exam. For example, a parrot may exhibit behaviors such as aggression or excessive vocalization that are not apparent when the veterinarian is in the room, or the caregiver may provide crucial information that was forgotten during the doctor's consultation. These observations are invaluable in fine-tuning the diagnosis and treatment plan. It is also important for veterinary technicians to advance their knowledge of avian behavior and disease processes.

As avian medicine and behavioral science evolve, staying up to date on the latest research and treatment modalities allows technicians to perform their job more effectively. This not only improves patient care but also empowers technicians to contribute meaningfully to the treatment process. Understanding the complex interplay between avian behavior, reproductive health, and disease processes enables technicians to better communicate with caregivers and advocate for the best interests of the bird. By being proactive in educating clients, providing clear explanations, and offering behavioral guidance, veterinary technicians play an integral role in managing the health and well-being of parrot patients. Whether advising on the prevention of chronic egg laying, addressing inappropriate sexual behaviors, or offering strategies for dealing with aggression, technicians can empower caregivers to take the right steps in providing long-term care. In addition, technicians are crucial in reinforcing plans that have been outlined by the veterinarian, offering additional resources or follow-up as needed. Veterinary technicians play a crucial, multifaceted role in avian healthcare, serving as the vital link between the veterinary care team and the parrot's caregiver. Their ability to facilitate meaningful conversations, provide educational resources, and clarify instructions not only enhances the quality of care but also fosters stronger relationships between the bird's owner and the veterinary practice.

References

1. Smith GE, Greene D, Hartsfield LA, et al. Initial evidence for eliciting contrafreeloading in grey parrots (*Psittacus erithacus*) via the opportunity for playful foraging. *J Comp Psychol.* 2021;135(4):516-533.

2. Toft CA, Wright TF. *Parrots of the Wild: A Natural History of the World's Most Captivating Birds*. Oakland, CA: University of California Press; 2015.

3. Samour J. *Avian Medicine*. 3rd Ed. St. Louis, MO: Elsevier; 2016.

4. Scagnelli AM, Tully TN, Jr. Reproductive disorders in parrots. *Vet Clin North Am Exot Anim Pract.* 2017 20(2):485-507.

5. Schmidt RE, Struthers JD, Phalen DN. Reproductive system. In: *Pathology of Pet and Aviary Birds*. 3rd Ed. Hoboken, NJ: Wiley-Blackwell; 2024:263-288.

6. Beaufrère H, Graham JE. *Blackwell's Five-Minute Veterinary Consult: Avian.* 2nd Ed. Hoboken, NJ: Wiley-Blackwell; 2025.

How to Place an Intraosseous Catheter: Anatomy, Technique and Tips

Kristina Palmer, RVT, VTS (Clinical Practice, Exotic Companion Animal)

Session #223

Affiliation: From UC Davis William R. Prichard Veterinary Medical Teaching Hospital, 1 Garrod Drive, Davis, CA 95616, USA.

Abstract: Fluid therapy is crucial for critically ill avian patients, particularly in cases of shock where vascular access may be difficult or impossible to obtain. In such instances, intraosseous (IO) catheterization serves as a vital alternative, allowing for the rapid absorption of fluids and medications into the bloodstream. This manuscript offers a comprehensive guide to performing IO catheterization in birds, covering key considerations, relevant anatomy, necessary equipment, and appropriate analgesia for the procedure.

Considerations for Use and Potential Risks

Intraosseous (IO) catheter placement is particularly valuable for small or critically ill birds, as well as during surgeries when intravenous catheterization is not feasible. Small avian species, such as cockatiels, budgerigars, canaries, and finches, present unique challenges for intravenous catheter placement, as even the smallest catheter sizes may be too large for their vessels. Additionally, attempting to place an IV catheter in these small birds can lead to complications of excessive blood loss. In such cases, IO catheterization is the preferred alternative. Birds are adept at hiding signs of illness, often masking symptoms until they are severely compromised. Subtle indicators, such as lethargy, regurgitation, anorexia, or simply a bird sitting at the bottom of its cage, may signal an emergency. When these signs are noticed and the bird is brought into the clinic, it may be severely dehydrated and in hypovolemic shock, making immediate access for fluids essential. IO catheterization provides a critical and effective solution in these situations.

Placing an IO catheter can be more challenging than an IV catheter; however, practicing the technique on cadavers can help build confidence for emergency situations. A solid technique also reduces the risk of complications, such as fluid extravasation, iatrogenic fractures, osteomyelitis, through-and-through penetration of the bone, and physeal plate injury. IO catheter placement should be avoided in septic patients or those with bone diseases. Additionally, reproductively active females may experience hyperostosis of the long bones, leading to thickened bone cortices and mineral deposition in the marrow cavity, making IO catheter placement more difficult.

Anatomy

Most species of birds seen in practice have pneumatized bones, which have evolved to aid in weight reduction for flight. These bones include the humerus, coracoid, pelvis, sternum, vertebrae, femur, scapula, and furcula, with the skull also containing honeycomb-like air spaces. Most of the distal bones are not pneumatized. In birds, the humerus and femur should be avoided for IO catheter placement because their medullary cavities connect to air sacs, risking fluid entry into the respiratory system and potential drowning, unlike in mammals, where these sites are commonly used. The two most utilized sites for IO catheter placement in birds are the distal ulna and proximal tibia (Fig 1).

Distal ulna

This is the most common site for IO catheterization for birds. In birds, the ulna is larger than the radius, which is the opposite of what we observe in mammals. This site is suitable for birds of all sizes and is often easy to access, secure, and maintain patency. A figure-of-8 bandage is commonly used to secure the catheter by immobilizing the wing and adding comfort for the patient.

Proximal tibia (tibial crest)

Typically used for medium to large birds, this method is generally easy to apply, secure, and maintain. Bandaging will be required to ensure the site remains clean.

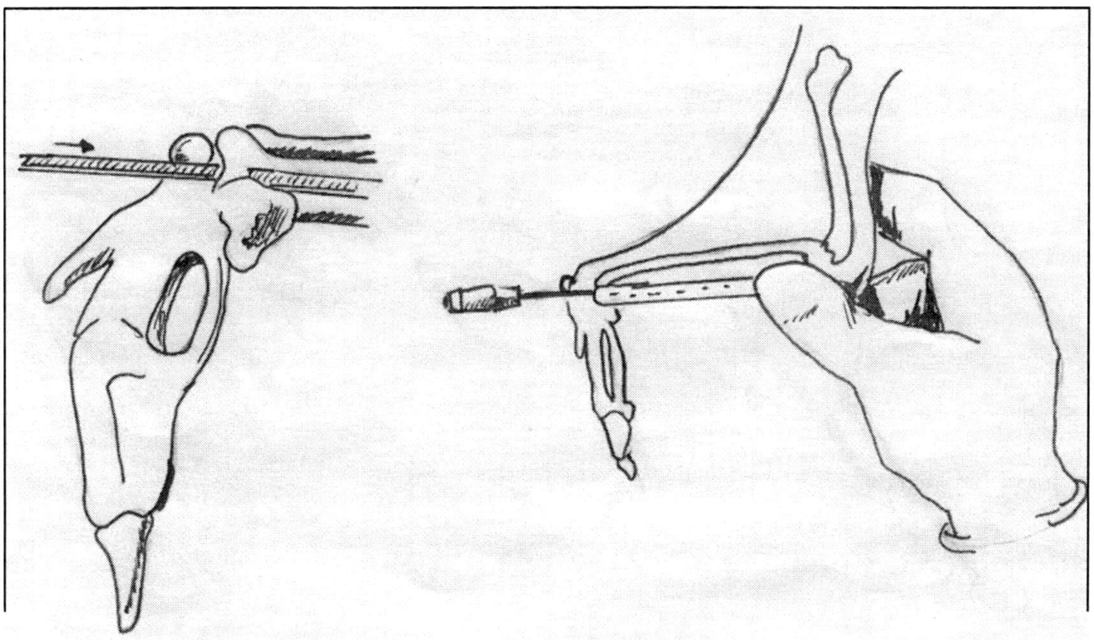

Figure 1. Intraosseous catheter placement.

Analgesia and Anesthesia

Placement of an IO catheter is both stressful and uncomfortable for the patient and appropriate analgesia should be used whenever possible. Placement of an IO catheter is ideally performed under general anesthesia or with heavy sedation with a local anesthetic, such as 2% lidocaine, at the insertion site prior to placement. If the patient is in crisis this may not be possible and appropriate analgesia should be given once the patient is stabilized. Once in place an IO catheter should be no more uncomfortable than an IV catheter.

Supplies Needed

Below is a list of supplies needed for IO catheter placement.

- Aseptic prep
- Sterile gloves
- Spinal or hypodermic needles
- Injection port
- Needle holders, forceps, scissors
- Suture
- Heparinized saline or fluids of choice
- Surgical tape
- Tape
- Glue
- Bandage material
- Syringe pump: if these are not available then alternatively slow boluses of fluid can be given manually
- Elizabethan collar (e-collar); the patient may need this to prevent destruction of the catheter

Step-by-Step Instructions for IO Catheter Placement in Birds (Distal Ulna or Proximal Tibia)

Either a spinal needle or hypodermic needle of an appropriate size can be used as an IO catheter. The spinal needle has the advantage of having a stylet, which will help prevent the lumen from becoming blocked with a bone core. For a small bird such as a cockatiel or smaller you can use a 26- to 25-gauge needle. For a medium bird such as an Amazon parrot you can use a 25- to -22-gauge needle or 22-gauge spinal needle. For large birds such as a macaw or a large raptor you can use a 22- to 20-gauge 1.5-inch needle or spinal needle.

1. Position the bird

- For distal ulna: Place the bird in lateral recumbency with the appropriate wing facing up.
- For proximal tibia: Place the bird in dorsal recumbency with the selected leg exposed.

2. Palpate landmarks

- Ulna: Locate the distal bony protuberance beneath the extensor carpi ulnaris and extensor digitorum communis ligaments. Feel for a plateau-like depression just distal to the condyle.
- Tibia: Flex the stifle and palpate the cranial cnemial (tibial) crest, identifying the patellar tendon to avoid during insertion.

3. Prepare the injection site

- Aseptically prepare the insertion site.
- Infuse 2% lidocaine to anesthetize the insertion site.
- Always identify anatomical landmarks before injecting lidocaine, as they may become obscured.

4. Stabilize the limb

- Ulna: Grasp the ulna for stability with one hand and approach the wing dorsally.
- Tibia: Use the non-dominant hand to hold the tibiotarsus stable.

5. Needle insertion

- Ulna: Insert the needle on the dorsal condyle of the distal ulna, parallel to its median plane.
- Tibia: Insert the needle through the tibial crest, avoiding the patellar ligament.

6. Advance the cannula

- Apply steady, firm pressure with a rotating motion, aiming:
  - Ulna: Toward the elbow, following the bone's ventral curvature.
  - Tibia: Toward the tarsus, keeping alignment with the tibial shaft.
- Think of this step like inserting an intramedullary pin.

7. Feel for the pop

- A distinct pop indicates entry into the marrow cavity.
- If resistance persists, you may be skimming along the lateral cortex.

8. Flush and confirm placement

- Ulna: Palpate to ensure the needle tip hasn't exited the bone.
- Blanching of the basilic vein during flushing suggests proper placement.
- Tibia: In large birds, palpation may not be possible; use radiographs to confirm placement.
- The cannula should flush easily. If blocked (e.g., by a bony plug), replace with a new needle using the existing guide hole.

9. Cap and secure the catheter

- Cap with a sterile infusion plug.
- Secure with butterfly tape and suture to surrounding soft tissue.
- Site-specific anchoring:
  - Ulna: Suture to soft tissue of the distal ulna.
  - Tibia: Suture to soft tissue of the proximal tibia.

10. Bandage the limb

- Ulna: Apply a figure-of-8 bandage around the wing.
- Tibia: Wrap the leg with cling gauze and cohesive bandage to protect the site and ensure comfort.

Conclusions

Securing IV access in a small or severely dehydrated bird can be a challenge. In these situations, knowing how to place an IO catheter in the distal ulna or proximal tibia becomes an essential skill. Practicing on cadavers helps build confidence and ensures proper placement, which ultimately reduces the risk of complications. It is important to perform the procedure aseptically, with the right anesthesia and analgesia. Always confirm proper catheter placement before use, and make sure it's securely fastened and maintained once in place.

References

1. Bowles H, Lichtenberger M, Lennox A. Emergency and critical care of pet birds. *Vet Clin North Am Exot Anim Pract*. 2007;10(2):355. doi: 10.1016/j.cvex.2007.04.001.

2. Chitty J. Basic techniques. In: Harcourt-Brown N, Chitty J, eds. *BSAVA Manual of Psittacine Birds*. Chapter 6. Gloucester, UK: British Small Animal Veterinary Association; 2005:55.

3. Dubé C, Dubois I, Struthers J. Intravenous and intraosseous fluid therapy in critically ill birds of prey. *J Exot Pet Med*. 2011;20(1):21-26. doi: 10.1053/j.jepm.2010.11.008.

4. Edis A. How to manage intraosseous catheters in exotic species. *Vet Nurse.* 2016;7(10):589-592. doi: 10.12968/vetn.2016.7.10.589.

5. Lennox A. Intraosseous catheterization of exotic animals. *J Exot Pet Med.* 2008;17(4):300-306. doi: 10.1053/j.jepm.2008.07.007.

6. Quesenberry KE, Hillyer EV. Supportive care and emergency therapy. In: Ritchie BW, Harrison GJ, Harrison LR, eds. *Avian Medicine: Principles and Application.* Lake Worth, FL: Wingers Publishing; 1994:382-416.

How Parrots Eat and How It Impacts Their Care

Jovanna Berls, CVA, LVT, FFCP (Veterinary-Avian)

Session #150

Affiliation: From ABC Animal & Bird Clinic, 11930 Highway 6 South, Sugar Land, TX 77498, USA.

Abstract: Monitoring patients' food intake is an important part of a veterinary technician's function. The mechanics of how companion parrots eat can have an important impact on whether they are eating and what they are eating. In addition, parrots are unusual in using their feet to prehend their food. Ensuring adequate nutrition can be crucial in assisting the healing process, especially in the case of injury or illness affecting the beak or the feet. This presentation covers topics including anatomy of the beak and the feet, conditions that can affect the ability to self-feed, techniques for assisting patients to consume appropriate nutrition, and counseling clients on appropriate feeding at home.

Introduction

Monitoring a psittacine patient's food intake is an important part of a veterinary technician's function. Much of that function naturally involves thinking about what parrots eat, because inadequate or inappropriate nutrition can lead to metabolic derangements with very serious consequences. The best nutrition in the world is no benefit to a patient unless the food reaches the digestive tract. Thus, it is also important to consider the mechanics of how parrots eat.

Even among species in the Psittaciformes order, there is considerable variety of food preferences, beak size and shape, and several differences in anatomy. This presentation focuses on psittacine birds commonly seen in clinical practice, including budgerigars (*Melopsittacus undulatus*), cockatiels (*Nymphicus hollandicus*), macaws (*Ara* spp.), Amazon parrots (*Amazona* spp.), conures (*Aratinga* spp., *Pyrrhura* spp.), Quaker parrots (*Myiopsitta monachus*), *Poicephalus* spp., most cockatoos (*Cacatua* spp.), African grey parrots (*Psittacus* spp.), *Pionus* spp., caiques (*Pionites* spp.), rose-ringed and Alexandrine parakeets (*Psittacula* spp.), and similar species. It excludes those with significantly different beak and tongue morphology, such as lorikeets (*Trichoglossus* spp.) and palm cockatoos (*Probosciger aterrimus*).

What Parrots Eat

The general feeding recommendations for captive parrots will be readily familiar to most veterinary technicians: at least 60% formulated diet and 40% vegetables, with seeds and fruit provided primarily as treats.[1] Parrots usually are considered granivores (grain-eaters), frugivores (fruit-eaters), florivores (flower-eaters), or some combination of those.[2] Recent research has shown that parrots in the wild are considerably more omnivorous than initially suspected. They have been reported eating what are called alternative foodstuffs, including insects, small fish and lizards, wood, roots, charcoal, algae, and tree sap.[3] Wild and feral parrots appear adept at switching from their usual preferred food of seeds, nuts, and fruits to different sources of nutrition, such as leaves and blossoms, depending on the season.[4,5]

The Amazing Parrot Beak

It may seem obvious that birds' beaks evolved to accommodate the types of food they typically eat. The long, thin beak of the hummingbird, for example, appears to have evolved to allow penetrating into blossoms to obtain a flower's nectar; the lamellae in a duck's bill seem designed for ease of filtering food from water. Recent research has shown the evolution of the beak is not quite that simple.[4] Birds use their beaks for much more than food intake: preening and feather maintenance, foraging, nest building, attack and defense, and, in the case of many psittacine birds, even locomotion, as when the bird uses its beak to assist in climbing.[4,6] Thus, parrots' beaks have evolved to be useful and versatile tools.

Parrots have a dramatically curved, pointed beak. Along with a muscular tongue and a complex combination of musculoskeletal structures, this beak is ideally suited for chewing and processing food items. Most species remove and discard the shell from seeds and nuts before consuming them.[7] The beak is renowned for being able to exert significant bite forces to crack seeds and even some of largest, hardest nuts (not to mention the fingers of veterinary technicians). Despite this shell-crushing strength, parrots' beaks also are capable of a wide variety of small, precise movements of the jaws and tongue to remove and manipulate the meat of the seeds and nuts they eat. Some species use their jaws and tongues to process their food against small ridges and edges on the underside of the upper beak and at the lateral aspects of the beak.[4]

Anatomically, the beak as a whole is described as the rostrum, which is made up of the maxillary rostrum (upper beak) and the mandibular rostrum (lower beak). The rostrum consists of epidermis, dermis, bone, and air spaces, and it is covered by a keratin sheath called the rhamphotheca, sometimes also known as the horny bill. The maxillary portion of the rhamphotheca is called the rhinotheca, and the mandibular portion of the rhamphotheca is called the gnathotheca.[8,9]

Fused premaxillary and nasal bones form the maxilla, which functions in parrots as a solid triangular block. At the junction of the upper jaw and skull, there is a synovial joint most often referred to as the craniofacial hinge. Movement of this hinge is called prokinesis.[10] Elevation and depression of the upper beak requires the movement of four pairs of bones: the jugal arch, the pterygoid, palatine, and quadrate bones. The main connection between the skull and the lower jaw is the pair of quadrate bones, which articulate not only with the skull and lower jaw but also the jugal arch and the pterygoid bone. (Parrots do not have a zygomatic arch or temporomandibular joint, as mammals do.)[9] When the quadrate bones are rotated rostrally, they push on the pterygoid and jugal arches, which push the palatine bone upward, and forces the upper jaw to open upward. When the quadrate bones rotate caudally, the upper jaw swings downward to close the jaws.[8]

Parrots do not have pseudotemporalis profundus muscles, which are common in other birds. They do have two unique pairs of muscles that act on the jaws, the ethmomandibularis and the pseudomasseter. To open the jaws requires action of the protractor pterygoidei et quadrati and depressor mandibulae muscles. Closing the jaws involves the ethmomandibularis and pterygoideus muscles.[8]

The primary motor innervation for the muscles of mastication comes from the mandibular nerve, except for the depressor mandibulae muscle, which is innervated by the facial nerve.[9] A bill tip organ, which plays a role in discrimination between food and non-food objects, has been reported in the Senegal parrot (*Poicephalus senegalus*) and has been theorized to exist in other parrot species, as well.[8] The bill tip organ contains Herbst corpuscles, which are very sensitive to vibration. Their function is related to the way the beak is used as an exploratory tool when feeding. Many species that crack and open seeds are well supplied with Herbst corpuscles at the tip and lateral ridges of the beak.[8] Watch a parrot with a nut or chunk of wood, and it will be apparent that the beak probes for the right spot to split or break the object.

The keratin that forms the rhamphotheca grows from the epidermis at the base of the beak and normally moves rostrally and laterally along the surface of the beak from its vascular bed. Under normal conditions, the keratin surface grows and is worn down continuously at the same rate as its growth. Damage to the keratin may regrow normally, but injury to deeper tissues or bone can lead to abnormal growth, potentially for the remainder of the parrot's life. Such damage may require regular, periodic trimming and/or reshaping of the rhamphotheca. The primary source of blood to the beak and oral cavity is the common carotid artery, which subdivides into the internal carotid and vertebral arteries.[9]

Craniofacial hinge

This synovial joint is present in many species of birds, offering an ability to widen the mouth gape by 5-15 degrees more than simply opening the jaw. In psittacine birds, there is additional mobility provided – a recent study measured the maximum excursion of the beak in peach-faced lovebirds (*Agapornis roseicollis*) at up to 40 degrees.[6] It has been hypothesized that this larger gape results from the fact that the craniofacial hinge of parrots arises from within the nasal capsule.[6] The greater mobility affords parrots the ability to adjust the gape as needed for larger and harder food items, as well as to produce larger biting forces.[11] Deeper, shorter beaks with a larger gape enhance bite force transmission and can withstand heavier stress without fracture. Because parrots evolved to eat relatively large, hard foods, it is thought that the craniofacial hinge and the unique ethmomandibularis and pseudomasseter muscles combine to provide the power needed to crack seeds and nuts – larger parrots can easily open nuts that a human could only break with pliers or a hammer.[12,13]

Tongue

Parrots have large, fleshy tongues. Only psittacine birds have intrinsic muscles of the tongue that insert on their hyoid apparatus, which imparts tremendous mobility to the tongue. Seeds and nuts are manipulated into position against the beak by the tongue and then held there while the jaws split the shells.[7]

Sense of taste

It is generally believed that birds do have a sense of taste but that it is not as acute as that of mammals. They do have taste buds located on the base of the tongue, but birds in general have many fewer taste buds than mammals.[9,13] For example, chickens have 767 taste buds, compared to 7,902 in humans. Birds have demonstrated the ability to distinguish nutrients including amino acids, fatty acids, salts, acids, bitterants, and possibly simple carbohydrates and calcium. The taste buds tend to be clustered in areas around salivary ducts.[13] Among other things, the sense of taste is important for aiding the bird to detect nutrient levels and avoid toxins;[2] however, a study of feeding habits of feral Amazon parrots in Europe showed they frequently ate plants containing toxic compounds such as tannins and the cyanogenic glycoside found in apple seeds without apparent adverse effects.[7]

The Amazing Parrot Foot

To hold food up to the beak with a foot, a parrot must move its whole leg – anything that impairs the mobility of the pelvic limb or foot can have an adverse impact on food intake. That said, parrots' feet have particular characteristics, and the birds certainly use them in some unusual ways.

Parrot feet are zygodactyl, which means they have two digits facing forward and two digits facing backward so that the first and fourth toes oppose the second and third toes when the foot is closed. This arrangement allows for a firm grasp of a perch or a food item.[15] The foot is well-adapted for climbing, as well as grasping.[16] Each digit has one more phalanx than the number of the digit; the first digit has two phalanges, the second digit has three phalanges, and so forth. The distal phalanx on each digit consists of a base and an apex that forms the underlying structure of the claw. Birds do not have the discrete tarsus familiar in mammals but rather a tarsometatarsus that is formed by a fusion of metatarsal bones with the distal end of the tarsus.[17]

When eating, parrots often use their feet to manipulate food items, especially larger, bulky items such as nuts or chunks of fruit. They are not the only birds to hold their food while eating; birds of prey, for example, often hold food between a foot and the ground or a perch. That forces the bird to look down to take a bite of food, lessening its ability to observe its surroundings for nearby predators. Lifting the food to the mouth with a foot, as parrots do, allows surveillance of the area at all times. Some sources have called this method prehensile-footed feeding. Further, parrots manipulate the food item with the foot holding it, turning it while at the same time gnawing or exploring it with the beak and tongue.

Smaller birds, especially ground-feeding birds like budgerigars and cockatiels, tend to grip food with their feet less often than larger birds. Whether there is a preference for using one foot or the other to hold food is controversial, but some sources have held that parrots often are left-footed, similar to the way a human may be left-handed.[15]

Diseases and Conditions Affecting Parrots' Food Intake

Conditions affecting the beak and oral cavity

Environmental, nutritional, infectious, and neoplastic factors can be involved in dysfunction of the beak. Developmental abnormalities also may be seen. Keep in mind that a parrot may present for a beak problem that is actually a symptom of an underlying disease process elsewhere in the body, such as a dramatically overgrown beak resulting from liver disease.

Environmental factors

These include trauma of any type, such as collision with a solid object, beak trimming accident, or attack by another bird or wild or pet animal. Trauma that interferes with normal motion of the jaws can result in abnormal beak growth for the future. Beak fractures can occur as a result of trauma or metabolic abnormalities; in a bird with metabolic bone disease, a collision can result in fracture of both the rhamphotheca and underlying bones. Simple fractures of the beak may cause a split in the rostral gnathotheca at the midline. Avulsion fractures are common with bite injuries to the beak from other birds.[17] Parrots that lose a large portion of the upper beak may lose the ability to crack nuts or seeds, so they will need a soft diet for the remainder of their lives.[9] Other environmental causes of beak injury are lacerations resulting from interaction with caging or toys.[18]

Nutritional factors

Inappropriate nutrition can result in softening or flaking of the rhamphotheca. Nutrition-related metabolic bone disease can cause abnormal bone growth, especially in younger birds. Hypovitaminosis A, as well as causing blunting of the choanal papillae, may lead to squamous metaplasia of tissues in the oral cavity, resulting in impaction of salivary ducts, whitish nodular or plaque like lesions. Secondary mycotic and bacterial infection is common.[9]

Developmental abnormalities

Congenital beak abnormalities may be seen, but the most common developmental beak deviations are acquired after hatching. Acquired developmental beak deviations can result from any condition that damages the musculoskeletal structure of the rhamphotheca. The most common abnormalities are lateral beak deformity (scissor beak) and mandibular prognathism. These most commonly affect young macaws and cockatoos that have been hand-fed.[18] Causes of scissor beak and mandibular prognathism are not known, but the conditions may be related to incubation issues, hand-feeding a young bird only on one side of the beak, infections such as sinusitis, mycotoxicosis, genetic influences, or inappropriate nutrition that leads to a softening of the growing beak.[9,18]

Infectious agents

Parasitic, viral, fungal, and bacterial agents can affect the beak. Among the most common of these conditions is infestation with scaly face and leg mites (*Knemidocoptes pilae*). These mites are most often seen on budgerigars and canaries (*Serinus canaria domestica*) but occasionally are seen in other species. In large numbers, the mites can severely deform the beak and feet. They also may be found on the legs and pericloacal area. Transmission is by direct contact. Severity of the clinical signs may depend on genetic factors and the immune status of the bird.[18]

The most common viral agent that can cause beak and claw deformities is parrot beak and feather disease (PBFD, formerly known as psittacine beak and feather disease). PBFD is caused by a virus in the Circoviridae family. It most commonly affects birds less than 2 years of age, especially Old World species such as cockatoos, eclectus parrots (*Eclectus* spp.), lovebirds (*Agapornis* spp.), and African grey parrots. Ingestion and inhalation of foor or air contaminated by feather and fecal dust is the most common transmission route. Fomites also may be involved. Abnormal feather development may be the most obvious symptom, but beak abnormalities that can develop include elongation, fractures, palatine necrosis, and oral ulceration; claw development also may be affected because the virus appears to disrupt keratin production.[18]

Chronic fungal infection with *Aspergillus* spp. can result in sinusitis that leads to drainage onto the beak from the nares. A groove-like indentation may form in the rhinotheca over time. Similarly, bacterial infection can cause chronic drainage from the sinuses onto the rhinotheca. Inappropriate nutrition leading to immunosuppression can be an important factor.[18] Sinusitis caused by *Cryptococcus neoformans* var *gatti* and *Penicillium cyclopium* has been reported as a cause of destruction of the rhinotheca in a Major Mitchell's cockatoo (*Cacatua leadbeteri*) and a blue and gold macaw (*Ara ararauna*), respectively.[9]

A bacterial condition known as lockjaw, in which a bird is unable to open its beak, is most often seen in juvenile cockatiels and is believed to associated with *Bordetella avium* and other bacterial infections. The resulting sinusitis and rhinitis can have involvement with nearby bone and musculature.[9]

Neoplasia

Tumors on the beak are not common, but squamous cell carcinomas of the rhamphotheca and oral cavity have been reported.[18] Masses in the oral cavity can adversely affect growth of the rhamphotheca, or may lead to hemorrhage or tissue necrosis, dyspnea, or result in difficulty prehending food. Fibrosarcoma is the most common primary beak tumor and is most common in budgerigars and cockatiels. If these tumors invade the nares, nasal turbinates, or bone, deformation of the beak can result. The most common tumor in the oral cavity of New World parrots is the mucosal papilloma, usually appearing as a raised, cauliflower-like mass, pink to white in color. These lesions can come and go, sometimes disappearing altogether, then recurring. Squamous cell carcinoma of the oral cavity has been reported in a number of psittacine species. Although these tumors rarely metastasize to other areas of the body, they can be locally invasive. Malignant melanoma of the beak has been reported in African grey parrots.[9]

Dysplasia

Beak overgrowth may have a variety of causes but can result from damage to the growth areas of the beak, nutritional factors, and underlying metabolic problems, such as liver disease. In addition, the beak is designed to be worn down in a normal process that may not be as available to captive birds as it is to wild birds. If normal wear does not occur, a real or perceived need to trim the keratin may be a consequence, to prevent progressively worsening malocclusion.[8]

Conditions affecting the feet

Similar to conditions affecting the beak, causes of problems with parrots' feet can include environmental, infectious, neoplastic, metabolic, and idiopathic factors.[18] Any condition that impairs the mobility of a parrot's pelvic limb or limbs can have an adverse effect on the bird's ability to eat.

Environmental factors

Trauma, lacerations, and claw trimming accidents are common. In addition, leg-band injuries involving constriction, thermal injuries and frostbite, fractures, pododermatitis (bumblefoot), and avascular necrosis of distal extremities (constricted toe syndrome) can occur.[18]

Fractures can result from trauma or metabolic bone diseases. Fractures of the feet or toes may require coaptation to immobilize the fracture site, with bandaging or with systems of wires or pins, depending on the fracture site, the fracture type and the size of the bird.[18] Complications following fracture repair, such as osteoarthrosis, may require long-term management.[19]

Pododermatitis is an inflammatory condition that initially affects the plantar aspect of the foot/feet but can progress to compromised blood supply and infection that can severely affect underlying soft tissue, bones and joints. Mobility may be severely reduced. The cause is multifactorial, often involving a trauma to the foot or toes. Inappropriate perches, inactivity, malnutrition, and obesity can be contributing factors.[18]

Avascular necrosis of distal extremities can be caused by heat, cold, low humidity, foreign-body constriction, aflatoxins, or unknown factors. It is more often seen in young birds but can be found in birds of any age or species, most commonly macaws and eclectus parrots.[18]

Infectious agents

PBFD and *Knemidocoptes pilae* mites also affect the feet, as mentioned previously. Papillomas of the feet can be caused by papillomavirus in African grey parrots and Cuban Amazon parrots (*Amazona leucocephala*), while papillomas found in cockatoos and macaws appear to be caused by a herpesvirus.[18] Both viruses result in wart-like growths on the feet.[18]

Idiopathic conditions

Self-mutilation of feet most commonly is found in Amazon parrots, especially yellow-naped Amazons (*Amazona auropalliata*) and double yellow-headed Amazons (*Amazona oratrix*), but it has been reported in Quaker parrots and conures.[20] The cause is unknown, but in some birds it appears to be seasonal, occurring primarily in spring. Other suggested causes are delayed hypersensitivity reaction, contact dermatitis following exposure to nicotine or tobacco smoke, exposure to crack cocaine or other toxins, nutritional deficiencies, or hormonal influences. These birds chew and bite at their feet and/or legs, leaving the skin reddened and dry or flaky. Bleeding and secondary infections can occur.[18]

Articular gout

This condition results in deposition of uric acid crystal deposits called tophi in the joints. Tophi can develop in the phalanges, making weight-bearing and manipulating objects with the feet painful and difficult. The cause is poorly understood and may be multifactorial. Hyperuricemia, diet, and genetics may be involved.[18]

Assisting Parrots that Have Difficulty Feeding Themselves

Anorexia in a parrot is a serious condition that can result in a variety of metabolic derangements and, if left unaddressed, can lead to death by starvation. While it is certainly possible to provide nutrition by syringe or gavage feeding when the bird's condition requires it, there are risks to force-feeding and the process can be stressful both for the bird and the human administering it. It is at best a temporary solution. Finding a method to encourage self-feeding as soon as possible will be advantageous for the bird, medical staff, and the client once the bird leaves the hospital.

It is important to know the type of diet the bird was eating before arriving at the hospital. A diet similar to the food provided at home is likely to be more readily accepted. Although a change of diet may be necessary because of the bird's condition, parrots can be neophobic and may not immediately recognize an unfamiliar diet as food. That can discourage self-feeding and adds the challenge of coaxing the patient to try the new diet. Parrots with beak injuries may not be able to return to a seed diet because of inability to crack the shells and extract the seeds; unlike many other species, parrots do not swallow seeds whole.

Leaving food in the cage with the patient is important even while providing supplemental syringe or gavage feeding. If there is nothing in the cage to eat, it will not be possible to know when the patient is ready to resume self-feeding.

Parrots with upper beak injuries often can learn to scoop food into their lower beak and may become very successful at self-feeding small pellets, soft fruits and vegetables, birdie bread, or mash. The same formula used for gavage feeding may be provided in a small bowl. If the injury involves the lower beak, the prognosis may be more guarded, especially if the injury is extensive, but the bird still may be able to consume soft foods such as moistened pellets. Providing a deep bowl that allows the parrot to push food against a hard surface or corner can be helpful.[21]

For parrots with leg or foot conditions, it is important to ensure the bird can reach the food. Cage furnishings may need rearrangement because of the mobility impairment related to the foot/leg issue – at that time, attention should be paid to the location of water and food bowls. Containers should be easily accessible and sized so that the bird has room to get its head into them easily. In-bowl foraging, in which a variety of healthy foods are mixed in the same bowl, is a good option for birds with mobility or visual impairments. Psittacine birds who are naturally ground feeders in the wild, such as budgerigars and cockatiels, may be enticed to eat by scattering food on the bottom of the cage. Note that fecal matter may build up on the feet and tail feathers of parrots who are unable to use a perch. This material will need to be removed regularly for adequate sanitation.

Every client's home environment and skill level for handling a parrot will be different, so it is important to query the client about how feeding will be accomplished at home. At a minimum, ask what the bird was eating before the beak or foot was impaired, how the cage was arranged to provide the food, and what foods are believed to be favored or likely to be rejected. Encourage the client to experiment with different foods or containers if the first approach is ineffective. For example, if birdie bread is rejected, try a mash or soft, cooked vegetables. Then, if the second attempt does not work, try a third, and so on. Offer several examples of possible food selections so the client is empowered to try a variety of approaches.

Clients should be advised to monitor the parrot's weight in grams daily and report to the clinic any weight loss in excess of 5%. Examples of appropriate scales, such as those used to weigh food, can be provided. Demonstrating how to place the bird in a container on a scale is a good idea. Explain that small fluctuations in weight (e.g., a few grams difference from morning to night) are not abnormal, but a large variation or a downward trend over a short period of time should raise concerns.

For a bird that has been hospitalized for a time, it will be very helpful to note the bird's feeding habits while in the hospital and relay that information to the client. The client should be advised that the patient's behavior may change after discharge. For example, a parrot that has been eating moistened pellets while hospitalized may fall back on previous habits once returned home. If syringe feeding must be continued, a demonstration of the technique will be required for any client who is unfamiliar with it.

Conclusions

Ensuring that a psittacine patient can self-feed nutritionally appropriate foods is vital to the patient's health and welfare. Fortunately, parrots are wonderfully versatile creatures, so despite impairments of the beak or foot, a wide variety of adaptations in the hospital and at home are possible to maintain their health and well-being.

References

1. Doneley B. Diets for Healthy Parrots. Veterinary Partner website. Available at: https://veterinarypartner.vin.com/default.aspx?pid=19239&id=10891591. Accessed March 16, 2025.

2. Matson K, Koutsos E. Captive parrot nutrition: interactions with anatomy, physiology, and behavior. In: Luescher A, ed. *Manual of Parrot Behavior.* Ames, IA: Blackwell Publishing; 2006:50.

3. Voltura E, Brightsmith D, Cornejo J, et al. Parrot dietary habits and consumption of alternate foodstuffs. *J Avian Med Surg.* 2023;37(4):297-313.

4. Bright J, Marugan-Lobon J, Rayfield E, et al. The multifactorial nature of beak and skull shape evolution in parrots and cockatoos (Psittaciformes). *BMC Evolutionary Biology.* 2019;19;1-9.

5. Martens J, Hoppe D, Woog F. Diet and feeding behavior of naturalized Amazon parrots in a European city. *Ardea.* 2013;101(1):71-76.

6. Young M, Wilken A, Manafzadeh A, et al. The dual function of prokinesis in the feeding and locomotor systems of parrots. *J Exp Biol.* 2023;226:1-10.

7. Homberger D. Classification and status of wild populations of parrots. In: Luescher A, ed. *Manual of Parrot Behavior.* Ames, IA: Blackwell Publishing; 2006:3-5.

8. Speer B. Beak deformities: form, function, and treatment methods. *Proc Annu Conf Assoc Avian Vet.* 2014;213-219.

9. Speer B, Powers L. Anatomy and disorders of the beak and oral cavity of birds. *Vet Clin North Am Exot Anim Pract.* 2016;19:707-736.

10. King A, McLelland J. *Birds: Their Structure and Function.* 2nd Ed. London, UK: Bailliere Tindall; 1984: 43-44.

11. Bout R, Zweers G. The role of cranial kinesis in birds. *Comp Biochem Physiol A.* 2001;131:194-205.

12. Tokita M. The skull development of parrots with special reference to the emergence of a morphologically unique cranio-facial hinge. *Zool Sci.* 2003;6:1-12.

13. Navalon G, Bright J, Marugan-Lobon J, Rayfield E. The evolutionary relationship among beak shape, mechanical advantage, and feeding ecology in modern birds. *Evolution.* 2018;73:3:422-435.

14. Niknafs S, Navarro M, Schneider E, et al. The avian taste system. *Front Physiol.* 2023;14:1-16.

15. Harris L. Footedness in parrots: three centuries of research, theory, and mere surmise. *Can J Psychol.* 1989;43(3):369-396.

16. King A, McLelland J. *Birds: Their Structure and Function.* 2nd Ed. London, UK: Bailliere Tindall; 1984: 18-22.

17. Konig H, Korbel R, Liebich H. *Avian Anatomy: Textbook and Colour Atlas.* 2nd Ed. Sheffield, UK: 5m Publishing; 2016:67-68.

18. Worell A. Dermatological conditions affecting the beak, claws, and feet of captive avian species. *Vet Clin North Am Exot Anim Pract.* 2013;16:777-799.

19. Harcourt-Brown N. Orthopedic conditions that affect the avian pelvic limb. *Vet Clin North Am Exot Anim Pract.* 2002;5(1):49-81.

20. Langlois I. Medical causes of feather damaging behavior. *Vet Clin North Am Exot Anim Pract.* 2021;24: 136-138.

21. Wheler C. Orthopedic conditions of the avian head. *Vet Clin North Am Exot Anim Pract.* 2002;5(1):83-95.

Section 6

Wildlife Session II

Stephany Lewis, DVM, CWR, DABVP (Avian Practice)
Session Chair

Fracture Management in Pediatric Raptors

Dana Franzen-Klein, DVM, MS, and
Annette Ahlmann-Garcia, DVM, MS

Session #139

Affiliation: From The Raptor Center, University of Minnesota College of Veterinary Medicine, St. Paul, MN 55108, USA.

Abstract: Young raptors heal quickly, which can be both a blessing and a curse when it comes to fracture management. Their rapid growth allows for faster healing than adults, while also necessitating adaptations to our treatment methodology to not create complications. Diagnosing orthopedic injuries is also very challenging in nestling aged birds owing to open growth centers and partially mineralized bones. This case-based lecture reviews the challenges of diagnosing and managing fractures in nestling, brancher, and fledgling aged raptors. We discuss how to evaluate prognosis, modify treatment plans for growing juveniles, and monitor healing, as well as the challenges and complications that may arise with common fracture management techniques when applied to pediatric raptors.

Introduction

Juvenile raptors are common patients at wildlife rehabilitation facilities. Occasionally these birds are normal, healthy juveniles that have been obtained by well-meaning members of the public, and they can be quickly returned to their recovery location. Traumatic injuries are also a common reason for admission. The raptor rehabilitation hospital at The Raptor Center at the University of Minnesota College of Veterinary Medicine (TRC) admits approximately 150 juvenile raptors each year. The clinicians at TRC have found that subtle orthopedic injuries secondary to trauma are quite common in nestling and brancher aged birds and are challenging to diagnose. This presentation discusses radiograph interpretation and treatment of orthopedic injuries in juvenile raptors in a wildlife rehabilitation setting.

Avian Bone Growth and Development

The bones of pediatric raptors grow in length through a process called endochondral elongation, where cartilage is replaced by bone.[1-3] Unlike mammals, the ends of the long bones (epiphyses) of pediatric raptors are cartilaginous rather than boney.[1-3]

Developmental stages of pediatric raptors

Nestling

This stage covers the age from hatching until most of the down on the bird's body has been replaced with contour feathers. Their bones and joints are in very early stages of development and the birds have limited mobility.

Brancher

Raptors at this stage have significant growth of the contour feathers on their wings, tail, and body. The flight and tail feathers are starting to grow. They have increased mobility compared to nestlings, and some species have started to move outside of the nest. They can stand and perch for longer periods of time and may have started making short flight attempts. Not all species have a brancher stage. This age is further along in the bone development process, but not all bones or joints are fully developed.

Fledgling

At this stage, raptors have started flying and have left the nest, but they are still dependent on their parents and gaining flight skills and strength. Their outermost primary flight feathers are still in blood and growing, but their inner primary flight feathers are fully grown (i.e., hard-penned and out of blood). At the fledgling age, the bones and joints are nearly fully developed but growth is not yet complete.

Post-fledgling juvenile (hatch year)

At this stage, all the flight feathers and tail feathers are fully grown (i.e., hard-penned and out of blood). The bird is fully flighted. Bone and joint development are complete.

Treatment of Pediatric Fractures

Key points for pediatric raptor fractures

- Pediatric raptors with fractures should be provided with analgesia for pain management.
- Nutrition is an important component of fracture healing and growth in pediatric raptors. A species appropriate, varied whole prey diet is essential. Calcium and vitamin D supplementation may be considered.
- Incomplete fractures, folding fractures, and fractures within the non-ossified, cartilaginous ends of the bone secondary to trauma are very common, even in chicks with appropriate nutrition and bone density (without metabolic bone disease).
- Radiographic interpretation of pediatric raptor fractures can be challenging if the damaged areas are unossified at the time of injury. Summation with blood feathers may also impair visibility and complicate radiographic interpretation.
- Pediatric raptor fractures start to heal very quickly, which can be a detriment if treatment is delayed. If fractures are displaced, the surrounding soft tissues may contract rapidly and dramatically, which can limit treatment options. A complete diaphyseal pediatric fracture may become irreparable within 72 hours due to the speed at which the bones and soft tissues are growing and changing.
- Damage to a growth plate can impact bone growth and development and has the potential to result in angular abnormalities or shortening of the affected bone.

Treatment

The appropriate fracture management technique will depend on the affected bone, degree of displacement, patient temperament, age, and species. Potential options for treatment include cage rest, coaptation (bandaging, splinting), surgical fixation, or a combination of methods.

Considerations for coaptation

In pediatric patients, coaptation with bandages and splints can sometimes cause more harm than good. The bandaged limb can develop restricted range of motion very quickly, and that restriction can be challenging or impossible to overcome. Additionally, blood feathers can be damaged by bandages or splints. If splinting or bandaging, use lightweight materials that do not add excess weight to the limb. When bandaging or splinting the leg, using tape to create hobbles between the tarsometatarsi can help prevent lateral rotation of the bandaged limb and associated strain on the coxofemoral joint. All bandages and splints placed on pediatric patients need to accommodate continued bone growth. Splint and bandage changes should be done frequently to adjust coaptation as the bone grows, perform physical therapy, and monitor for pressure sores or other complications. The increased frequency of bandage changes in juvenile raptors compared to adults must be balanced with the need to keep the fracture as stable as possible during the healing process. In general, bandages and splints may need to be changed as frequently as every 48 hours for nestling aged birds that are growing rapidly. Comparatively, fledgling aged birds may be able to wait up to 4 days between bandage changes due to their slower growth rate. The attending clinician must weigh the pros and cons associated with the bandage change, reassess the patient for possible complications at each exam, and adjust the treatment plan accordingly.

Considerations for surgical fixation

Soft (unossified) bones cannot hold pins well, which may limit surgical options in pediatric patients. All methods of surgical fixation must accommodate continued growth. Accommodation for continued bone growth may affect the type of surgical fixation that is chosen, where the pins are placed, and the pin size and length. For example, if an intra-medullary pin is used as a component of surgical fixation for a tibiotarsus fracture, extra length must be left where the pin emerges from the stifle joint so that the bone and soft tissues may continue to grow around the intramedullary pin. Partial surgical fixation combined with splinting or coaptation is a common treatment method for complete, displaced fractures. For example, clinicians at The Raptor Center have had success treating complete, diaphyseal tibiotarsus fractures in brancher and fledgling owls, including barred owls (*Strix varia*), great horned owls (*Bubo virginianus*), and an eastern screech owl (*Megascops asio*), by placing an intramedullary pin and creating a spoon splint with thermoplastic material (Rolyan® Aquaplast-T thermoplastic splinting material, 1/8 inch or 1/16 inch; JorVet™ Marq-Easy thermoplastic splint sheets) on the caudal aspect of the leg extending from the tarsometatarsus, over the hock joint, to the proximal aspect of the tibiotarsus. The stifle joint is not immobilized in this treatment method due to close anatomic relationship of the stifle joint with the coelomic wall, especially when the leg is flexed in a normal perching position, making it impossible to effectively immobilize this joint with coaptation without creating severe pressure sores on the coelomic wall and inguinal region.

When may cage rest be appropriate

Consider the age, species, activity level, and temperament of your patient. For some fractures, cage rest may be more appropriate for a nestling than a fledgling due to their low activity levels. Incomplete long bone fractures in very young patients may be suitably managed by cage rest. Shoulder girdle, pelvic, or keel fractures are often suitably managed by cage rest. Complete long bone fractures should be assessed on a case-by-case basis with regards to the patient age and fracture location (e.g., mineralized versus unmineralized area of the bone, wing versus leg, severity of displacement) to determine the best course of treatment. The lecture provides case examples to discuss the decision-making process used by the clinicians at The Raptor Center.

Rate of healing

Fractures in nestling and brancher aged raptors heal remarkably fast, sometimes healing in a third to half the time it would take an adult raptor to heal the same fracture. Based on the subjective experience of the clinicians at The Raptor Center, the younger the bird is, the faster the fracture will heal and remodel, as they are mineralizing and growing their bones at a remarkably fast rate. For example, a brancher barred owl (approximately 4-5 weeks old) with an incomplete proximal humerus fracture in the unossified area of the bone was treated with cage rest and analgesics and completely healed and remodeled the area within 7 days. An adult barred owl with an incomplete proximal humerus fracture would take approximately 3 to 4 weeks to fully heal and remodel with coaptation and analgesics. This rapid rate of healing requires an adjustment to the clinician's treatment plan; often requiring an expedited timeline for initiating treatment, necessitating changes in the normal frequency of rechecks and bandage changes, and destabilizing or removing surgical fixation and/or coaptation earlier than what would be typically considered for a fully grown bird.

References

1. Atterholt J, Woodward HN. A histological survey of avian post-natal skeletal ontogeny. *PeerJ.* 2021;9:e12160. doi:10.7717/peerj.12160.

2. Leach RM Jr, Gay CV. Role of epiphyseal cartilage in endochondral bone formation. *J Nutr.* 1987;117(4): 784-790. doi:10.1093/jn/117.4.784.

3. Tully TN Jr. Basic avian bone growth and healing. *Vet Clin North Am Exot Anim Pract.* 2002;5(1):23-30. doi:10.1016/s1094-9194(03)00044-6.

Use of Readily Available Materials to Provide Tensile Force for Repair of Bilateral Leg Fractures in a Peregrine Falcon

Crystal Matt, DVM

Session #153

Affiliation: From the Avian and Exotic Animal Clinic of Indianapolis, Indianapolis, IN 46268, USA.

Abstract: This report describes the treatment of a free-ranging, juvenile peregrine falcon presented to a veterinary hospital with bilateral tarsometatarsal fractures from a reported window collision injury. Both fractures were proximal, transverse, comminuted, and minimally displaced, with an additional midshaft lateral greenstick fracture on the right limb. A modified version of a Type II external skeletal fixation (ESF) system was applied to each limb using two, 0.9-mm, positive-profile, Kirschner-wire pins. To counteract the compressive forces from the flexor tendon attachment at the foot and the bird's bodyweight while standing, tongue depressors were wedged between the K-wires to replace the connecting bars. The slight flexibility of the K-wires, combined with tongue depressors cut to be slightly longer than the spacing between the wires when the fractured limbs were unsupported, caused the wires to behave like springs and create tensile forces. Additional tongue depressors attached perpendicular to the K-wires with flat sides adhered to the K-wires with white tape were added on the lateral aspects of each fixator and served to counteract the rotational forces generated by the tension from the K-wires. The bird was able to stand the following day and used both legs throughout recovery. The proximal K-wire on the left leg broke at the medial cortex 2 weeks postoperatively, suspected to be from excess tension. The K-wire fragment was removed, and the remainder of the K-wire was able to be advanced to reconstruct the fixation. This method resulted in successful healing and full functional recovery of both limbs.

Therapeutic Botulinum Toxin Type A in an American Barn Owl (*Tyto furcata*) with Propatagial Tendon Contracture

Mei Schultz, DVM,
Hollie Chiles, CVT,
Lauren Savoy, BS,
Alexandra Goe, DVM, Dipl ACZM

Session #190

Affiliation: From Midwestern University College of Veterinary Medicine, 19555 North 59th Avenue, Glendale, AZ 85308, USA.

Abstract: A wild, adult American barn owl (*Tyto furcata*) of unknown sex developed unilateral contracture of the patagium and propatagial tendon, associated with a malunion humerus fracture. Physical examination findings included contracture of the right propatagial tendon (right side 13 cm; left side 24 cm) and decreased range of motion of the elbow (right side 65°; left side 130°). Patagial contracture is a widespread issue affecting birds in rehabilitation settings, the wild, and in managed care.[1] Therapeutic botulinum toxin type A (BoNT A) injections are routinely documented in human medicine, and rarely in veterinary medicine, to treat a variety of tendon contracture disorders.[(2-4)] OnabotulinumtoxinA (4.24 U/kg IM once; BOTOX, AbbVie Inc., North Chicago, IL, USA) was administered in two aliquots into the muscle *Deltoideus propatagialis* and the muscle *Pectoralis propatagialis* under inhalant anesthesia. After a single treatment, the propatagial tendon length improved by 7.6% (14 cm) and elbow range of motion improved by 30% (85°) when evaluated 4 days post-treatment. The bird was monitored daily for 10 days, and no adverse effects were noted. At 2 weeks post-treatment, propatagial tendon length improved an additional 7.4% (15 cm) and elbow range of motion improved an additional 16% (95°). Recheck examination 3 months later showed no additional improvement. While the tendon contracture did not resolve, the degree of improvement along with absence of adverse effects make this a promising therapy to further investigate in cases of avian tendon contracture.

References

1. Jato-Schez OS, Otero I, Lopez I, Mendoza JL. Patagium rehabilitation treatment in wild birds following long-term wing immobilization. *Wildl Rehabil Bull.* 2011;29(2):33-41.

2. Dressler D, Altavista MC, Altenmueller E, et al. Consensus guidelines for botulinum toxin therapy: general algorithms and dosing tables for dystonia and spasticity. *J Neural Transm (Vienna).* 2021;128(3):321-335.

3. McGeachan RI, Schwarz T, Gunn-Moore DA, Marioni-Henry K. Botulinum toxin type A for the treatment of muscle contractures secondary to acute spinal cord injury in a young cat. *JFMS Open Rep.* 2020;6(1):2055116920922648.

4. Wijnberg ID, Hardeman LC, van der Meij BR, et al. The effect of *Clostridium botulinum* toxin type A injections on motor unit activity of the deep digital flexor muscle in healthy sound Royal Dutch sport horses. *Vet J.* 2013;198(Suppl 1):e147-51. doi: 10.1016/j.tvjl.2013.09.050.

Successful Outcomes Following Conservative Management of Complicated Bilateral Radius and Ulna Fractures in Rock Pigeons (*Columba livia*)

Mei Schultz, DVM,
Hollie Chiles, CVT,
Alexandra Goe, DVM, Dipl ACZM

Session #191

2025 House Officer Manuscript Competition Winner

Affiliation: From Midwestern University College of Veterinary Medicine, 19555 North 59th Avenue, Glendale, AZ 85308, USA.

Abstract: This case series describes largely successful outcomes following conservative management of complicated bilateral radius and ulna fractures in wild rock pigeons (*Columba livia*). A group of 22 wild, adult, rock pigeons was admitted to a wildlife rehabilitation center (Liberty Wildlife, Phoenix, AZ, USA) with bilateral radius and ulna fractures, verified radiographically upon intake. All birds were placed in bilateral figure-of-eight wing wraps and treated with a 7-day course of an oral nonsteroidal anti-inflammatory medication. Birds with open fractures also received a 7-day course of an oral antibiotic. From the 22 birds initially admitted, six were euthanized due to comorbidities and two died. Recheck radiographs of the remaining birds obtained 4 weeks post-intake showed healing complications consisting of unilateral or bilateral partial nonunion and/or malunion fractures in 6/14 birds (42.8%). Over the following 1-5 months, 13/14 birds (92.8%) were released. At the time of this submission, 1/14 (7.2%) was under active rehabilitation, showing promising flight ability. A prior study reported that wild raptors with concomitant radioulnar fractures were less likely to have a successful outcome than other types of antebrachial fractures.[1] Synostosis is the most frequent complication of antebrachial fractures and occurs due to insufficient stabilization.[2] This series suggests that concomitant radioulnar fractures in rock pigeons may carry a good prognosis for release even in the presence of healing complications like partial nonunion fractures or radioulnar synostosis.

References

1. Vergneau-Grosset C, Dubé C, Fitzgerald G, Lair S. Characteristics of antebrachial fractures associated with a successful outcome among free-ranging birds of prey that received treatment in a rehabilitation program. *J Am Vet Med Assoc*. 2020;256(5):580-589.

2. Redig PT. Orthopedic Surgery. In: Samour J, ed. *Avian Medicine*. 3rd Ed. Philadelphia, PA: Mosby; 2016:320.

Range of Motion Measurement in the Thoracic Limbs of Raptors at a Wildlife Rehabilitation Center in Chile

Carolina Sánchez-Saavedra, DMV, MSc,
Ariela Cultrera Rozowski, DMV,
Nicole Sallaberry-Pincheira, DMV, MSc

Session #146

Affiliation: From Unidad De Rehabilitacion De Fauna Silvestre, Escuela De Medicina Veterinaria, Facultad De Ciencias De La Vida, Universidad Andrés Bello, Santiago, Chile (Sánchez-Saavedra, Cultrera Rozowski), Department of Small Animal Medicine and Surgery, Veterinary Teaching Hospital, University of Georgia, Athens, GA 30605, USA (Sallaberry-Pincheira).

Abstract: Wildlife rehabilitation centers (WRRCs) integrate veterinary medicine, animal physiotherapy, and conservation biology to reintroduce animals into the wild.[1,2] (1) (2)Range of motion (ROM) is a key clinical parameter for assessing limb function;[3,4] (3) however, in raptors, standardized ROM values are lacking in some species,[5] (5) so clinicians rely on empirical data and extrapolations. This study aimed to establish reference ROM values for raptor forelimbs using a universal goniometer. Measurements were taken on various Accipitriformes, Falconiformes, and Strigiformes treated at La Unidad de Rehabilitación de Fauna Silvestre (UFAS) Universidad Andrés Bello (UNAB) and Ñamku UNAB WRRC in 2024 and 2025. Flexion and extension angles of the shoulder, elbow, and wrist were assessed. Measurements were obtained from 152 individuals. The mean thoracic limb ranges for the following orders for shoulder (S), elbow (E), and wrist (W) were: Strigiformes: S: 20-160°, E: 25-170°, W: 20-180°; Accipitriformes: S: 15-140°, E: 25-150°, W: 20-150°; Falconiformes: S: 25-140°, E: 20-150°, W: 20-175°. Measurements were consistent among individuals of the same species. Accipitriformes and Falconiformes showed similar shoulder and elbow ROM, but the wrist joint exhibited a greater ROM in Falconiformes. In addition, Strigiformes had a wider elbow ROM compared to diurnal species such as Accipitriformes and Falconiformes. This work provides a preliminary basis for the standardization of joint parameters in Chilean raptors, with potential clinical application in functional assessment during rehabilitation processes. Studies with larger sample sizes are recommended to validate and expand upon these findings.

References

1. Hernández S, Barron HW, Miller EA, et al. *Medical Management of Wildlife Species: A Guide for Practitioners.* Ames, IA: Wiley-Blackwell; 2020.

2. Sallaberry-Pincheira N, Vera C. *Manual Básico Operacional para Rescate y Rehabilitación de Fauna Silvestre en Situaiones de Desastres y Consideraciones para Incorporar el Componente Fauna en Proyectos de Restauración Ecológica.* Santiago, Chile: Corporación Nacional Forestal; 2018.

3. Scott D. *Raptor Medicine, Surgery and Rehabilitation.* Boston, MA: CABI; 2020.

4. Biewener A. Biomechanics of avian flight. *Curr Biol.* 2022;R1042-R1172.

5. Gjeltema JL, Degernes LA, Buckanoff HD, Marcellin-Little DJ. Evaluation of goniometry and electrogoniometry of carpus and elbow joints in the barred owl (*Strix varia*). *J Avian Med Surg.* 2018;32(4):267-278.

Survival of a Wedge-Tailed Eagle After Hallux Talon Amputation: Case Report

Ellen K. Rasidi, BBiomedSc, BSc (Vet) (Hons), BVSc,
Charlie I.O. Carter, BVSc (Hons), Dip Pract Mgt,
Peg J. McDonald, OAM, CF

Session #215

Affiliation: From Australian Raptor Care and Conservation Inc., 48 Berrima Rd, Moss Vale NSW, Australia 2577 (Rasidi, Carter, McDonald), Beecroft Avian and Exotics Veterinary Clinic, 991 E Alexandra Rd, #01-27, Singapore 119973 (Rasidi), Southern Highlands Veterinary Centre, 48 Berrima Rd, Moss Vale NSW, Australia 2577 (Carter).

Abstract: Hallux amputation deems an eagle non-releasable because digit one is necessary for successful hunting and carrying of live prey.[1] North American eagles—the bald eagle (*Haliaeetus leucocephalus*), golden eagle (*Aquila chrysaetos*), Stellar's sea eagle (*Haliaeetus pelagicus*), and the white-tailed eagle (*Haliaeetus albicilla*)—are predominantly hunters of live prey, but will scavenge carrion if available, which makes up a relatively small proportion of the diet. The wedge-tailed eagle (*Aquila audax*), Australia's largest, is a predominantly carrion-feeder, filling the ecologic niche occupied by vultures elsewhere. Since vultures are reported to survive well if released without the hallux talon,[1] it is hypothesized that so would wedge-tailed eagles. A fledgling wedge-tailed eagle was presented with radiographic and clinical pathology findings consistent with traumatic amputation of the distal phalanx of the right hallux. The amputation site was debrided and closed, and the bird released at the appropriate dispersal time after 4 months of rehabilitation.[2] A satellite telemetry device was mounted to the base of the rectrices[3] to monitor post-release survival. The bird was monitored remotely for 10 months after release, and travelled an estimated distance of 10,000 km, over an area of 450,000 km^2. This post-release behavior is consistent with the recorded behavior of intact, healthy juvenile and subadult wedge-tailed eagles.[4] We conclude that in this case, the loss of a hallux talon did not adversely impact the survival of this bird, and that hallux amputation may not preclude the release of individuals of this species in the future.

Acknowledgments: The authors thank the rehabilitation and veterinary volunteers of Australian Raptor Care and Conservation, Inc., for their assistance in the care and tracking of this bird.

References

1. Scott DE. *Raptor Medicine, Surgery, and Rehabilitation.* 3rd Ed. Wallingford, Oxfordshire, UK: CABI; 2021.

2. State of NSW and Office of Environment and Heritage. Code of Practice for Injured, Sick and Orphaned Birds of Prey. Available at: https://www.environment.nsw.gov.au/resources/wildlifelicences/code-of-practice-birds-of-prey-160665.pdf. Accessed January 30, 2025.

3. Groom C, Warren K, Le Souef A, Dawson R. Attachment and performance of Argos satellite tracking devices fitted to black cockatoos (*Calyptorhynchus* spp.). *Wildl Res.* 2015;41(7)571-583.

4. Cherriman SC. Movement ecology of Australia's avian apex predator: the Waalitj/Warlawurru (wedge-tailed eagle *Aquila audax*). Doctoral thesis. Murdoch University. 2024. Available at: https://researchportal.murdoch.edu.au/esploro/outputs/doctoral/Movement-ecology-of-Australias-avian-apex/991005726584507891. Accessed January 30, 2025.

Seasonal Facial Dermatitis in Captive Arizona Harris's Hawks and Treatment Responses: A Case Series

Erica Giles, DVM,
Emily Mocarski, DVM,
Rachel Raguin, DVM,
Stephanie Lamb, DVM, DABVP (Avian Practice)

Session #136

Affiliation: From Arizona Sonora Desert Museum, 2021 N Kinney Rd, Tucson, AZ 85743, USA (Giles), Stanford University Department of Comparative Medicine, 300 Pasteur Drive Edwards Building RoomR350, Stanford, CA 94305, USA (Mocarski), Avian and Exotic Animal Hospital, 1276 Morena Blvd, San Diego, CA 92110, USA (Raguin), Arizona Exotic Animal Hospital, 2340 E Beardsley Rd Ste 100, Phoenix, AZ 85024, USA (Lamb).

Abstract: While skin disease is commonly noted in captive psittacine birds there are few cases describing dermatitis in raptors.[1-2] Furthermore, there are uncertainties around whether birds truly experience allergic skin disease, but we believe that to be the underlying cause in this case series.[2] Skin disease in birds is often multifactorial and difficult to diagnose.[1] Common underlying causes for avian skin disease include nutritional factors, parasitic disease, bacterial disease, fungal disease, viral disease, metabolic causes, referred pain response and behavioral disease. With so many potential factors contributing to skin disease, and often more than one factor in a single patient, a thorough history and full diagnostic workup—including radiographs, bloodwork, skin cultures and skin biopsies—are critical to determining the cause and treatment plan for the avian patient. This case series describes 12 Harris's hawks from three different institutions in Arizona experiencing seasonal facial pruritus, dermatitis and xeroderma, with suspected underlying allergic etiology. This case series reviews varying treatments completed and perceived responses to those treatments. While varying treatments were used (i.e., topicals, NSAIDs, steroids, antibiotics, antihistamines), antihistamines showed the most consistent response. Skin disease was also seasonal, with the clinical signs and need for treatment, starting in late spring and resolving in the fall. The antihistamines and doses used were as follows: diphenhydramine 1-2 mg/kg PO q12h and cetirizine 2-4 mg/kg PO q24h.[3]

References

1. Girling S, Fraser M. Chapters 1-5. In: Paterson S, ed. *Skin Diseases of Exotic Pets.* Oxford, UK: Blackwell Science Ltd; 2006:3-72.

2. Pendl H, Speer BL. Immune-mediated disease in avian species: hypersensitivity and autoimmune disease. *Proc Annu Conf Assoc Avian Vet.* 2012;169-180.

3. Sanchez-Migallon Guzman D, Beaufrere H, Welle KR., et al. Birds. In: Carpenter JW, Harms CA, eds. *Carpenter's Exotic Animal Formulary.* 6th Ed. St. Louis, MO: Elsevier; 2023:222-443.

Trombiculosis in Two Strigidae Species from Arizona: Mite We Be Missing Them?

Jason D. Struthers, DVM, MVetSc, DACVP,
Anne Justice-Allen, DVM, MS,
Alexandra Brower, DVM, DACVP,
Cal Welbourn, PhD,
Ashley P.G. Dowling, PhD

Session #127

Affiliation: From Department of Pathology, College of Veterinary Medicine, Midwestern University, 19555 N 59th Ave, Glendale, AZ 85308, USA (Struthers), Arizona Game and Fish Department, 5000 W. Carefree Highway, Phoenix, AZ 85086, USA (Justice-Allen), Wyoming State Veterinary Laboratory, University of Wyoming, 1174 Snowy Range Road, Laramie, WY 82070, USA (Brower), Department of Entomology & Plant Pathology, University of Arkansas, 312 Agriculture Bldg, Fayetteville, AR 72701, USA (Welbourn, Dowling).

Abstract: Thousands of trombiculid mite species exist globally and are diverse in tropical and southern temperate regions. The life cycle involves free-living, non-parasitic nymphs and adults. Adults lay eggs in the soil to produce parasitic larvae that require a vertebrate host. Larvae feed on a host for 3-5 days, sometimes 30 days or more, before becoming nymphs. Feeding larvae may cause chigger dermatitis or trombiculosis. Between 2022 and 2024, chigger dermatitis was diagnosed in two euthanized great horned owls (*Bubo virginianus*) (GHOW) and multiple deceased burrowing owls (*Athene cunicularia*) (BUOW) from Arizona, USA. In all owls the ventral skin of the sternum and coelom had multifocal raised flaky, tan to light brown crusts and raised, < 5 mm diameter, centrally depressed tan papules. Both GHOWs had lesions with ovoid, < 0.5-mm long orange chiggers. All owls had microscopic necrotizing dermatitis with characteristic intralesional chigger stylostomes. Chiggers analyzed from one GHOW were *Neoschoengastia* spp. In North America, *Neoschoengastia* sp. infections occur in turkeys, southeastern Northern bobwhite, and southeastern Arizona Passeriformes without reports in Strigidae. Trombiculosis of the GHOWs was likely facilitated by recumbency since trauma prevented flight in both. Comparatively, perhaps BUOWs are a more typical host for chiggers given their burrowing behavior. The significance and the incidence of chigger dermatitis in Strigidae is unclear, but, at minimum, the skin lesions could be a portal of entry for bacteria. Owls admitted for veterinary care following prolonged recumbency should receive an examination of the ventral skin to detect chiggers that may warrant treatment.

Bilateral Elbow Synovial Chondromatosis in a Great Horned Owl (*Bubo virginianus*)

Evanna Lai, DVM,
Amanda Day, DVM,
Allison M. Watson, DVM, DACVP,
Alex Ohlendorf, DVM, DACVR,
Gretel Tovar-Lopez, MVZ, DABVP (Avian Practice), Dipl ACZM,
Miranda J. Sadar, DVM, Dipl ACZM

Session #111

Affiliation: From the Western College of Veterinary Medicine, University of Saskatchewan, 52 Campus Drive, Saskatoon, Saskatchewan S7N 5B4, Canada (Lai), James L. Voss Veterinary Teaching Hospital, Colorado State University, 300 West Drake Road, Fort Collins, CO 80523, USA (Day, Tovar-Lopez), Department of Microbiology, Immunology, and Pathology, College of Veterinary Medicine and Biomedical Sciences, Colorado State University, 300 West Drake Road, Fort Collins, CO 80523, USA (Watson), Department of Environmental and Radiological Health Sciences, College of Veterinary Medicine and Biomedical Sciences, Colorado State University, 300 West Drake Road, Fort Collins, CO 80523, USA (Ohlendorf, Yoon), Department of Clinical Sciences, College of Veterinary Medicine and Biomedical Sciences, Colorado State University, 300 West Drake Road, Fort Collins, CO 80523, USA (Sadar).

Current affiliation: From Department of Clinical Sciences, College of Veterinary Medicine, North Carolina State University, Raleigh, NC 27606, USA (Day).

Current affiliation: From Department of Clinical Sciences, College of Veterinary Medicine, Kansas State University, Manhattan, KS 66506, USA (Tovar-Lopez).

Abstract: Synovial chondromatosis is an uncommon condition involving metaplasia of the synovium in joints.[1,2] Cases have been seen in dogs, cats, equids, and varying species of birds.[1-18] Among these, the most common avian species is the great horned owl (*Bubo virginianus*).[5-7] The lesion has been reported in the sternocoracoid, scapulohumeral, metacarpophalangeal, femoroacetabular, and femorotibiotarsal joints in avian species.[5-7] In previous reports, morbidity associated with this disease process has been high, and birds have generally been euthanized upon presentation or shortly thereafter due to quality of life concerns.[5-7] This case is the first reported instance of synovial chondromatosis appearing in the elbow joints bilaterally, as well as the first documented long-term monitoring and medical management of a captive adult great horned owl with chondromatosis. Cases of synovial chondromatosis may be amenable to long-term management for avian species under managed care.

References

1. Żylińska B, Sobczyńska-Rak A, Lisiecka U, Stodolak-Zych E, Jarosz Ł, Szponder T. Structure and pathologies of articular cartilage. *In Vivo*. 2021;35(3):1355-1363. doi: 10.21873/invivo.12388.

2. Meuten DJ, ed. *Tumors in Domestic Animals*. 5th Ed. Hoboken, NJ: Wiley Blackwell; 2017.

3. Freire M, Meuten D, Lascelles D. Pathology of articular cartilage and synovial membrane from elbow joints with and without degenerative joint disease in domestic cats. *Vet Pathol*. 2014;51(5):968-978. doi: 10.1177/0300985813516648.

4. Smith MRW. Synovial osteochondromatosis. *Equine Vet Educ.* 2014;26(8):400-401. doi: 10.1111/eve.12205.

5. Howard MO, Nieves MA, Miles KG. Synovial chondromatosis in a great horned owl (*Bubo virginianus*). *J Wildl Dis.* 1996;32(2):370-372. doi: 10.7589/0090-3558-32.2.370.

6. Stone EG, Walser MM, Redig PT, Rings B, Howard DJ. Synovial chondromatosis in raptors. *J Wildl Dis.* 1999;35(1):137-140. doi: 10.7589/0090-3558-35.1.137.

7. Xie S, Nevis J, Lezmi S. Pathology in practice. *J Am Vet Med Assoc.* 2014;245(7):767-769. doi: 10.2460/javma.245.7.767.

8. Díaz-Bertrana C, Durall I, Rial MRial, et al. Extra- and intra-articular synovial chondromatosis and malignant transformation to chondrosarcoma. *Vet Comp Orthop Traumatol.* 2010;23(04):277-283. doi: 10.3415/VCOT-09-03-0036.

9. Edinger DT, Manley PA. Arthrodesis of the shoulder for synovial osteochondromatosis. *J Small Anim Pract.* 1998;39(8):397-400. doi: 10.1111/j.1748-5827.1998.tb03740.x.

10. Aeffner F, Weeren R, Morrison S, et al. Synovial osteochondromatosis with malignant transformation to chondrosarcoma in a dog. *Vet Pathol.* 2012;49(6):1036-1039. doi: 10.1177/0300985811432347.

11. Smith TJ, Baltzer WI, Löhr C, Stieger-Vanegas SM. Primary synovial osteochondromatosis of the stifle in an English Mastiff. *Vet Comp Orthop Traumatol.* 2012;25(02):160-166. doi: 10.3415/VCOT-11-04-0049.

12. Gregory SP, Pearson GR. Synovial osteochondromatosis in a labrador retriever bitch. *J Small Anim Pract.* 1990;31(11):580-583. doi: 10.1111/j.1748-5827.1990.tb00694.x.

13. Levitin B, Aroch I, Aizenberg I, et al. Linear osteochondromatosis in a cat. *Vet Radiol Ultrasound.* 2003;44(6):660-664. doi: 10.1111/j.1740-8261.2003.tb00529.x.

14. Tas O, De Cock H, Lemmens P, Pool RR. Synovial osteochondromatosis and sclerosing osteosarcoma in a cat. *Vet Comp Orthop Traumatol.* 2013;26(02):160-164. doi: 10.3415/VCOT-12-04-0059.

15. Weissengruber G, Loupal G. Osteochondroma of the tracheal wall in a Fischer's lovebird (*Agapornis fischeri*, Reichenow 1887). *Avian Dis.* 1999;43(1):155. doi: 10.2307/1592776.

16. Smith RKW, Coumbe A, Schramme MC. Bilateral synovial chondromatosis of the metatarsophalangeal joints in a pony. *Equine Vet J.* 1995;27(3):234-238. doi: 10.1111/j.2042-3306.1995.tb03068.x.

17. Kirk MD. Radiographic and histologic appearance of synovial osteochondromatosis of the femorotibial bursae in a horse: a case history report. *Vet Radiol.* 1982;23(4):167-170. doi: 10.1111/j.1740-8261.1982.tb01100.x.

18. Rothschild BM, Panza RK. Epidemiologic assessment of trauma-independent skeletal pathology in non-passerine birds from museum collections. *Avian Pathol.* 2005;34(3):212-219. doi: 10.1080/03079450500096455.

Definition of Intraocular Pressure Values in Raptors at a Native Wildlife Rehabilitation Center in Chile

Ariela Cultrera Rozowski, DMV,
Carolina Sánchez-Saavedra, DMV, MSc,
Nicole Sallaberry-Pincheira, DMV, MSc

Session #149

Affiliation: From the Unidad De Rehabilitacion De Fauna Silvestre, Escuela De Medicina Veterinaria, Facultad De Ciencias De La Vida, Universidad Andres Bello, Santiago, Chile (Cultrera Rozowski, Sánchez-Saavedra), Department of Small Animal Medicine and Surgery, Veterinary Teaching Hospital, University of Georgia, Athens, GA 30605, USA (Sallaberry-Pincheira).

Abstract: The primary cause of admission for avian patients in wildlife rehabilitation centers (WRRC) is trauma,[1-6] which can lead to ophthalmologic injuries given their anatomic features.[7] Rebound tonometry is a practical and minimally invasive diagnostic method for measuring intraocular pressure (IOP), facilitating the detection of abnormalities resulting from ocular trauma. Accurate diagnosis is often hindered by the lack of published IOP reference values for several species or subspecies extant in Chile.

Between 2024 and 2025, IOP measurements were obtained via rebound tonometry from birds of prey admitted to the Wildlife Rehabilitation Unit of Andrés Bello University (UFAS-UNAB) and Ñamku UNAB WRRC. The samples included individuals from the orders Accipitriformes, Falconiformes, and Strigiformes, with no ophthalmologic alterations during admission or rehabilitation, with the following average results in the following species: *Accipiter bicolor* (18.66 mmHg), *Athene cunicularia* (10.16 mmHg), *Bubo virginianus* (19.91 mmHg), *Falco peregrinus* (23.6 mmHg), *Falco sparverius* (16.0 mmHg), *Geranoaetus melanoleucus* (44.87 mmHg), *Geranoaetus polyosoma* (35.04 mmHg), *Glaucidium nana* (20.5 mmHg), *Milvago chimango* (20.03 mmHg), *Parabuteo unicinctus* (30.19 mmHg), *Strix rufipes* (11.5 mmHg) and *Tyto furcata* (11.16 mmHg). The results were heterogeneous; while certain species, such as *Tyto furcata*, exhibited IOP values consistent with existing literature[1], other species, namely *Falco sparverius* and *Falco peregrinus*, displayed significant deviations from previously reported data.[7] This study underscores the paucity of reference parameters for raptor species commonly admitted to rehabilitation centers in South America. It contributes new reference ranges for species with previously undocumented values and advocates for further research aimed at standardizing IOP measurements to improve the diagnostic and therapeutic approaches for these birds.

References

1. Cococcetta C, Collarile T, Masi M, et al. Antemortem diagnosis of congenital glaucoma in a white-bellied caique (*Pionites leucogaster*). *Vet Ophthalmol.* 2021;24(1):93-96. doi:10.1111/vop.12821.

2. Hernandez CL, Oster SC, Newbrey JL. Retrospective study of raptors treated at the southeastern raptor center in Auburn, Alabama. *J Raptor Res.* 2018;52(3):379-388.

3. Maphalala M, Monadjem A, Bildstein KL, et al. Causes of admission to a raptor rehabilitation centre and factors that can be used to predict the likelihood of release. *Afr J Ecol.* 2021;59(2):510-517.

4. Molina-López RA, Casal J, Darwich L. Causes of morbidity in wild raptor populations admitted at a wildlife rehabilitation centre in Spain from 1995-2007: a long term retrospective study. *PLOS One.* 2011;6(9):e24603.

5. Pahl A. Raptor rehabilitation in Florida: admission reasons, care duration, and release rates during the last fifteen years. *Slovak Raptor J.* 2024;18(1):85-98.

6. Panter CT, Allen S, Backhouse N, et al. Causes, temporal trends, and the effects of urbanization on admissions of wild raptors to rehabilitation centers in England and Wales. *Ecol Evol.* 2022;12(4):e8856.

7. Carter RT, Lewin AC. Ophthalmic evaluation of raptors suffering from ocular trauma. *J Avian Med Surg.* 2021;35(1):2-27.

Bird Strike: Causal Background, Prevention and Therapeutic Approaches to the Treatment of Intraocular Hemorrhage

Rüdiger Korbel, DipECZM (Avian), Prof, Dr med vet habil, and
Isabella Gruenberger, med vet

Session #204

Affiliation: From Center for Clinical Veterinary Medicine, Clinic for Birds, Small Mammals, Reptiles and Ornamental Fish, University Ludwig Maximilian of Munich, Germany.

Abstract: Bird strike refers to the collision of birds with objects. There are approximately 250,000 bird strikes daily in Europe and approximately 70-100 million cases in Germany annually.[1] Causes of bird strikes include collisions with vehicles, overhead power lines, glass panes and wind turbines.[2-4] Public safety considerations are also important, with two officially recorded bird collisions in air traffic and one bird strike per 0.21-61 kilometers of rail travel in Germany every day. The Federal Aviation Administration and the U.S. Department of Agriculture reported 276,846 wildlife strikes, with 272,016 (98.3%) occurring in the United States. In 2023, about 19,400 bird strikes were reported at 713 U.S. airports.[3,4] The International Civil Aviation Organization reports approximately 34,000 bird strikes annually worldwide, but the actual number is likely higher due to underreporting. Due to the expansion of renewable energies and urban infrastructure, collision trauma in birds is being recorded with increasing frequency, making the issue more relevant in terms of animal welfare law, medical care and politics. From a medical and therapeutic perspective, in addition to deaths, fractures and soft tissue trauma, intraocular lesions are of particular relevance, with an incidence of around 35% in all avian trauma victims.[5-8] Intravitreal or retinal hemorrhages lead to consecutive visual impairments and even loss of vision, which due to the visual lifestyle of wild birds, especially birds of prey, has far-reaching consequences for individual animal welfare and species conservation. Preventive measures regarding bird strikes at wind turbines include specifications for nationwide distances to breeding sites of birds of prey when operating onshore wind turbines and innovative research on AI-controlled anti-collision systems.[9] Furthermore, the Brandenburg Paper offers guidelines for documenting the impacts of wind energy use on birds and corresponding raptor-oriented documentation in Southern Germany/Bavaria.[2] Against the background of a study funded by the Bavarian State Ministry for Economic Affairs, Regional Development and Energy and conducted in cooperation with the Bavarian Chapter of the German Falconry Association/Bavarian Hunting Association, the causes, prevention strategies, diagnostics (using electroretinography and optical coherence tomography)[7] and aspects of an ongoing project on therapeutic approaches for the treatment of bird strike-related ocular lesions are outlined.

References

1. Naturschutzbund Deutschland e.V. Available at: https://www.nabu.de/wir-ueber-uns/infothek/publikationen/index.html. Accessed July 14, 2025.

2. State Office for Environment. Informations on influences on wind power plants on birds [Brandenburger Papier; German]. Brandenburg, Germany: State Office for Environment; 2025.

3. Diaz J. What are bird strikes and should you be worried about them during your next flight? Texas Public Radio. Available at: https://www.tpr.org/news/2024-12-31/what-are-bird-strikes-and-should-you-be-worried-about-them-during-your-next-flight. Accessed July 14, 2025.

4. Gard K, Groszos MS, Brevik EC, Lee GW. Spatial analysis of bird-aircraft strike hazard for Moody Air Force Base aircraft in the state of Georgia. *Georgia J Sci.* 2007;65(4):161-169.

5. Korbel RT. Avian ophthalmology: principles, application and practical cases. *Proc ExoticsCon.* 2022; 97-107.

6. Korbel RT. Eye. In: Liebich HG, Korbel RT, Koenig H, eds. *Avian Anatomy: Textbook and Colour Atlas.* 2nd Ed. Translated and revised by C. Klupiec. Sheffield, UK: 5M Publishing Ltd; 2016.

7. Korbel RT. Advanced ocular imaging and optical coherence tomography (OCT) in avian and reptiles. *Proc Annu Conf Eur Coll Vet Ophthalmol.* 2024.

8. Carter RT, Lewin AC. Ophthalmic evaluation of raptors suffering from ocular trauma. J Avian Med Surg. 2021;35(1):2-27.

9. ProTecBird. Available at: https://www.protecbird.com/. Accessed July 14, 2025.

Adequate Hunting Performance Despite Unilateral Evisceration in a Harris's Hawk (*Parabuteo unicinctus*) with Bilateral Anterior Uveitis and Unilateral Secondary Glaucoma

Clara Y. Verdenius,
Zoë van der Plaats, DVM, MSc,
IJM Slenter,
Yvonne van Zeeland, DVM, MVR, PhD, DipECZM (Avian),
DipECZM (Small Mammal), CPBC,
Sylvia Djajadiningrat-Laanen, DVM, DipECVO

Session #189

Affiliation: From Department of Clinical Sciences of Companion Animals, Division of Zoological Medicine and Surgery of Companion Animals, Ophthalmology Section at Utrecht University, 3584 CM Utrecht, The Netherlands.

Abstract: A 6-year-old female Harris's hawk (*Parabuteo unicinctus*), professionally trained to repel and catch seagulls, presented with acute blepharospasm and corneal edema on the right eye (OD) for one day. Complete physical and ophthalmic examination revealed signs of bilateral anterior uveitis (OD > OS) and markedly increased intraocular pressure (IOP OD 99 mmHg) without further abnormalities. As work-up into potential underlying causes was declined, treatment was initiated using dorzolamide 2% (1gtt q12h OD) and meloxicam (0.5 mg/kg q12h PO). Three days later, no improvement was seen, and CBC revealed marked leukocytosis. Hence, anti-inflammatory therapy was intensified using dexamethasone 0.1% (1gtt q4h OU) and prednisolone (1 mg/kg q24h PO) with additional antimycotic (itraconazole, 10 mg/kg q24h PO) and antibiotic prophylaxis (doxycycline 35 mg/kg q12h PO). As uveitis improved and right-sided IOP normalized (28 mmHg), medication was gradually tapered, but acute flare-up of right anterior uveitis occurred 5 weeks post-presentation. Due to persistent leukocytosis, doxycycline was replaced by enrofloxacin (15 mg/kg q12h PO). Additionally, due to poor prognosis for the right eye, evisceration was performed using the technique described by Murray et al.[1] The hawk recovered well post-surgery, and following leukocyte counts returned within normal limits without signs of anterior uveitis or glaucoma 6 weeks later. All medication except dorzolamide (1gtt q12h OS) was discontinued or gradually tapered. Three years after initial presentation, the hawk is doing well and has fully resumed its professional hunting activities, illustrating that, contrary to common belief, trained hawks may effectively hunt without restrictions after unilateral evisceration.

References

1. Murray M, Pizzirani S, Tseng F. A technique for evisceration as an alternative to enucleation in birds of prey: 19 cases. *J Avian Med Surg.* 2013;27:120-127.

Determination of Reference Intervals for Plasma Bile Acids in Bald Eagles (*Haliaeetus leucocephalus*)

Michael P. Jones, DVM, DABVP (Avian Practice),
Fabiano Montiani-Ferreira, DVM, PhD,
Carolyn Cray, PhD

Session #217

Affiliation: From American Eagle Foundation, 510 Kodak Rd., Kodak, TN 37764, USA (Jones), Division of Comparative Pathology, Department of Pathology & Laboratory Medicine, University of Miami, Miami, P.O. Box 016960 R-46, Miami, FL 33101, USA (Cray, Montiani-Ferreira), Federal University of Parana, Curitiba, Brazil (Montiani-Ferreira).

Abstract: While hematologic and biochemical reference intervals have been established for bald eagles[1] and other raptors, there are still significant gaps in our knowledge base that need to be filled. Once such gap is the determination of plasma bile acids in bald eagles. The objective of this study was to determine reference intervals for bile acids in a collection of bald eagles (*Haliaeetus leucocephalus*) in managed care. Reference intervals were determined from blood samples collected during annual wellness examinations for 39 bald eagles (19 male and 20 female) with ages ranging between 3 and 39 years (15 median, 9-21 IQR) at the time of sample collection. All eagles were fasted for approximately 24 hours prior to sample collection. Blood samples were collected from either the right or left basilic vein using a 22-gauge needle on a 3-cc syringe and immediately placed in sample collection tubes containing lithium heparin. All samples were centrifuged to obtain plasma samples prior to overnight shipment to the University of Miami Comparative Pathology laboratory for analysis. Samples were analyzed on the Ortho Vitros 5600 chemistry analyzer (Rochester, NY) using a colorimetric assay validated for use in eagles. The reference interval of 1.6-32.9 μmol/L (median 9.3) was determined in accordance with the American Society for Veterinary Clinical Pathology Quality Assurance and Laboratory Standards Guidelines[2] and robust method. There was no effect of age or sex on bile acids values.

References

1. Jones MP, Arheart KL, Cray C. Reference intervals, longitudinal analyses, and index of individuality of commonly measured laboratory variables in captive bald eagles (*Haliaeetus leucocephalus*). *J Avian Med Surg*. 2014:28(2),118-126.

2. Friedrichs KR, Harr KE, Freeman KP, et al. ASCVP reference interval guidelines: determination of de novo reference intervals in veterinary species and other related topics. *Vet Clin Pathol*. 2012;41(4):441-453.

Section 7

Computed Tomography, Avian Bornavirus, and Reproductive Topics Session

Jess Magnotti, DVM, DABVP (Avian Practice)
Session Chair

Computed Tomography Imaging in Birds: Clinical Applications

Minh Huynh, DVM, MRCVS, DipECZM (Avian), Dipl ACZM

Session #176

Affiliation: From Centre Hospitalier Vétérinaire Frégis, 9 Rue de Verdun 94250 Gentilly.

Abstract: Computed tomography (CT) has revolutionized avian diagnostic imaging by enabling high-resolution, cross-sectional visualization of complex anatomical structures. This presentation explores the multifaceted clinical applications of CT imaging in birds, with an emphasis on anatomic-functional relationships and their relevance to diagnosis.

Respiratory System: Air Sacs, Lung Architecture, and Pneumatization

Birds possess a unique respiratory system consisting of rigid, non-expansible lungs and a series of paired and unpaired air sacs. These air sacs are intimately connected to the lung parenchyma and extend into various pneumatic bones, including the vertebrae, sternum, humerus, and femur. Understanding the anatomic connections between air sacs and lungs is essential to identifying the potential spread of infectious (e.g., fungal pneumonia) or neoplastic disease.

Skull pneumatization involves the quadrate, parasphenoid complex, pterygoid, and frontal bone, with extensions originating from the tympanic cavity;[1] however, structures like the mandible are rarely pneumatized. Infections invading one of the bones such as otitis can extend into these pneumatic structures, complicating treatment. It is important to note that pneumatization can vary significantly both within and between avian species.[2,3]

Cervical, clavicular, and cranial thoracic air sacs arise from the medioventral secondary bronchi. Caudal air sacs—comprising the caudal thoracic and abdominal sacs—originate from the lateroventral secondary bronchi or the primary bronchus and receive air from the trachea via the medioventral secondary bronchi.[4] Consequently, infections in the bronchial tree frequently involve the caudal thoracic and abdominal air sacs.

Birds possess one pericardial cavity, two pleural cavities, and five distinct peritoneal cavities. Four hepatic peritoneal cavities (HPCs) surround the right and left liver lobes. Ascites caused by portal hypertension or right heart failure typically occurs first in the HPCs and only later spills into the intestinal peritoneal cavity (IPC) once a critical volume is reached. Effusion specifically in the IPC may also result from internal ovulation, or from ovarian, testicular, splenic disease, or lymphoid neoplasia.[5]

Vascular Anatomy

Major vessels such as the aorta, brachiocephalic trunks, carotids, and pulmonary arteries are readily identifiable on CT. Reference values for arterial diameters have been previously published.[6]

Vascular lesions such as atherosclerosis are prevalent in captive birds—especially Amazon parrots and African grey parrots—and are commonly identified on CT by calcified plaques within the arterial walls.[7] While lesions in the aorta and brachiocephalic trunk are routinely observed, advances in CT imaging have enabled the identification of vascular abnormalities in other arteries, including the renal and femoral arteries. Both atherosclerosis and aneurysms have been described in avian patients, and a potential association between these conditions has been suggested, paralleling findings in human medicine.

Skull Abnormalities: Trauma, Luxation, Deformities

CT imaging provides invaluable insight into beak pathologies and facial trauma. In psittacine species, cranial kinesis depends on a complex array of articulations and syndesmoses, which can easily luxate or fracture.[8] Accurate identification of luxation or fracture patterns is essential for surgical planning and optimal clinical management. Palatine bone luxation, choanal atresia, quadrate bone luxation and mandible shortening are common abnormalities which have been recently described by CT-scan evaluation.[9,10]

Visceral Organ Abnormalities

Coelomic evaluation in birds is inherently limited by low natural contrast. CT imaging can improve detection of organomegaly, such as hepatomegaly, splenomegaly, and renomegaly. It also aids in the assessment of reproductive pathologies such as ectopic egg, salpingitis and coelomitis. Mass characterization can be enhanced by identifying the organ of origin and analyzing contrast enhancement patterns.

Skeletal Abnormalities

CT imaging provides unparalleled three-dimensional visualization of the avian skeleton, overcoming the superimposition limitations of traditional radiography. This is particularly valuable for identifying the following conditions.

Angular limb deformities

These deformities may result from nutritional imbalances, metabolic bone disease, or congenital malformations. CT helps quantify the degree of angulation, assess the long bone deviation and potentially plan corrective osteotomies or orthopedic interventions.

Osteodystrophic processes

CT can reveal changes in bone density and architecture consistent with fibrous osteodystrophy, osteosclerosis or osteopenia. This is especially useful in polyostotic disease processes or systemic metabolic conditions affecting bone integrity.

Fractures and nonunions

Subtle or complex fractures, particularly in the thoracic girdle, skull, spine, or pelvic girdle, may go undetected on radiographs. CT imaging offers precise localization and characterization, aiding in surgical planning and follow-up.

Neoplastic bone lesions

CT is instrumental in evaluating primary or metastatic bone tumors, their margins, and the extent of cortical destruction or periosteal reaction.

Oncologic Applications and Neoplastic Staging

Neoplastic diseases are relatively common in certain species, such as chickens with ovarian carcinoma. Spontaneous ovarian carcinoma in chickens may serve as a comparative model for human ovarian cancer.[11] CT is useful for disease staging, assessing coelomic effusion, and detecting potential metastases.

It also allows evaluation of hepatic, renal, and intestinal involvement, all of which are essential for long-term case management.

Conclusions

CT has become an indispensable tool in avian medicine, offering unparalleled detail in the evaluation of both soft tissues and bony structures. From diagnosing complex respiratory and vascular conditions to guiding surgical interventions for trauma and neoplasia, CT imaging enhances clinical decision-making and patient outcomes. Its non-invasive nature and rapid acquisition time make it particularly well-suited to exotic species, where minimizing handling and stress is paramount. As technology advances and accessibility improves, the integration of CT into routine avian diagnostics is likely to expand, supporting early detection, accurate staging, and more tailored therapeutic strategies.

Understanding species-specific anatomy, anatomic variation, and the functional relevance of imaging findings is key to maximizing the diagnostic utility of CT scans in birds. When combined with clinical history, physical examination, and complementary diagnostics, CT emerges not only as a diagnostic tool, but as a window into the intricate and often enigmatic physiology of avian species.

Disclosure statement: The author runs a telediagnostic company, Telenacologie.

References

1. Hogg DA. The development of pneumatisation in the skull of the domestic fowl (*Gallus gallus domesticus*). *J Anat.* 1990;169:139-151.

2. Hogg DA. The distribution of pneumatisation in the skeleton of the adult domestic fowl. *J Anat.* 1984;138(4):617-629.

3. Lawson AB, Martinez A, Hedrick BP, et al. Variation in air sac morphology and postcranial skeletal pneumatization patterns in the African grey parrot. *J Anat.* 2025;246(1):1-19.

4. King AS, McLelland J. *Birds: Their Structure and Function.* 2nd Ed. Sussex, EN: Baillière Tindall; 1984.

5. Taylor WM. Pleura, pericardium, and peritoneum: the coelomic cavities of birds and their relationship to the lung-air sac system. In Speer B, ed. *Current Therapy in Avian Medicine and Surgery.* St. Louis, MO: Elsevier; 2016:345-362.

6. Beaufrère H, Rodriguez D, Pariaut R, et al. Estimation of intrathoracic arterial diameter by means of computed tomographic angiography in Hispaniolan Amazon parrots. *Am J Vet Res.* 2011;72(2):210-218.

7. Beaufrère H, Pariaut R, Rodriguez D, Tully TN. Avian vascular imaging: a review. *J Avian Med Surg.* 2010;24(3):174-184.

8. Huynh M, González MS, Beaufrére H. Avian skull orthopedics. *Vet Clin North Am Exot Anim Pract.* 2019;22(2):253-283.

9. Pérez AH, García MA, Barceló AM, Nadeu CB. Diagnostic approach using computerized tomography and successful surgical resolution of a palatine luxation and entrapment in a blue and yellow macaw (*Ara ararauna*). *J Avian Med Surg.* 2025;39(1):46-53.

10. Cococcetta C, Collarile T, Vetere A, et al. Successful conservative management of multiple skull fracture and quadrate-mandibular medial luxation in a blue-and-yellow macaw (*Ara ararauna*). *BMC Vet Res.* 2025;21(1):123.

11. Hawkridge AM. The chicken model of spontaneous ovarian cancer. *Proteomics Clin Appl.* 2014;8(9-10): 689-699.

Avian Cervical Spinal Diseases: A Review of Computed Tomography Findings, 2017-2025

M. Scott Echols, DVM, DABVP (Avian Practice)

Session #192

Affiliation: From Parrish Creek Veterinary Hospital and Diagnostic Center, 86 N 70 W St, Centerville, UT 84014, USA, Scarlet Imaging, LLC, PO Box 57515, Murray, UT 84157-0515, USA, Medical Center for Birds, 3807 Main St A, Oakley, CA 94561, USA.

Abstract: Avian intervertebral diseases are poorly documented. This study seeks to better understand the disease incidence, specifically in the cervical region as defined as C1-T1. Computed tomography (CT) records were reviewed from 2017 through June 2025. Of the 405 CT cases that met study inclusion criteria, 107 had demonstrable spinal column disease (26.4% of total cases). Fifty-one cases demonstrated cervical disease represented by Psittaciformes (45, 88.2% of cases), Falconiformes (2, 3.9%), Anseriformes (2, 3.9%) and Pelecaniformes (2, 3.9%). Lesions were categorized and divided into thirds based on the number of cervical vertebrae reported for the order: proximal (C1-C3 for parrots), mid (C4-C7 for parrots) and distal (C8+ for parrots) cervical regions. While there were 51 birds, a total of 59 contiguous lesions were found as several birds had more than one cervical lesion. Four birds (6.8% of the lesions and all parrots) had proximal cervical lesions. Twenty-three birds (39% of the lesions and all parrots) had mid cervical lesions. Thirty-two birds (54.2% of the lesions and 26 parrots, 2 pelicans, 2 ducks and 2 birds of prey [eagle and a hawk]) had distal cervical lesions. Of the 51 birds, 16 (31.4%) had some degree of spinal cord compression or swelling; 33 (64.7%) had non-spinal osteoarthritis; 8 (15.7%) had intervertebral tissue mineralization (of which 7 [13.7%] had mineralized intervertebral tissue bulging dorsally into the intervertebral canal); 1 (2.0%) had intervertebral tissue bulging without mineralization; and 1 (2.0%) had a cervical spinal column fracture.

Avian Notarial, or Thoracic, Spinal Diseases: A Review of Computed Tomography Findings, 2017-2025

M. Scott Echols, DVM, DABVP (Avian Practice)

Session #194

Affiliation: From Parrish Creek Veterinary Hospital and Diagnostic Center, 86 N 70 W St, Centerville, UT 84014, USA, Scarlet Imaging, LLC, PO Box 57515, Murray, UT 84157-0515, USA, Medical Center for Birds, 3807 Main St A, Oakley, CA 94561, USA.

Abstract: Avian intervertebral diseases are poorly documented. This study seeks to better understand the disease incidence, specifically in the notarial (i.e., thoracic) region as defined as T1-synsacrum. Computed tomography (CT) records were reviewed from 2017 through June 2025. Of the 405 CT cases that met study inclusion criteria, 107 had demonstrable spinal column disease (26.4% of total cases). Thirty-nine cases demonstrated notarial disease represented by Psittaciformes (28, 71.8% of cases), Falconiformes (3, 7.7%), Anseriformes (7, 17.9%) and Galliformes (1, 2.6%). Lesions were categorized and divided into thirds based on the number of thoracic vertebrae reported for the order: proximal (T1-T3 for parrots), mid (T4-T5 for parrots) and distal (T6+ for parrots) notarial/thoracic regions. While there were 39 birds, a total of 44 contiguous lesions were found as several birds had more than one thoracic lesion. One bird (2.5% of lesions, parrot) had its entire notarium affected. Eleven birds (25% of lesions, 10 parrots, 1 duck) had proximal lesions. Three birds (6.8% of lesions, 2 parrots and 1 hawk) had mid lesions. Twenty-nine birds (65.9% of lesions and 17 parrots, 4 birds of prey, 7 ducks and 1 chicken) had distal lesions. Of the 39 birds, 7 (17.9%) had some degree of spinal cord compression or degeneration; 20 (51.3%) had non-spinal osteoarthritis; 2 (5.1%) had intervertebral tissue mineralization; 1 (2.6%) had intervertebral tissue bulging into the intervertebral canal with no spinal cord compression; 2 (5.1%) had a notarial (i.e., thoracic) spinal column fracture; and 3 (7.7%) had compression of the nerves laterally exiting the spinal column.

Avian Synsacral Spinal Diseases: A Review of Computed Tomography Findings, 2017-2025

M. Scott Echols, DVM, DABVP (Avian Practice)

Session #195

Affiliation: From Parrish Creek Veterinary Hospital and Diagnostic Center, 86 N 70 W St, Centerville, UT 84014, USA, Scarlet Imaging, LLC, PO Box 57515, Murray, UT 84157-0515, USA, Medical Center for Birds, 3807 Main St A, Oakley, CA 94561, USA.

Abstract: Avian intervertebral diseases are poorly documented. This study seeks to better understand the disease incidence, specifically of the synsacrum. Computed tomography (CT) records were reviewed from 2017 through June 2025. Of the 405 CT cases that met study inclusion criteria, 107 had demonstrable spinal column disease (26.4% of total cases). Ten cases demonstrated synsacral disease represented by Psittaciformes (5, 50.0% of cases), Falconiformes (1, 10.0%), Anseriformes (2, 20.0%) and Galliformes (2, 20.0%). Lesions were defined as: reactive/destructive synsacral mass (5, 50% of the cases), non-reactive/non-destructive synsacral mass (1, 10%), significant synsacral deformity (3, 30%) and synsacral bone marrow soft tissue infiltration (1, 10%). Of the 10 birds, 5 (50.0%) had spinal cord compression and/or sacral nerve plexus compression; 5 (50.0%) had non-spinal osteoarthritis; 1 (10.0%) had a synsacral fracture; and 6 had reactive bone changes (2 histologically confirmed cancers, 2 probable cancers, 1 confirmed *Aspergillus* sp. fungal infection and 1 suspected other infection).

Avian Caudal Vertebral Spinal Diseases: A Review of Computed Tomography Findings, 2017-2025

M. Scott Echols, DVM, DABVP (Avian Practice)

Session #197

Affiliation: From Parrish Creek Veterinary Hospital and Diagnostic Center, 86 N 70 W St, Centerville, UT 84014, USA, Scarlet Imaging, LLC, PO Box 57515, Murray, UT 84157-0515, USA, Medical Center for Birds, 3807 Main St A, Oakley, CA 94561, USA.

Abstract: Avian intervertebral diseases are poorly documented. This study seeks to better understand the disease incidence, specifically in the caudal region as defined as synsacrum-Cd-pygostyle. Computed tomography (CT) records were reviewed from 2017 through June 2025. Of the 405 CT cases that met study inclusion criteria, 107 had demonstrable spinal column disease (26.4% of total cases). Forty cases demonstrated caudal disease represented by Psittaciformes (33, 82.5% of cases), Galliformes (1, 2.5%), and Anseriformes (6, 15.0%). Due to the significant variation in number of caudal vertebrae, lesions were divided as follows: synsacral-caudal joint, proximal (Cd1-Cd3), and distal (Cd4+). While there were 40 birds, a total of 48 contiguous lesions were found as several birds had more than one caudal lesion. Nine birds (18.8% of lesions and all parrots) had synsacral-caudal joint lesions (two of which were S-Cd2). Twenty-three birds (47.9% of the lesions, 17 parrots, 5 ducks and 1 curassow) had proximal lesions. Sixteen birds (33.3% of the lesions and 15 parrots and 1 duck) had distal lesions. Of the 40 birds, 14 (35.0%) had some degree of spinal cord compression or significant spinal canal narrowing; 20 (50.0%) had non-spinal osteoarthritis; 1 (2.5%) had intervertebral tissue mineralization; 2 (5.0%) had a fracture; and 1 (2.5%) had a diagnosed cancer (squamous cell carcinoma) all in the caudal spinal column.

Frantic Mutilative Behavior Associated with Avian Ganglioneuritis in a Young Blue and Gold Macaw

Jared S. Zion, DVM, and
Stephanie K. Lamb, DVM, DABVP (Avian Practice)

Session #161

Affiliation: From Arizona Exotic Animal Hospital, Phoenix, AZ 85024, USA.

Abstract: Feather destructive behavior (FDB) is a commonly observed phenomena that companion bird owners often report to their veterinarians. Signalment varies across the Aves class with no defined sex or age predilections. Despite some psittacine birds reported to have suspected higher incidences than others, the etiology of the condition is multifactorial. Psychotropic, hormonal, seasonal, and environmental factors are proposed as well as wide-ranging pathologic conditions such as infectious agents, inflammatory conditions, heavy metal toxins, auto-immune conditions, and neoplastic conditions, to name a few.[1] An extensive diagnostic workup is required in FDB cases for the avian clinician to best direct treatment. The clinical presentation varies widely among patients with some at higher risks for secondary infections as a result of disruption of the epithelial barrier. Management may involve a mixture of behavioral and environmental modifications, pharmacologic therapies and in severe cases, the use of physical barrier devices (e.g., Elizabethan collars, cloth vests). The use of psychogenic medications is anecdotal and remains controversial.[2] Avian ganglioneuritis is an inflammatory condition affecting the autonomic nervous system, particularly the enteric (i.e., gastrointestinal) ganglia, and may also involve other peripheral or central nervous system components. Diagnosis involves histologic confirmation of lymphoplasmacytic infiltration of the impacted ganglia often at necropsy. The condition has been associated with bornaviral infection in multiple psittacine species but is not solely correlated as a sequela to the infection and therefore may be complicated by the results of bornavirus diagnostic tests.[3] This case presents the management of a 4-year-old blue and gold macaw with frantic, repeatedly localized, severe mutilative behavior. In addition to detailing the diagnostic workup involved such as computed tomography, heavy metal testing, bornavirus antigen testing, anti-ganglioside antibody titers and serology testing, as well as aspergillus titers, the presentation outlines the use of varied pharmacologic agents involved in the management of this case. Histopathology on necropsy confirmed the diagnosis of ganglioneuritis as a comorbidity involved.

References

1. Langlois I. Medical causes of feather damaging behavior. *Vet Clin North Am Exot Anim Pract.* 2021;24(1): 119-152.

2. Rubinstein J, Lightfoot T. Feather loss and feather destructive behavior in pet birds. *Vet Clin North Am Exot Anim Pract.* 2014;17(1):77-101.

3. McCready JE. Treatment of gastrointestinal infectious diseases in exotic animals. *Vet Clin North Am Exot Anim Pract.* 2024;28(2):485-501.

Establishment of the Intranasal Transmission Pathway to Advance Avian Bornavirus Experimental Research

Caitlin P. Mencio, PhD,
Dallas Clontz, DVM,
Tyler Tortella, DVM,
Jill Heatley, MS, DVM, DABVP (Avian Practice), DABVP (Reptile & Amphibian Practice), Dipl ACZM,
Raquel Rech, MS, PhD, DVM,
Donald Brightsmith, PhD,
Sarah Hamer, MS, PhD, DVM, Dipl ACVPM (Epidemiology)

Session #185

Affiliation: From Schubot Center for Avian Health, College of Veterinary Medicine and Biomedical Sciences, Raymond Stotzer Pkwy, College Station, TX 77843, USA.

Abstract: Parrot bornavirus-4 (PaBV-4) is one of several avian bornaviruses (ABVs) that are an etiologic agent for avian ganglioneuritis (AG), also known as proventricular dilatation disease (PDD).[1] There is no cure or treatment for AG, nor vaccine against ABV infection. ABV transmission and disease progression is poorly understood. While the modes of naturally occurring transmission are not confirmed, evidence of shed virus in the feces[2] and urine[3] supports a role for ABV horizontal transmission through the fecal-oral route. Current experimental infection protocols rely on ABV inoculation via intramuscular (IM)[4,5] or combination inoculation routes.[6,7] We therefore aimed to develop an intranasal (IN) transmission pathway using cockatiels (*Nymphicus hollandicus*) as a model species. Ten cockatiels received inoculations of 1×10^5 FFU of parrot bornavirus-4 (PaBV-4) through the nares while under anesthesia to ensure uptake. Three cockatiels, serving as a control, received an IM injection of the same amount of virus. Four of the 10 IN inoculated developed severe clinical symptoms between 1-2 months post inoculation and were euthanized. All other birds showed no clinical symptoms to the end of the study at day 130 post inoculation (PI). Upon necropsy, all cockatiels, both IN and IM, showed similar cellular disease progression and pathology. Tissue collected at necropsy showed the presence of viral RNA present in at least one organ of every bird; however, distribution was not uniform. As a less invasive and likely natural transmission route, the IN inoculation pathway will facilitate our ongoing work on ABV vaccination and diagnostics to improve bird health.

References

1. Rubbenstroth D. Avian bornavirus research-a comprehensive review. *Viruses.* 2022;14(7):1513. doi: 10.3390/v14071513.

2. Rinder M, Ackermann A, Kempf H, et al. Broad tissue and cell tropism of avian bornavirus in parrots with proventricular dilatation disease. *J Virol.* 2009;83(11):5401-5407.

3. Heatley JJ, Villalobos AR. Avian bornavirus in the urine of infected birds. *Vet Med* (Auckl). 2012;3:19-23.

4. Runge S, Olbert M, Herden C, et al. Viral vector vaccines protect cockatiels from inflammatory lesions after heterologous parrot bornavirus 2 challenge infection. *Vaccine.* 2017;35(4):557-563.

5. Leal de Araujo J, Rech RR, Heatley JJ, et al. From nerves to brain to gastrointestinal tract: a time-based study of parrot bornavirus 2 (PaBV-2) pathogenesis in cockatiels (*Nymphicus hollandicus*). *PLOS One.* 2017;12(11):e0187797.

6. Gancz AY, Kistler AL, Greninger AL, et al. Experimental induction of proventricular dilatation disease in cockatiels (*Nymphicus hollandicus*) inoculated with brain homogenates containing avian bornavirus 4. *Virol J.* 2009;6:100.

7. Escandon P, Heatley JJ, Tizard I, et al. Treatment with nonsteroidal anti-inflammatory drugs fails to ameliorate pathology in cockatiels experimentally infected with parrot bornavirus-2. *Vet Med (Auckl).* 2019;10: 185-195.

Establishing the Pharmacokinetics of an Oral Dose of Robenacoxib in Hispaniolan Amazon Parrots (*Amazona ventralis*)

Hailey B. Penticoff, DVM,
Thomas N. Tully, Jr., DVM, MS, DABVP (Avian Practice), DipECZM (Avian),
Dharmikkumar Vora, MS,
Levent Dirikolu, DVM, MVSc, PhD

Session #169

Affiliation: From the Louisiana State University School of Veterinary Medicine, Skip Bertman Dr, Baton Rouge, LA 70803, USA.

Abstract: Robenacoxib is a coxib class nonsteroidal anti-inflammatory drug (NSAID) and its mechanism of action involves selectively inhibiting cyclooxygenase-2 (COX-2) to reduce inflammation via reduction of prostaglandin synthesis in dogs, cats, and rats.[1-3] Robenacoxib has been used anecdotally in avian species at doses ranging from 2-10 mg/kg PO every 3-7 days to manage pain associated with inflammation.[4] Robenacoxib concentrates in inflamed tissues, as demonstrated in cats, dogs, and rats, and its clinical effects persist longer than its short terminal half-life in blood.[1-3,5] In this study, the pharmacokinetic parameters of a single oral dose of robenacoxib were evaluated in a group of seven healthy, adult Hispaniolan Amazon parrots ranging from 10-15 years of age. Four birds were male and three were female. The results indicated that a single 4 mg/kg oral dose of robenacoxib was well tolerated; appetite, thirst, fecal output, activity levels and mentation remained normal. The mean peak plasma concentrations were reached at 0.45 hours, with a peak concentration of 88.75 ng/ml. The mean half-life was 1.3 ± 0.37 hours. The mean total body clearance per fraction of dose absorbed was 24.9 L/kg/hr. The study suggested that robenacoxib remains in plasma for up to 6 hours post-administration of an oral dose, which is higher than a comparable pharmacokinetics study performed in geese.[6] Further pharmacodynamics studies of this drug in this species would provide stronger support for dose determination and effectiveness of robenacoxib in psittacine birds.

References

1. Pelligand L, King JN, Toutain PL, et al. Pharmacokinetics/pharmacodynamic modelling of robenacoxib in a feline tissue cage model of inflammation. *J Vet Pharmacol Ther.* 2012;35(1):19-32.

2. Kongara K, Chambers J. Robenacoxib in the treatment of pain in cats and dogs: safety, efficacy, and place in therapy. *Vet Med (Auckl).* 2018;9:53-61.

3. King JN, Dawson J, Esser RE, et. al. Preclinical pharmacology of robenacoxib: a novel selective inhibitor of cyclooxygenase-2. *J Vet Pharmacol Ther.* 2009;32(1):1-17.

4. Orosz SE, Antinoff N. Clinical avian neurology and neuroanatomy. In: Speer BL, ed. *Current Therapy in Avian Medicine and Surgery.* St. Louis, MO: Elsevier; 2016:363-377.

5. Lees P, Toutain PL, Elliott J. Pharmacology, safety, efficacy, and clinical uses of the COX-2 inhibitor robenacoxib. *J Vet Pharmacol Ther.* 2022;45(4):325-351.

6. Fadel C, Lebkowska-Wieruszewska, Lisowski A, et al. Disposition kinetics of robenacoxib following intravenous and oral administration in geese (*Anser anser domesticus*). *J Vet Pharmacol Ther.* 2023;46(6):413-420.

Resolution of Egg Binding is Possible in Most Client-Owned Parrots When Multiple Treatment Strategies Are Considered

Ali Vavlas, DVM,
Holly Galusha, DVM,
Joerg Mayer, DVM, MS, DABVP (Exotic Companion Mammal Practice), DipECZM (Small Mammal), Dipl ACZM,
Brian Speer, DVM, DABVP (Avian Practice), DipECZM (Avian),
Nicola Di Girolamo, DVM, PhD, DipECZM (Herpetology), Dipl ACZM, Dipl ACVPM

Session #167

Affiliation: From College of Veterinary Medicine, University of Georgia, Athens, GA 30602, USA (Vavlas, Mayer), Medical Center for Birds, Oakley, CA 94561, USA (Galusha, Speer), Department of Clinical Sciences, College of Veterinary Medicine, Cornell University, Ithaca, NY 14853, USA (Di Girolamo).

Abstract: Egg binding is a common and potentially life-threatening disorder in avian species.[1-7] The aim of this study was to analyze the success rate and associated complications of different treatment approaches for egg binding in parrots. All parrots treated in an avian species-exclusive veterinary practice for egg binding during an 11-year period (2009 to 2020) were eligible for inclusion in this retrospective study. The primary targeted outcome of the study was resolution of egg binding, defined as removal or laying of the egg and survival at 7 days from presentation. Of the 150 events that had complete follow-up, 109 (72.7%) resulted in a successful outcome. A successful outcome was observed in 44 of 133 events (33.1%) that were initially managed medically without resorting to additional treatment strategies, in 31 events (86.1%) managed with mechanical assistance, in 20 events (60.6%) managed surgically, and in 12 events (85.7%) managed with ovocentesis. Time from presentation to resolution was a median of 36 hours (range, 1 to 240 hours; n = 119). Multivariable logistic regression demonstrated associations between higher body weight, administration of medical treatment and mechanical assistance, absence of administration of sedation or anesthesia, and absence of obtundation or stupor on presentation with a positive outcome.[8] Overall, egg binding has a favorable outcome if multiple treatment strategies in addition to medical management are considered. We identified factors associated with resolution and outlined complications of egg binding in parrots, laying the foundation for additional prospective research on this complex condition.

References

1. Jenkins JR. Surgery of the avian reproductive and gastrointestinal systems. *Vet Clin North Am Exot Anim Pract.* 2000;3(3):673-692. doi: 10.1016/S1094-9194(17)30069-5.

2. Stout JD. Common emergencies in pet birds. *Vet Clin North Am Exot Anim Pract.* 2016;19(2):513-541. doi: 10.1016/j.cvex.2016.01.002.

3. Scagnelli AM, Tully TN Jr. Reproductive disorders in parrots. *Vet Clin North Am Exot Anim Pract.* 2017;20(2):485-507. doi: 10.1016/j.cvex.2016.11.012.

4. Harrison GJ, Lightfoot T. *Clinical Avian Medicine.* Palm Beach, FL: Spix Publishing; 2006.

5. Brown C. Clinical approach to egg binding: clinical management of female psittacine birds in reproductive distress. *Proc Annu Conf Assoc Avian Vet.* 2007;7-14.

6. Clayton LA, Ritzman TK. Egg binding in a cockatiel (*Nymphicus hollandicus*). *Vet Clin North Am Exot Anim Pract.* 2006;9(3):511-518. doi:10.1016/j.cvex.2006.05.004.

7. Joyner KL. Avian reproductive emergency: medical management. *Vet Med Rep.* 1990;2(3):246-249.

8. Vavlas A, Galusha H, Mayer J, et al. Resolution of egg binding is posible in most client-owned parrots when multiple treatment strategies are considered. *J Am Vet Med Assoc.* 2025;263(5):628-634.

Yolk Embolism in Birds: A Retrospective Review of Twenty-Four Cases

Anna Sickmueller, DVM,
Stamatios A. Tahas, DVM, MSc, Dipl ACZM, DipECZM (Zoo Health Management), MRCVS,
Mads Frost Bertelsen, DVM, DVSc, Dipl ACZM, DipECZM (Zoo Health Management),
Michael M. Garner, DVM, DACVP

Session #110

Affiliation: From Ahornstraße 3, 86459 Gessertshausen, Germany (Sickmueller), Copenhagen Zoo, Roskildevej 38, Frederiksberg, 2000, Denmark (Tahas, Bertelsen), Northwest ZooPath, 654 West Main Street, Monroe, WA 98272, USA (Garner).

Abstract: Reports of fatal yolk embolism exist in reptiles, but this condition has only anecdotally been referred to in avian patients, primarily cockatiels, in sources found online. A retrospective review of post-mortem records from The Copenhagen Zoo, in Denmark and Northwest ZooPath in the United States revealed 24 cases of yolk embolism in 19 zoo-housed and five privately-owned birds, spanning the last 26 years. Cases were reviewed histologically by a board-certified veterinary pathologist (MMG). Birds from 12 avian orders and 23 avian species were affected. All animals were female and age varied from 4 days to 33 years. Gross lesions accompanying histologic evidence of yolk in aberrant locations were infrequently reported but included a prominent yolk sac or active ovary. Histologically, globular homogeneous basophilic material, indistinguishable from yolk, was identified in vasculature of organs such as the ovary (50%), lung (33%), brain (33%), kidney (29%) and liver (17%) and sometimes associated with thrombosis or mild hemorrhage. Occasionally, yolk droplets were also distinguishable in the ventricles of the heart and in the oviduct. Yolk embolism was most associated with yolk coelomitis (50%) or blunt trauma (29%). Yolk embolism was suspected to be the primary cause of death in 15 (63%) individuals whereas was considered an incidental finding in nine (37%). This study reveals that, although rare, yolk embolism should be considered as a differential diagnosis for acute death or central nervous system signs in female birds of multiple orders that have a history of coelomitis or blunt trauma.

Radiation Therapy and Orchidectomy in Two Pekin Ducks (*Anas platyrhynchos domesticus*) with Testicular Tumors

Sarah Dickerson, DVM,
Sue Chen, DVM, DABVP (Avian Practice),
M. Scott Echols, DVM, DABVP (Avian Practice),
Dustin Lewis, DVM, DACVR (Radiation Oncology)

Session #158

Affiliation: From Gulf Coast Veterinary Specialists, 8042 Katy Fwy, Houston, TX 77024, USA (Dickerson, Chen), Medical Center for Birds, 3807 Main Street, Oakley, CA 94561, USA (Echols), Red Bank Veterinary Hospital, 100 Schulz Drive, Red Bank, NJ 07701, USA (Lewis).

Abstract: The presence of avian testicular tumors is well documented in a variety of species; however, there is little in the literature regarding the successful treatment of these tumors.[1-3] This report follows the cases of two Pekin ducks (*Anas platyrhynchos domesticus*) that underwent radiation therapy and orchidectomies for unilateral testicular tumors. An 8-year-old male Pekin duck presented with respiratory signs. A CT scan and fine-needle aspirate confirmed a right testicular seminoma. A Varian Linear Accelerator was used to deliver hypofractionated radiation of 6 Gy x 5 fractions, once weekly, totaling 30 Gy. Following radiation therapy, the size of the right testicular tumor was markedly reduced by approximately 80%. Approximately 19 weeks after radiation therapy, the patient was anesthetized for orchidectomy. Surgery was successful with an uneventful recovery; however, significant adhesions between the neoplastic testicle and surrounding soft tissue were noted. Another case, a 9-year-old male Pekin duck, presented with a left leg lameness. A large testicular tumor was identified on CT and the patient received 8 Gy x 2 fractions, 1 week apart, with reduction in tumor size demonstrated. Two weeks after radiation therapy, the patient was anesthetized for orchidectomy. During surgery, severe adhesions were present between the testicular tumor and the common iliac vein, vena cava, and caudal aorta leading to significant hemorrhage. During recovery, the patient experienced cardiac arrest, and extensive resuscitative efforts were unsuccessful. These results highlight radiation therapy protocols that were well tolerated and produced a reduction in tumor size during treatment in two Pekin ducks.

References

1. Wernick MB, Tobias JR, Moeller RB, et al. Bilateral malignant seminomas in two unrelated, aged trumpeter hornbills (*Bycanistes buccinator*). *J Avian Med Surg.* 2015;29(2):125-129.

2. Kirk N, Echols MS, Reavill D, et al. Diagnosis of seminoma with metastasis to the brain and kidneys in a Moluccan eclectus (*Eclectus roratus*) using advanced contrast based imaging. *J Avian Med Surg.* 2022;36(3):325-332.

3. Sullivan PJ, Fasina OO, Cushing AC. Bilateral malignant seminoma with ventricular metastasis in a bald eagle (*Haliaeetus leucocephalus*). *J Avian Med Surg.* 2018;32(3):240-245.

Section 8

Analgesia and Anesthesia Session

Michelle Guarin, DVM
Session Chair

Advances in Pain Recognition, Assessment and Treatment of Pain in Birds

David Sanchez-Migallon Guzman, LV, MS, DipECZM (Avian), DipECZM (Small Mammal), Dipl ACZM

Session #121

Affiliation: From Department of Medicine and Epidemiology, School of Veterinary, University of California Davis, Davis, CA 95616, USA.

Abstract: Pain management is one of the areas of avian medicine that has advanced substantially in recent years. These advances include information regarding the anatomy and physiology of pain in birds, the recognition of pain behaviors as well as their assessment in addition to information regarding newer studies on opioids, nonsteroidal anti-inflammatory drugs, local anesthetics, and other drugs like gabapentin, amantadine, cannabinoids, acetaminophen and polysulfated glycosaminoglycans in birds.

Pain Recognition

An integrative literature review on pain recognition was performed in 2020 over five online databases: MEDLINE/PubMed, CAB Direct, Biosis, Zoological Record, and Scopus.[1] Additional snowballing was incorporated by looking at the references and articles that cited the 126 articles from the initial abstract and full-text screening. Of the 194 full-text articles reviewed, 132 sources of literature were included in the final analysis. From these 132 sources of literature, 31.8% were general review articles in which avian pain behaviors were described irrespective of species, with others being specific to a particular species (chickens 47.8%, turkeys 7.6%, parrots 3.8%, pigeons 3%, raptors 3%, and other 3%, including two on ducks, one on emus, and one on Eurasian blue tits). Pain stimulus varied depending on species, although the vast majority of the pain stimuli involved welfare issues such as beak trimming, limb abnormalities, and keel bone fractures in chickens. Although information regarding this topic remains limited for many avian species, this review provides a more thorough understanding of behavioral indicators of pain in species such as chickens, turkeys, psittacine birds, pigeons, raptors, and select others. More recently, the behaviors associated with inflammatory pain induced by carrageenan injection in the cockatiel was described.[2] Cockatiels injected with carrageenan in the plantar surface exhibited significantly increased focal preening and increased burst preening, while control cockatiels spent significantly more time in an upright stance. Although the remainder of behaviors observed were not statistically significant between groups, additional variables of interest seen more frequently in treatment cockatiels included non–weight-bearing stance, holding of the body low, and being nonvigilant. The level of agreement between observers was variable based on the specific behaviors; nevertheless, the dynamic behaviors were substantial to strong.

Pain Assessment

Using pain scales requires understanding normal and pain-related behavior for the species and individual. Pain score sheets can help improve uniformity and efficacy of pain scoring using behavioral analysis. Score-sheet descriptions of behavior must be validated and terms must be clearly defined to reduce observer bias and interobserver variability. Accurate assessment of avian pain can be confounded by the effect of the observer on the bird's behavior (i.e., especially in terms of wildlife species), effects of some analgesics or sedatives that can alter behavior, and concurrent physiologic conditions (e.g., anesthetic recovery, hypothermia, and others). Although a numeric rating pain scale was created in one study that exhibited construct validity in relation to orthopedic pain in pigeons and in another study that described a scored behavioral ethogram in Hispaniolan Amazon parrots, these scales may be of limited use for other avian species and clinical presentations.[1] Ideally, scales more specific to individual species and common clinical presentations and procedures (e.g., coeliotomy, enucleation, cloacopexy, limb or tail amputation, pelvic limb osteoarthritis) would be developed to increase sensitivity and specificity in pain recognition and assessment for various situations and is particularly critical for psittacine birds, in which there is a true lack of studies evaluating pain scales in a clinical setting. Although this specificity is achievable with time, more broad scales that include general pain-related behaviors could be developed and used while considering the possible variations between different avian species, types of pain, and severity. A series of behavioral and postural parameters that could be considered includes mentation, environmental interaction, perching, maintenance behaviors, locomotion, focal preening of singular area, body posture and appetite.[3] The need for tools such as those described above among a wide range of evaluators, including veterinary students, veterinary clinicians, veterinary technicians, wildlife rehabilitators, and household caregivers, is critical for the advancement of pain management in birds. It is possible that behavioral indicators of pain can vary drastically between avian species. Therefore, species-specific algorithms are essential, and extrapolations made to other avian species may not be accurate. In lieu of species-specific pain score sheets, a generic pain scale of 1 to 10 could be useful for assessing bird's pain and the response to treatment and recovery from a painful condition. Assessing a patient both before and after drug administration can help to evaluate the effectiveness of analgesics provided and determine if any new or altered behaviors are more likely to be a consequence of the drug administered.

Treatment of Pain

Effective pain management addresses the underlying cause of pain, utilizing both pharmacologic and non-pharmacologic methods to reduce nociceptor activation, signal transmission, and modulation in the central nervous system. Multimodal analgesia, which involves a combination of drugs acting at different stages of pain pathways with diverse mechanisms, is recommended. This approach considers the type and severity of pain, allowing for a balanced strategy. Commonly used drugs include opioids, nonsteroidal anti-inflammatory drugs (NSAIDs), local anesthetics, and others, aiming to enhance pain relief while minimizing adverse effects.

Opioid drugs

Opioids are used for treating moderate to severe pain, acting on μ, κ, δ, and nociceptin receptors in the central and peripheral nervous system. While opioid receptors in the brain stem and spinal cord are conserved across species, variations exist in other brain regions, affecting clinical effects. Recent studies looking at opioid receptor in pigeons and cockatiels,[4] together with a large number of recent studies on opioids, challenge the traditional view favoring κ-opioid receptor agonists (e.g., butorphanol, nalbuphine) in birds, suggesting that μ-opioid receptor agonists (e.g., morphine, hydromorphone, methadone, fentanyl, buprenorphine) may be more effective in some species. Peregrine falcons, snowy owls, and blue-fronted Amazon parrots[5] exhibit nucleotide homologies with humans (77-82%) in opioid receptor structures, possibly contributing to efficacy differences. Variability in pharmacokinetics and pharmacodynamics among species, ages, and sexes emphasizes caution in generalizations and extrapolations within the avian class. Morphine, hydromorphone, methadone, fentanyl, buprenorphine, and butorphanol are common opioids studied and used in birds.

Morphine

Morphine, a μ-opioid receptor agonist, has been rarely used in avian medicine due to conflicting results from early studies in domestic fowl. A recent study assessed the analgesic efficacy of morphine in lame broiler chickens.[6] The chickens were subjected to an obstacle course and latency-to-lie test before injection, and at 30 minutes and 2 hours after injection. The injection of 2 mg/kg IV of morphine led to increased sedation, causing chicks to sit rather than walk, making it challenging to evaluate its analgesic effects. The study suggested that lower doses of morphine might produce analgesia without sedation but would require more frequent administration due to its shorter elimination half-life. Further research is needed to explore this hypothesis.

Hydromorphone

Hydromorphone, a potent μ-opioid receptor agonist used for acute and chronic pain, was studied in orange-winged Amazon parrots and great horned owls. In parrots,[7] hydromorphone at 1 and 2 mg/kg increased thermal foot withdrawal response significantly, with positive pharmacokinetics after intravenous and intramuscular administration; however, side effects like agitation and nausea-like behavior were observed, suggesting consideration for lower doses in clinical use. In great horned owls,[8] 0.6 mg/kg showed high bioavailability and rapid elimination, with a documented metabolite H3G. Subsequent studies[9] demonstrated thermal antinociceptive effects and mild sedation at lower doses, supporting hydromorphone use in great horned owls for pain management.

Methadone

Methadone, a μ-opioid receptor agonist, is comparable to morphine in analgesia but has additional mechanisms, acting as a non-competitive NMDA receptor inhibitor and affecting norepinephrine and serotonin re-uptake. It is more lipid-soluble and potent than morphine, with minimal histamine release and less excitation or vomiting. In hens,[10] a 6 mg/kg IM dose decreased isoflurane minimum anesthetic concentrations but produced mild respiratory acidosis, reduced heart rate, and increased blood pressure. In orange-winged Amazon parrots, methadone showed similar pharmacokinetics requiring frequent dosing due to short half-life (Guzman D S-M, unpublished data). A 6 mg/kg IM dose exhibited desired antinociceptive effects but caused agitation, ataxia, and nausea-like behavior, suggesting lower doses for clinical use (Guzman D S-M, unpublished data). While pharmacokinetic properties were comparable between chickens and parrots, species differences in drug metabolism and response necessitate further studies in avian species.

Fentanyl

Fentanyl, a short-acting μ-opioid receptor agonist, is commonly administered through continuous rate infusion or transdermal patches for severe pain. Transdermal patch delivery, particularly in chickens,[11] was studied following application of 25 μg/hour with variations in individual absorption leading to differences in maximum plasma fentanyl concentrations. Despite this variability, all chickens reached human target plasma concentrations within 2-4 hours, maintaining therapeutic levels for 72 hours. The study suggests that fentanyl transdermal patches for chickens targeting 8-10 μg/kg per hour can achieve human therapeutic plasma concentrations; however, due to substantial individual variation, further studies are needed to establish effective and safe plasma concentrations in various bird species.

In another study involving helmeted guineafowl,[12] a transdermal fentanyl solution at 5 mg/kg applied to the interscapular region resulted in mean plasma concentrations above analgesic levels for dogs for at least 7 days, with no observed adverse effects. Unfortunately, the commercial formulation used in the study had been discontinued at the time of writing.

Buprenorphine

Buprenorphine hydrochloride, known for its slow onset and prolonged duration, acts as a μ-receptor agonist with less defined κ-receptor activities. In American kestrels,[13] a sustained-release formulation (Buprenorphine SR-LABTM) administered subcutaneously (SC) and intramuscularly (IM) at 1.8 mg/kg exhibited rapid absorption, maintaining plasma concentrations above 1 ng/ml for 48 hours. Subcutaneous hematomas were observed in some birds but resolved without intervention. Thermal antinociceptive effects indicated a potential need for administration every 24 hours to manage pain.[14] In red-tailed hawks,[15] a concentrated buprenorphine formulation (Simbadol®) achieved maximum concentration rapidly after subcutaneous administration at 0.3 mg/kg and 1.8 mg/kg, maintaining therapeutic levels for 24 and 48-72 hours, respectively, with mild to moderate sedation. No adverse effects were noted. This formulation is cost-effective and beneficial for raptor species. In American flamingos, the same formulation showed a longer terminal half-life of 12.6 hours, maintaining therapeutic concentrations for other species for 48-72 hours.[16] Further studies are needed to assess efficacy in flamingos. Orange-winged Amazon parrots[17] exhibited significant effect at lower doses (0.1 and 1 mg/kg) of Simbadol®, while 2 mg/kg resulted in a small but significant increase in thermal threshold. These findings, consistent with previous studies in cockatiels, emphasize species-specific differences in buprenorphine efficacy, requiring further evaluation in psittacine birds and other species.

Butorphanol

Butorphanol, a mixed κ-opioid receptor agonist and μ-opioid receptor antagonist, has been widely used in birds for moderate to severe pain, despite studies suggesting the need for frequent dosing. In lame broiler chickens,[6] a study found that 2 mg/kg IV of butorphanol provided analgesia for up to 2 hours, indicating its potential effectiveness in chickens.

In Hispaniolan Amazon parrots,[18] a formulation containing butorphanol in a 25% poloxamer 407 base (But-P407 25%) was evaluated for thermal antinociceptive effects following SC administration. The formulation exhibited sustained release, with plasma concentrations of butorphanol remaining above 100 ng/ml for 3 to 8 hours; however, later studies in orange-winged Amazon parrots did not confirm the theoretical analgesic effects, and only mild sedation was noted.[19] The standard butorphanol tartrate formulation also showed a short-lived increase in thermal threshold, confirming its historically known short duration of action in psittacine birds.

In common peafowl,[20] the pharmacokinetics of butorphanol administered via an osmotic pump were assessed. The osmotic pumps provided sustained release, reaching plasma concentrations of 60 ng/ml after 24 hours in most birds and maintaining therapeutic concentrations for seven days. No evidence of sedation or adverse effects was observed. Osmotic pumps present a promising technique for pain management in large birds, although deep sedation or general anesthesia may be required for implantation and removal. The historical paradigm of κ-opioid receptor agonists-μ-opioid receptor antagonists as the opioids of choice for acute pain management in psittacine birds may need reevaluation based on recent findings with butorphanol and other opioids in avian species.

Tramadol

Tramadol hydrochloride, a synthetic μ-opioid receptor agonist, is commonly used in avian medicine for moderate pain. Its efficacy varies among species and is believed to be related to the production of the O-desmethyltramadol metabolite (M1), which has a much higher affinity for the opioid receptor than the parent drug. Tramadol also affects κ and δ-opioid receptors, inhibits norepinephrine and serotonin reuptake, and is effective in treating moderate pain. In Muscovy ducks,[21] tramadol at 30 mg/kg PO was effective in improving various variables in a temporary arthritis model induced by monosodium urate solution. A pharmacokinetic study in the same species[22] showed an elimination half-life of 3.95 hours, with a Cmax of 780 ng/ml at 1.16 hours. The ducks readily produced the M1 metabolite, maintaining concentrations above therapeutic levels for humans and other avian species. African penguins[23] administered 10 mg/kg PO showed a longer elimination half-life of 7.3 hours, with a Cmax of 815 ng/ml at 4.6 hours. Tramadol and M1 concentrations were consistent with therapeutic levels in humans for up to 12 hours in 9/15 birds for tramadol and 36 hours in 14/15 birds for M1. Authors suggested an oral dose of 10 mg/kg tramadol once daily based on plasma concentrations.

In Harris hawks,[24] electrical antinociceptive effects of tramadol were evaluated. No effects were observed at 5 mg/kg IM, but 15 and 30 mg/kg IM resulted in electrical nociceptive effects and mild to moderate sedation. The study highlights the variable response to tramadol in different avian species.

Tapentadol

Tapentadol, an analgesic agent acting as both a μ-opioid receptor agonist and a norepinephrine reuptake inhibitor, is commonly used in human medicine for pain management. Its potential use in veterinary medicine is under investigation, and the pharmacokinetics of tapentadol were assessed in Hispaniolan Amazon parrots.[25] A single 30 mg/kg PO dose of a compounded 5 mg/ml tapentadol suspension resulted in detectable plasma concentrations in only two out of seven birds. The bird with the highest plasma levels reached a maximum concentration of 143 ng/ml with a half-life of 24.8 minutes. The low oral bioavailability and short half-life in Hispaniolan Amazon parrots suggest limited clinical use in this species. Further studies in other avian species are recommended to evaluate the analgesic potential of tapentadol.

Nonsteroidal anti-inflammatory drugs

Nonsteroidal anti-inflammatory drugs (NSAIDs) have different mechanisms of action, primarily through the inhibition of the enzyme cyclo-oxygenase 1 and/or 2 (COX-1 and COX-2). These drugs help relieve mild to moderate pain associated with chronic inflammation. Newer COX-inhibitors specifically target COX-2 inhibition while sparing COX-1, limiting the inhibition of physiologically important prostaglandins. Another class of NSAIDs has emerged with a different mechanism of action, involving prostaglandin E2 (PGE2) receptor antagonism. NSAIDs are commonly used in the post-operative period and for patients experiencing trauma or inflammation in musculoskeletal tissues. In addition to their effects on acute pain, NSAIDs are useful in alleviating chronic pain, such as that seen in patients with osteoarthritis. Commonly used NSAIDs in birds include meloxicam, carprofen, ketoprofen, celecoxib, and piroxicam. Each of these drugs has a distinct COX-1/COX-2 ratio, leading to differences in efficacy and toxicity.

Meloxicam

Meloxicam is a widely used anti-inflammatory medication in avian practice, particularly in birds. It is a COX-2-selective NSAID that may show more COX-1 selectivity at higher dosages. Meloxicam is available in various formulations, including oral tablets, oral suspension, and injectable forms. Pharmacokinetic studies of meloxicam have been conducted in a variety of avian species, revealing differences in dosage, bioavailability, and elimination half-lives. Some of the more recent pharmacokinetic studies include species like zebra finches,[26] chukar partridge,[27] Japanese quail,[28] pigeons[29] or Bilgorajska geese.[30] The elimination half-life of meloxicam varies significantly among species, contributing to clinical differences in its use. For example, chickens and pigeons[31] have three times longer elimination half-life compared to mallard ducks, ostriches, and turkeys. African penguin[32] and brown pelican[33] have shown recently to have a longer elimination half-life than most avian species studied. Within psittacine species, recent studies have highlighted differences in the pharmacokinetics of meloxicam that have a direct effect on dosage recommendations. The elimination half-life and bioavailability of meloxicam in Hispaniolan Amazon parrots was 15.8 ± 8.6 hours and 49-75%, respectively;[34] however, in grey parrots elimination half-life was twice as long as that with oral bioavailability of 38.1% ± 3.6%.[35] In cockatiels, an elimination half-life of less than one hour and an oral bioavailability of 11% was found.[36] Within duck species, even more significant changes have been found. Pekin ducks have 20 times longer elimination half-life of meloxicam than in mallard ducks.[31,37] Even within chicken breeds, there are differences in the meloxicam pharmacokinetics that affect the dosage recommendations. White Leghorn chickens have a shorter terminal half-life of meloxicam compared to Bantam Cochin and Wyandotte breeds.[38,39] Also, as has been demonstrated with emus, significant differences in some pharmacokinetic parameters of meloxicam can be found in individuals of different ages.[40] Dose extrapolation may not be suitable, and careful consideration of each species and/or breeds and/or age is necessary for safe and effective use.

The sustained-release formulation of meloxicam (Meloxicam SR) was assessed in Hispaniolan Amazon parrots[41] and American flamingos.[42] In Hispaniolan Amazon parrots, a single subcutaneous dose of 3 mg/kg resulted in a rapid increase to peak concentrations, maintaining above target levels for 12 to 96 hours; however, due to significant interindividual variability, the formulation's use in this species is not recommended until further evaluations in pharmacokinetics, safety, and pharmacogenomics are conducted to establish accurate dosing recommendations and understand the variability. In American flamingos, the sustained-release formulation exhibited a swift rise to peak plasma concentration, but it only sustained target levels for 0.5 to 4 hours after a single subcutaneous dose of 3 mg/kg. Consequently, this formulation may not extend the interval between treatments compared to the regular formulation of meloxicam in this species. The osmotic pump evaluated in pigeons was reported to maintain targeted plasma concentrations from 12 hours to up to 6 days.[43]

Meloxicam has generally shown minimal adverse effects in various avian species, with studies rarely reporting no significant renal, gastrointestinal, or hemostatic adverse effects in several birds at dosages used clinically. In a study involving grey parrots,[44] various doses of meloxicam (1 mg/kg IM q24h; 1 mg/kg PO q24h; and 1.6 mg/kg PO q24h) were administered in a multiple-dose pharmacokinetic analysis. Across all routes and doses, plasma concentrations consistently exceeded effective analgesic levels observed in other avian species. There was moderate evidence of drug accumulation with both intramuscular (IM) and oral (PO) administrations, and while there were no adverse effects observed in parrots receiving 1 mg/kg by either route, an increased N-acetyl-β-d-glucosaminidase activities were noted in birds receiving 1.6 mg/kg PO. In chickens, previous doses of 5 mg/kg considered to be effective in pharmacodynamic studies, were found to result in varying levels of renal acute tubular injury and gout.[45] This underscores the significance of assessing pharmacokinetics, pharmacodynamics and safety tailored to each avian species in meloxicam studies.

Meloxicam residues in eggs have been observed in chickens, requiring withdrawal times for eggs meant for consumption. The withdrawal time is influenced by factors such as breed and individual variability. In chickens, dosing of meloxicam 1 mg/kg PO q12 hours for 5 days resulted in meloxicam residues in eggs in the yolks and whites after 8 days and 3 days, respectively.[46] On the basis of these results, a 2-week withdrawal time should be adequate to avoid drug residues in eggs meant for consumption.[46,47]

Celecoxib, mavacoxib, robenacoxib

Celecoxib, mavacoxib, and robenacoxib are COX-2 selective NSAIDs selective as opposed to meloxicam which is COX-2 preferential, inhibiting both COX-1 and COX-2. In a study evaluating the pharmacokinetics of celecoxib (10 mg/kg PO) and mavacoxib (4 mg/kg PO) in cockatiels,[36] mavacoxib exhibited complete absorption with a bioavailability of 111%, while celecoxib had a mean oral bioavailability of 56%. Plasma concentrations of 0.4 μg/ml mavacoxib, effective in osteoarthritic dogs, were maintained for approximately 12 days in cockatiels, and Caribbean flamingos maintained concentrations for 120 hours after a single dose of 6 mg/kg. Elimination half-lives in cockatiels (135 hours) and Caribbean flamingos (74.5 hours) were longer than most NSAIDs in birds, requiring further studies for dosing frequencies, efficacy, and safety. A further study in cockatiels with ad libitum feeding and additional gavage feedings during the study showed that changes in body temperature, mentation, and posture were less pronounced with a single dose of mavacoxib 4 mg/kg PO compared to celecoxib 10 mg/kg at the tested dosages.[48] Robenacoxib, a long-acting NSAID used anecdotally in birds, showed a short elimination half-life in a single PK study in domestic geese,[49] suggesting caution in its use in this species.

Carprofen

Carprofen, considered a weak COX inhibitor at therapeutic doses, exhibits good anti-inflammatory activity despite its weak COX inhibition. A recent pharmacokinetic (PK) and toxicokinetic study in white-backed vultures[50] showed that carprofen at 5 mg/kg, administered via various routes, was well tolerated, but its oral bioavailability was only 42%. In a toxicology study with a subset of birds receiving a higher dose (64 mg/kg), one vulture died with elevated uric acid and potassium levels, similar to toxicities associated with diclofenac and ketoprofen in Gyps vultures. The study suggests caution in using carprofen in domesticated ungulates in areas accessible to vultures or at supplementary feeding sites.

Piroxicam

Piroxicam, a nonselective NSAID known for its anti-inflammatory properties and potential as a chemo-preventive and antitumor agent, exhibits higher potency against COX-1 than COX-2. Limited new studies on its avian use exist. Cranes have been treated with piroxicam for over 20 years for trauma, inflammation, and postoperative pain with minimal observed side effects. A pharmacokinetic study in brolga cranes[51] suggested daily administration might be appropriate for this species based on plasma concentrations.

Grapiprant

Grapiprant, a prostaglandin E2 (PGE2) receptor antagonist NSAID approved for dogs in the US, binds to the EP4 receptor, blocking sensitization of sensory neurons and inflammation stimulation. In red-tailed hawks,[52] oral administration of grapiprant at 30 mg/kg achieved plasma concentrations approximately 100 times canine therapeutic concentrations. While adverse effects were not identified in the main study, regurgitation occurred in two birds at 45 mg/kg in a pilot study. Non-fasted red-tailed hawks[53] showed delayed and decreased absorption, with only 50% reaching targeted concentrations for a shorter time. Further studies on higher doses, increased frequency with food, and multi-dose administration are needed.

Local anesthetics

Local anesthetics reversibly bind to Na+ channels and block impulse conduction, thereby making them favorable for use during surgical procedures to prevent pain conduction and sensitization of the central nervous system. The duration of action depends on the molecular properties and lipid solubility of the drug. Following local application, the anesthetic will be quickly absorbed into the blood stream, increasing the potential for toxic reactions. Typically, local anesthetics are administered via local line or splash blocks.

Lidocaine

Lidocaine, an amide local anesthetic, reduces neuronal excitability by blocking sodium channels, preventing nociception conduction. In broiler chickens,[54] lidocaine administered intrathecally provided analgesia, with 2 mg/kg resulting in a longer duration (21.3 minutes) compared to 0.5 or 1 mg/kg. The spinal anesthesia technique in chickens[55] may be suitable for providing local anesthesia in species with disorders affecting the cloaca, but caution is advised in other species where the technique has not been evaluated.

Bupivacaine

Bupivacaine, a long-acting amide local anesthetic, reduces neuronal excitability by blocking sodium channels, preventing nociception conduction. Recent studies in chickens[56] under isoflurane anesthesia determined that 1.96 mg/kg IV bupivacaine is associated with a 50% probability of a clinically relevant change in MAP and HR, while 1.33 mg/kg IV bupivacaine has a 1% probability of causing the same effect. In broiler chickens,[54] spinal anesthesia with bupivacaine at doses of 0.1, 0.25, and 0.5 mg/kg showed varying onset times (9, 4.33, and 3.33 minutes) and durations of action (54, 11, and 18 minutes), with 0.5 mg/kg having the longest duration. Bupivacaine may be preferred over lidocaine in chickens for this technique.

Other drugs

Gabapentin

Gabapentin, an analog of the neurotransmitter gamma-aminobutyric acid (GABA), binds to voltage-gated calcium channels, acting presynaptically to reduce excitatory neurotransmitter release and treat neuropathic pain. Recent pharmacokinetic studies in avian species, like great horned owls, Hispaniolan Amazon parrots, and Caribbean flamingos, have utilized a compounded formulation to avoid potential risks associated with xylitol in commercially available gabapentin oral suspension for humans. In great horned owls,[57] an oral dose of 11 mg/kg resulted in a Cmax of 6.17 µg/ml, Tmax of 51.43 minutes, and an elimination half-life of 4.4 hours. For Hispaniolan Amazon parrots,[58] oral bioavailability was estimated at 80-89%, and doses of 10 and 30 mg/kg showed varying Cmax, Tmax, and elimination half-life. In Caribbean flamingos,[59] 15 mg/kg was likely to maintain therapeutic plasma concentrations for about 12 hours, with higher doses up to 37 mg/kg every 12 hours suggested for some individuals.

Amantadine

Amantadine, an antiviral medication, increases dopamine in the central nervous system and weakly inhibits NMDA receptors contributing to chronic pain. It synergizes with NSAIDs, opioids, and gabapentin/pregabalin. In dogs, it improves osteoarthritis pain management when combined with meloxicam. In orange-winged Amazon parrots,[60] 10 mg/kg PO once and 5 mg/kg PO q24h for 7 days resulted in plasma concentrations supporting once-daily 5 mg/kg oral administration with weak accumulation. No obvious adverse effects were observed. Amantadine use in chickens, ducks, and turkeys is prohibited in the United States by the FDA. Further studies on safety and efficacy in various avian species are recommended.

Cannabidiol and other cannabinoids

Cannabidiol (CBD) oil, a non-psychoactive cannabinoid from hemp, interacts with the endocannabinoid system. In Hispaniolan Amazon parrots, a CBD oral formulation (50 mg/ml) showed variable results with a short half-life, suggesting a need for 3-4 daily administrations at 120 mg/kg. A commercial hemp extract (100 mg/ml) containing various cannabinoids was studied in orange-winged Amazon parrots[61] at 30/32.5 mg/kg PO of CBD/cannabidiolic acid (CBDA). The multiple dose study indicated a half-life of 8.6 and 6.3 hours for CBD and CBDA, supporting an every 12-hour administration. Encouraging results suggest further research on different formulations and bird species to understand the analgesic potential of cannabinoids.

Acetaminophen

Acetaminophen (N-acetyl-para-aminophenol or paracetamol, APAP) is an antipyretic and analgesic drug commonly used in human and veterinary medicine. The primary mechanism of analgesia is through the metabolite AM404's action on the brain and spinal dorsal horn. AM404 acts directly on the transient receptor potential vanilloid 1 (TRPV1) receptors, which are important for pain modulation, and the cannabinoid 1 receptors (CB1). Administration of a single dose of APAP 10 mg/kg orally to domestic geese[62] resulted in maximum plasma concentration of 5.31 µg/ml at 0.25 hours, with an elimination half-life of 5 hours. No adverse effects were noted during the study; however, minor changes to the liver consistent with fatty degeneration of hepatocytes were noted on histopathology 24 hours after APAP administration. In orange-winged Amazon parrots[63] receiving APAP (100 mg/kg) with silymarin (50 mg/kg, twice, q 12 h); (2) APAP (100 mg/kg) with NAC (400 mg/kg); or (3) APAP (100 mg/kg) alone, APAP plasma concentrations were lower than in other avian species despite a relatively high dose. APAP showed fast absorption, short half-life, and marked glucuronidation and concluded that a single oral dose administration of APAP at 100 mg/kg, alone or with NAC, appears safe based on plasma biochemistries, but multidose and pharmacodynamic studies are needed in this and other psittacines to determine if can be an effective and safe component of multimodal analgesia in psittacines

Polysulfated glycosaminoglycans

Polysulfated glycosaminoglycans have also been used anecdotally in the management of degenerative joint disease in birds, but fatal coagulopathies following IM doses from 0.5 mg/kg to 100 mg/kg in different avian species (i.e., one coraciiform, two raptors, and one psittacine) have been reported.[64] A recent *in vitro* study evaluating the hypocoagulability effects the same specific formulation (100 mg/ml; Adequan, Elanco, Luitpold Animal Health, Greenfield, USA) previously associated with adverse effects, paired with the anecdotal reports of one of the authors of use in a wide range of species over 20 years, supports its safe and effective use in birds at 1 mg/kg SC once weekly for four doses, followed by once every other week for two doses, then once monthly until the problem is resolved, or chronically at this dose for long-term maintenance.[65]

Acknowledgments: Much of this masterclass is based on publications by Mikoni et al.[3] and Guzman and Hawkins.[66] I acknowledge my co-authors in those publications Dr. Mikoni, Dr. Paul-Murphy and Dr. Hawkins.

References

1. Mikoni NA, Guzman DS, Fausak E, et al. Recognition and assessment of pain-related behaviors in avian species: an integrative review. *J Avian Med Surg.* 2022;36:153-172.

2. Mikoni NA, Sanchez-Migallon Guzman D, Beaufrere H, et al. Carrageenan-induced inflammation elicits behavioral changes in cockatiels (*Nymphicus hollandicus*) for potential pain scale development. *Am J Vet Res.* 2023;84:1-11.

3. Mikoni NA, Guzman DS, Paul-Murphy J. Pain recognition and assessment in birds. *Vet Clin North Am Exot Anim Pract.* 2023;26:65-81.

4. Fousse SL, Golsen BM, Sanchez-Migallon Guzman D, et al. Varying expression of mu and kappa opioid receptors in cockatiels (*Nymphicus hollandicus*) and domestic pigeons (*Columba livia domestica*). *Front Genet.* 2020:1278.

5. Duhamelle A, Raiwet DL, Langlois I, et al. Preliminary findings of structure and expression of opioid receptor genes in a peregrine falcon (*Falco peregrinus*), a snowy owl (*Bubo scandiacus*), and a blue-fronted Amazon parrot (*Amazona aestiva*). *J Avian Med Surg.* 2018;32:173-184.

6. Singh PM, Johnson CB, Gartrell B, et al. Analgesic effects of morphine and butorphanol in broiler chickens. *Vet Anaesth Analg.* 2017;44:538-545.

7. Sanchez-Migallon Guzman D, Douglas JM, Beaufrere H, et al. Evaluation of the thermal antinociceptive effects of hydromorphone hydrochloride after intramuscular administration to orange-winged Amazon parrots (*Amazona amazonica*). *Am J Vet Res.* 2020;81:775-782.

8. Sosa-Higareda M, Sanchez-Migallon Guzman D, Knych HK, et al. High bioavailability, short half-life, and metabolism into hydromorphone-3-glucuronide following single intramuscular and intravenous administration of hydromorphone hydrochloride to great horned owls (*Bubo virginianus*). *Am J Vet Res.* 2023;84.

9. Monopoli MR, Guzman DS, Paul-Murphy J, et al. Evaluation of thermal antinociceptive effects of intramuscular hydromorphone hydrochloride in great horned owls (*Bubo virginianus*). *J Avian Med Surg.* 2023;37:209-216.

10. Escobar A, da Rocha RW, Pypendop BH, et al. Effects of methadone on the minimum anesthetic concentration of isoflurane, and its effects on heart rate, blood pressure and ventilation during isoflurane anesthesia in hens (*Gallus gallus domesticus*). *PLOS One.* 2016;11:e0152546.

11. Delaski KM, Gehring R, Heffron BT, et al. Plasma concentrations of fentanyl achieved with transdermal application in chickens. *J Avian Med Surg.* 2017;31:6-15.

12. Waugh L, Knych H, Cole G, et al. Pharmacokinetic evaluation of a long-acting fentanyl solution after transdermal administration in helmeted guineafowl (*Numida meleagris*). *J Zoo Wildl Med.* 2016;47:468-473.

13. Guzman DS-M, Knych HK, Olsen GH, et al. Pharmacokinetics of a sustained release formulation of buprenorphine after intramuscular and subcutaneous administration to American kestrels (*Falco sparverius*). *J Avian Med Surg.* 2017;31:102-107.

14. Guzman DS, Ceulemans SM, Beaufrere H, et al. Evaluation of the thermal antinociceptive effects of a sustained-release buprenorphine formulation after intramuscular administration to American kestrels (*Falco sparverius*). *J Avian Med Surg.* 2018;32:1-7.

15. Gleeson MD, Guzman DS, Knych HK, et al. Pharmacokinetics of a concentrated buprenorphine formulation in red-tailed hawks (*Buteo jamaicensis*). *Am J Vet Res.* 2018;79:13-20.

16. Sarvi JY, Gardhouse SM, Knych HK, et al. High-concentration buprenorphine exceeds target plasma concentrations for extended periods following subcutaneous administration in American flamingos (*Phoenicopterus ruber*). *Am J Vet Res.* 2023;84:1-6.

17. Douglas JM, Guzman DS, Beaufrere H, et al. High doses of buprenorphine hydrochloride are well tolerated and produce a mild and prolonged thermal antinociceptive effect in orange-winged Amazon parrots (*Amazona amazonica*). *Am J Vet Res.* 2025:1-6.

18. Laniesse D, Guzman DS, Knych HK, et al. Pharmacokinetics of butorphanol tartrate in a long-acting poloxamer 407 gel formulation administered to Hispaniolan Amazon parrots (*Amazona ventralis*). *Am J Vet Res.* 2017;78:688-694.

19. Laniesse D, Sanchez-Migallon Guzman D, Smith DA, et al. Evaluation of the thermal antinociceptive effects of subcutaneous administration of butorphanol tartrate or butorphanol tartrate in a sustained-release poloxamer 407 gel formulation to orange-winged Amazon parrots (*Amazona amazonica*). *Am J Vet Res.* 2020;81:543-550.

20. Clancy MM, KuKanich B, Sykes IV JM. Pharmacokinetics of butorphanol delivered with an osmotic pump during a seven-day period in common peafowl (*Pavo cristatus*). *Am J Vet Res.* 2015;76:1070-1076.

21. Bailey RS, Sheldon JD, Allender MC, et al. Analgesic efficacy of tramadol compared with meloxicam in ducks (*Cairina moschata domestica*) evaluated by ground-reactive forces. *J Avian Med Surg.* 2019;33: 133-140.

22. Bailey RS, Sheldon JD, Allender MC, et al. Pharmacokinetics of orally administered tramadol in Muscovy ducks (*Cairina moschata domestica*). *J Vet Pharmacol Ther.* 2019;42:380-384.

23. Kilburn JJ, Cox SK, Kottyan J, et al. Pharmacokinetics of tramadol and its primary metabolite O-desmethyltramadol in African penguins (*Spheniscus demersus*). *J Zoo Wildl Med.* 2014;45:93-99.

24. Hernandez-Millan CL, Quezada Tristan T, Ortiz Martinez R, et al. Use of tramadol as analgesic alternative in Harris hawk (*Parabuteo unicinctus*). *Vet Med Sci.* 2023;9:2686-2692.

25. Duvall A, Tully TN, Carpenter JW, et al. Pilot study of a single dose of orally administered tapentadol suspension in Hispaniolan Amazon parrots (*Amazona ventralis*). *J Avian Med Surg.* 2021;35:45-50.

26. Miller KA, Hill NJ, Carrasco SE, et al. Pharmacokinetics and safety of intramuscular meloxicam in zebra finches (*Taeniopygia guttata*). *J Am Assoc Lab Anim Sci.* 2019;58:589-593.

27. Cetin G, Corum O, Corum D, et al. Pharmacokinetics of intravenous meloxicam, ketoprofen and tolfenamic acid in chukar partridge (*Alectoris chukar*). *Br Poult Sci.* 2021:1-7.

28. Turk E, Tekeli IO, Corum O, et al. Pharmacokinetics of meloxicam, carprofen, and tolfenamic acid after intramuscular and oral administration in Japanese quails (*Coturnix coturnix japonica*). *J Vet Pharmacol Ther.* 2021;44:388-396.

29. Chen JC, Yang F, Duan MH, et al. Pharmacokinetics of meloxicam in pigeons after single intravenous, oral, and intramuscular administration. *Poult Sci.* 2023;102:102869.

30. Sartini I, Łebkowska-Wieruszewska B, Lisowski A, et al. Pharmacokinetic profiles of meloxicam after single IV and PO administration in Bilgorajska geese. *J Vet Pharmacol Ther.* 2020;43:26-32.

31. Baert K, De Backer P. Comparative pharmacokinetics of three non-steroidal anti-inflammatory drugs in five bird species. *Comp Biochem Physiol C Toxicol Pharmacol.* 2003;134:25-33.

32. Morrison J, Greenacre CB, George R, et al. Pharmacokinetics of a single dose of oral and intramuscular meloxicam in African penguins (*Spheniscus demersus*). *J Avian Med Surg.* 2018;32:102-108.

33. Horgan MD, Knych HK, Siksay SE, et al. Pharmacokinetics of a single dose of oral meloxicam in rehabilitated wild brown pelicans (*Pelecanus occidentalis*). *J Avian Med Surg.* 2020;34:329-337.

34. Molter CM, Court MH, Cole GA, et al. Pharmacokinetics of meloxicam after intravenous, intramuscular, and oral administration of a single dose to Hispaniolan Amazon parrots (*Amazona ventralis*). *Am J Vet Res.* 2013;74:375-380.

35. Montesinos A, Ardiaca M, Gilabert JA, et al. Pharmacokinetics of meloxicam after intravenous, intramuscular and oral administration of a single dose to African grey parrots (*Psittacus erithacus*). *J Vet Pharmacol Ther.* 2017;40:279-284.

36. Dhondt L, Devreese M, Croubels S, et al. Comparative population pharmacokinetics and absolute oral bioavailability of COX-2 selective inhibitors celecoxib, mavacoxib and meloxicam in cockatiels (*Nymphicus hollandicus*). *Sci Rep.* 2017;7:12043.

37. Coskun D, Corum O, Durna Corum D, et al. Pharmacokinetics and bioavailability of meloxicam in Pekin ducks following intravenous, intramuscular and oral administration. *Vet Anaesth Analg.* 2023.

38. Stilz CR, Cox S, Bergman J, et al. Pharmacokinetics and egg residues of oral meloxicam in Bantam Cochin chickens. *J Avian Med Surg.* 2022;36:140-144.

39. Souza MJ, Gerhardt LE, Shannon L, et al. Breed differences in the pharmacokinetics of orally administered meloxicam in domestic chickens (*Gallus domesticus*). *J Am Vet Med Assoc.* 2021;259:84-87.

40. Castineiras D, Armitage L, Lamas LP, et al. Perioperative pharmacokinetics and pharmacodynamics of meloxicam in emus (*Dromaius novaehollandiae*) of different age groups using nonlinear mixed effect modelling. *J Vet Pharmacol Ther.* 2021;44(4):603-618.

41. Guzman DS, Court MH, Zhu Z, et al. Pharmacokinetics of a sustained-release formulation of meloxicam after subcutaneous administration to Hispaniolan Amazon parrots (*Amazona ventralis*). *J Avian Med Surg.* 2017;31:219-224.

42. Sim RR, Cox SK. Pharmacokinetics of a sustained-release formulation of meloxicam after subcutaneous administration to American flamingos (*Phoenicopterus ruber*). *J Zoo Wildl Med.* 2018;49:839-843.

43. Coutant T, Cococcetta C, Prouillac C, et al. Osmotic pump maintains plasma concentrations of meloxicam for 6 days after orthopedic surgery in pigeons (*Columba livia*). *Am J Vet Res.* 2023;84.

44. Montesinos A, Encinas T, Ardiaca M, et al. Pharmacokinetics of meloxicam during multiple oral or intramuscular dose administration to African grey parrots (*Psittacus erithacus*). *Am J Vet Res.* 2019;80:201-207.

45. Houck EL, Petritz OA, Chen LR, et al. Clinicopathologic, gross necropsy, and histopathologic effects of high-dose, repeated meloxicam administration in Rhode Island red chickens (*Gallus gallus domesticus*). *J Avian Med Surg.* 2022;36:128-139.

46. Souza MJ, Bailey J, White M, et al. Pharmacokinetics and egg residues of meloxicam after multiple day oral dosing in domestic chickens. *J Avian Med Surg.* 2018;32:8-12.

47. Richards ED, Dutch RS, Burmas NC, et al. Pharmacokinetic parameters and estimating extra-label tissue withdrawal intervals using three approaches and various matrices for domestic laying chickens following meloxicam administration. *Front Vet Sci.* 2022;9:826367.

48. Gasthuys E, Houben R, Haesendonck R, et al. Development of an in vivo lipopolysaccharide inflammation model to study the pharmacodynamics of COX-2 inhibitors celecoxib, mavacoxib, and meloxicam in cockatiels (*Nymphicus hollandicus*). *J Avian Med Surg.* 2019;33:349-360.

49. Fadel C, Łebkowska-Wieruszewska B, Lisowski A, et al. Disposition kinetics of robenacoxib following intravenous and oral administration in geese (*Anser anser domesticus*). *J Vet Pharmacol Ther.* 2023;46(6): 413-420.

50. Naidoo V, Taggart MA, Duncan N, et al. The use of toxicokinetics and exposure studies to show that carprofen in cattle tissue could lead to secondary toxicity and death in wild vultures. *Chemosphere.* 2018;190:80-89.

51. Keiper NL, Cox SK, Doss GA, et al. Pharmacokinetics of piroxicam in cranes (Family Gruidae). *J Zoo Wildl Med.* 2017;48:886-890.

52. Rodriguez P, Paul-Murphy JR, Knych HK, et al. Pharmacokinetics of grapiprant administered to red-tailed hawks (*Buteo jamaicensis*) after food was withheld for 24 hours. *Am J Vet Res.* 2021;82:912-919.

53. Rodriguez P, Paul-Murphy JR, Knych HK, et al. Absorption of grapiprant in red-tailed hawks (*Buteo jamaicensis*) is decreased when administered with food. *Am J Vet Res.* 2022;83(6):ajvr.21.10.0170.

54. Khamisabadi A, Kazemi-Darabadi S, Akbari G. Comparison of anesthetic efficacy of lidocaine and bupivacaine in spinal anesthesia in chickens. *J Avian Med Surg.* 2021;35:60-67.

55. Kazemi-Darabadi S, Akbari G, Shokrollahi S. Development and evaluation of a technique for spinal anaesthesia in broiler chickens. *N Z Vet J.* 2019;67:241-248.

56. DiGeronimo PM, da Cunha AF, Pypendop B, et al. Cardiovascular tolerance of intravenous bupivacaine in broiler chickens (*Gallus gallus domesticus*) anesthetized with isoflurane. *Vet Anaesth Analg.* 2017;44: 287-294.

57. Yaw TJ, Zaffarano BA, Gall A, et al. Pharmacokinetic properties of a single administration of oral gabapentin in the great horned owl (*Bubo virginianus*). *J Zoo Wildl Med.* 2015;46:547-552.

58. Baine K, Jones MP, Cox S, et al. Pharmacokinetics of compounded intravenous and oral gabapentin in Hispaniolan Amazon parrots (*Amazona ventralis*). *J Avian Med Surg.* 2015;29:165-173.

59. Browning GR, Carpenter JW, Magnin GC, et al. Pharmacokinetics of oral gabapentin in Caribbean flamingos (*Phoenicopterus ruber ruber*). *J Zoo Wildl Med.* 2018;49:609-616.

60. Berg KJ, Sanchez-Migallon Guzman D, Knych HK, et al. Pharmacokinetics of amantadine after oral administration of single and multiple doses to orange-winged Amazon parrots (*Amazona amazonica*). *Am J Vet Res.* 2020;81:651-655.

61. Sosa-Higareda M, Guzman DS-M, Knych H, et al. Twice-daily oral administration of a cannabidiol and cannabidiolic acid–rich hemp extract was well tolerated in orange-winged Amazon parrots (*Amazona amazonica*) and has a favorable pharmacokinetic profile. *Am J Vet Res.* 2023;1:1-11.

62. Pisack EK, Kleine SA, Hampton CE, et al. Evaluation of the analgesic efficacy of grapiprant compared with robenacoxib in cats undergoing elective ovariohysterectomy in a prospective, randomized, masked, non-inferiority clinical trial. *J Feline Med Surg.* 2024;26:1098612X241230941.

63. Megan L, Sanchez-Migallon Guzman D, Knych H, et al. Pharmacokinetics of single-dose oral acetaminophen with and without concurrent administration of silymarin or N-acetylcysteine in orange-winged Amazon parrots (*Amazona amazonica*). *Am J Vet Res.* 2025:1-11.

64. Anderson K, Garner MM, Reed HH, et al. Hemorrhagic diathesis in avian species following intramuscular administration of polysulfated glycosaminoglycan. *J Zoo Wildl Med.* 2013;44:93-99.

65. Wonn AM, Brooks MB, Hu H, et al. Hypocoagulability effect of Adequan in domestic chickens (*Gallus gallus*) and Chilean flamingos (*Phoenicopterus chilensis*). *J Zoo Wildl Med.* 2022;53:126-132.

66. Guzman DS-M, Hawkins MG. Advances in management of pain in birds. In: Speer BL, van Zeeland YRA, eds. *Current Therapy in Avian Medicine and Surgery.* Vol. 2. St. Louis, MO: Elsevier; 2025:578-589.

Ex-Vivo Pharmacologic Characterization of Grapiprant and a Multi-Dose Pharmacokinetic Study in Red-Tailed Hawks (*Buteo jamaicensis*)

Adam A. Moreno, DVM,
Michelle Hawkins, VMD, DABVP (Avian Practice),
Heather Knych, DVM, PhD, DACVCP

Session #183

Affiliation: From Veterinary Medical Teaching Hospital (Moreno), Department of Medicine and Epidemiology (Hawkins), California Animal Health and Food Safety Lab Molecular Biosciences, K.L. Maddy Equine Analytical Chemistry Laboratory (Knych), School of Veterinary Medicine, University of California Davis, Davis, CA 95616, USA.

Abstract: Pain management in birds, including raptors, often relies on nonsteroidal anti-inflammatory drugs (NSAIDs) that inhibit COX-1 and COX-2 enzymes. Grapiprant, a novel NSAID targeting the EP4 receptor, represents an alternative by bypassing COX inhibition, preserving COX physiologic functions and blocking PGE_2-mediated neuronal sensitization and inflammatory responses.[1] Recent studies show grapiprant is safe for red-tailed hawks (*Buteo jamaicensis*, RTHAs) and achieves plasma concentrations similar to therapeutic concentrations in dogs.[2,3] This study aims to evaluate grapiprant's pharmacokinetics and ex-vivo pharmacodynamic characteristics after 14 days of administration in RTHAs. Five non-releasable RTHAs were utilized for this study. For the multi-dose pharmacokinetic study, RTHAs received grapiprant 30 mg/kg PO with food q12h for 14 days. Blood was collected on days 1, 3, 10, and 14 to evaluate bioaccumulation and on day 15 to assess multidose pharmacokinetics. The mean maximum plasma concentration was 3,127 ng/ml, the time to maximum plasma concentration was 2.79 hours and the mean terminal half-life was 8.6 hours. No adverse effects were noted on physical examination, complete blood count or biochemistry panel. Minimal bioaccumulation was noted and plasma concentrations exceeded targeted therapeutic concentrations for more than 24 hours. The results of this study suggest that grapiprant at 30 mg/kg administered with food q12h is a safe dosing regimen in RTHA. For the ex-vivo study, the inhibitory role of PGE_2 on lipopolysaccharide-induced TNF-α production by macrophages and monocytes will be measured in RTHA whole blood spiked with differing concentrations of grapiprant as a minimally invasive way to study grapiprant's anti-inflammatory potential.[4]

References

1. Sartini I, Giorgi M. Grapiprant: a snapshot of the current knowledge. *J Vet Pharmacol Ther.* 2021;44(5): 679-688.

2. Rodriguez P, Paul-Murphy JR, Knych HK, et al. Pharmacokinetics of grapiprant administered to red-tailed hawks (*Buteo jamaicensis*) after food was withheld for 24 hours. *Am J Vet Res.* 2021;82(11):912-919.

3. Rodriguez P, Paul-Murphy JR, Knych HK, et al. Absorption of grapiprant in red-tailed hawks (*Buteo jamaicensis*) is decreased when administered with food. *Am J Vet Res.* 2022;83(6):ajvr.21.10.0170.

4. MacKenzie S, Planas JV, Goetz FW. LPS-stimulated expression of a tumor necrosis factor-alpha mRNA in primary trout monocytes and in vitro differentiated macrophages. *Dev Comp Immunol.* 2003;27(5):393-400.

Pharmacokinetics of an Extended-Release Liposomal Formulation of Buprenorphine in Hispaniolan Amazon Parrots (*Amazona ventralis*) and Its Effects on Body Temperature After a Single Subcutaneous Administration

Hugo A. Gonzalez-Jassi, MVZ,
Levent Dirikolu, DVM, MVSc, PhD,
Chin-Chi Liu, MapStat, PhD,
Thomas N. Tully Jr., DVM, MS, DABVP (Avian Practice), DipECZM (Avian)

Session #173

Affiliation: From Louisiana State University School of Veterinary Medicine, Skip Bertman Dr, Baton Rouge, LA 70803, USA.

Abstract: Pain management in birds often requires multimodal analgesia.[1] Buprenorphine has been used as an analgesic in birds, due to its longer half-life.[2-7] New formulations have shown analgesic concentrations for more than 12 hours in kestrels and mild antinociception in Amazon parrots.[6,7] The goals of this study were to describe the pharmacokinetic parameters (PK) of a single dose, liposomal, extended-release buprenorphine (ER-B) (2 mg/kg SC, 1.3 mg/ml, Fidelis Animal Health) and to describe its effects on body temperature (BT) in healthy Hispaniolan Amazon parrots (*Amazona ventralis*). Fifteen parrots were used for this study. BT was measured using subcutaneous transponders (TP1000, BMDS-Bio Medic Data system). One week before the PK study, BT was measured at 0.25, 1, 2, 3, 6, 12, 24, 48 and 72 hours, so each bird could serve as their control (CTR). A single dose of ER-B was administered, blood collections and BT measurements were obtained at the same 72-hour time points. Signs of sedation or adverse effects were recorded. A standard PK analysis was carried out. Statistical analysis for the BT of the CTR and ER-B groups was performed. An ANOVA with a mixed effect model and a post-hoc Tukey comparison was used. ER-B mean time to maximum plasmatic concentration was 2.1 hours, ER-B plasma concentration was > 1 ng/ml for more than 48 hours. Some birds showed mild sedation, and BT significantly decreased at 3 hours (ER-B 103.4 ± 0.6 °F CTR 105.3 ± 0.6 °F, $P = 0.013$) and 6 hours (ER-B 103.4 ± 0.6 °F CTR 105.6 ± 0.6 °F, $P = 0.001$) post-administration, providing foundational data on its PK and BT effects in Amazon parrots.

References

1. Sanchez-Migallon Guzman D, Hawkins MG. Treatment of pain in birds. *Vet Clin North Am Exot Anim Pract*. 2023;26(1):83-120.

2. Aliansyah E, Chng HT, Xie S. A critical review of the pharmacokinetics and pharmacodynamics of opioid medications used in avian patients. *Birds*. 2022;3(1):1-28.

3. Paul-Murphy J, Hes, JC, Fialkowski JP. Pharmacokinetic properties of a single intramuscular dose of buprenorphine in African grey parrots (*Psittacus erithacus erithacus*). *J Avian Med Surg*. 2004;18(4):224-228.

4. Guzman DS, Houck EL, Knych HKD, et al. Evaluation of the thermal antinociceptive effects and pharmacokinetics after intramuscular administration of buprenorphine hydrochloride to cockatiels (*Nymphicus hollandicus*). *Am J Vet Res*. 2018;79(12):1239-1245.

5. Guzman DS, Knych HK, Olsen GH, Paul-Murphy JR. Pharmacokinetics of a sustained release formulation of buprenorphine after intramuscular and subcutaneous administration to American kestrels (*Falco sparverius*). *J Avian Med Surg*. 2017;31(2):102-107.

6. Sarvi JY, Gardhouse SM, Knych HK, et al. High-concentration buprenorphine exceeds target plasma concentrations for extended periods following subcutaneous administration in American flamingos (*Phoenicopterus ruber*). *Am J Vet Res*. 2023;84(10):1-6.

7. Douglas JM, Guzman DS, Beaufrère H, Paul-Murphy JR. High doses of buprenorphine hydrochloride are well tolerated and produce a mild and prolonged thermal antinociceptive effect in orange-winged Amazon parrots (*Amazona amazonica*). *Am J Vet Res*. 2025:1-6. PMID: 39965328.

Pharmacologic Assessment of Intravenous versus Intramuscular Medetomidine-Zoletil Protocols for Precision Immobilization in Ostriches (*Struthio camelus*)

Boon Allwin,
Ramakrishnan Angappan,
Jayvin Hitendra Kelaiya,
Pa. Kalaignan

Session #165

Affiliation: From Greens Zoological, Rescue & Rehabilitation Centre, SSO Annexe-5, At & Post- R Greens, Village: Moti Khavdi, District: Jamnagar 361142, Gujarat, India.

Abstract: Effective immobilization of ostriches (*Struthio camelus*) is essential for veterinary, conservation, and management procedures. This study compared intravenous (IV; Group 1, n = 6) and intramuscular (IM; Group 2, n = 6) administration of medetomidine-Zoletil in clinically healthy adult ostriches to assess induction time, recovery, physiologic stability, reflex suppression, and safety outcomes. Intravenous administration achieved significantly faster induction (36.3 ± 3.1 seconds) and recovery (6.2 ± 1.5 minutes) compared to intramuscular (487.2 ± 35 seconds; 10.4 ± 2.5 minutes). Physiologic parameters (i.e., heart rate, respiratory rate, rectal temperature, SpO_2) remained stable in both groups, but reflex suppression was more rapid and complete in the intravenous group. These findings support intravenous medetomidine-Zoletil as a superior immobilization protocol for ostriches, offering improved procedural efficiency and reduced stress.

Introduction

The immobilization of large ratites such as ostriches poses unique challenges due to their body mass, strength, and heightened stress responses.[1,2] Various chemical restraint protocols have been explored, but few provide rapid, reliable sedation with smooth recovery. Medetomidine, an α_2-adrenergic agonist, combined with Zoletil (tiletamine-zolazepam), offers balanced sedative, analgesic, and muscle-relaxant effects, widely applied in mammals and avian species;[3,4] however, its comparative use in ostriches by intravenous (IV) versus intramuscular (IM) routes remains undocumented. This study aims to fill that gap by providing a controlled pharmacologic evaluation of the two administration routes.

Materials and Methods

Clinically healthy adult ostriches (n = 12) were divided into two groups: Group 1 (n = 6) received IV medetomidine-Zoletil via the basilic vein; Group 2 (n = 6) received IM medetomidine-Zoletil into the thigh or pectoral muscles. Physiologic monitoring included induction time, recovery time, heart rate, respiratory rate, rectal temperature, pulse oximetry (SpO_2), ocular and pedal reflexes, and ocular positioning. Data were collected at regular intervals and analyzed statistically (mean ± SD) following established avian anesthesia monitoring protocols.[5,6]

Results and Discussion

The results of this study demonstrate that intravenous (IV) administration of medetomidine-Zoletil produces significantly faster induction and recovery compared to intramuscular (IM) delivery in ostriches (Table 1). This aligns with the known pharmacokinetics of IV administration, which allows immediate systemic distribution, whereas IM absorption is dependent on local perfusion and tissue uptake.[1,2] Comparable studies in other avian species, including red-tailed hawks (*Buteo jamaicensis*), African grey parrots (*Psittacus erithacus*), and Amazon parrots (*Amazona* spp.), have similarly shown that IV routes provide faster onset and smoother sedation compared to IM, particularly when using α_2-agonists in combination with dissociative anesthetics.[3,5,6] In ratites, prior immobilization attempts have often relied on ketamine, xylazine, or diazepam combinations, which sometimes resulted in unpredictable depth, prolonged recovery, or excitation phases during emergence.[4,7,8] In this study, the IV group displayed rapid suppression of corneal, palpebral, and pedal reflexes, providing consistent sedation with minimal struggling, thereby reducing risks of trauma to the bird and handlers. This is particularly important in ostriches, where excessive pre-induction restraint or delayed sedation can result in severe self-injury or capture myopathy.[1] Physiologic parameters remained within acceptable ranges in both groups, confirming that medetomidine-Zoletil, when dosed appropriately, provides cardiovascular stability.[2,3,8] Although pulse oximetry readings were slightly higher in the IV group, no hypoxemia or adverse cardiopulmonary events were observed, consistent with findings from anesthetic monitoring studies in large avian species.[6]

Table 1. Anesthesia monitoring parameters for ostriches administered intravenous versus intramuscular medetomidine-Zoletil.

Monitoring parameter	Intravenous (Group 1, n = 6)	Intramuscular (Group 2, n = 6)
Induction time (seconds)	36.3 ± 3.1	487.2 ± 35
Recovery time (minutes)	6.2 ± 1.5	10.4 ± 2.5
Heart rate (beats/min)	55 ± 3	61 ± 5
Respiratory rate (breaths/min)	8 ± 2	10 ± 2
Rectal temperature (°F)	101.4 ± 3.1	101.6 ± 2.4
Capillary refill time (seconds)	< 2	< 2
Corneal reflex	Absent	Initially sluggish, then absent
Palpebral reflex	Absent	Initially reduced, then absent
Pedal reflex	Absent	Initially reduced, then absent
Ocular position	Central to medioventral	Central to medioventral
Pulse oximetry (SpO_2, %)	94 ± 3	92 ± 4

Importantly, we recognize that this was a small-scale study (n = 12), and caution should be taken when extrapolating these findings broadly. Factors such as individual variation, environmental conditions, and long-term outcomes require further investigation. Additionally, while this represents, to our knowledge, the first comparative evaluation of IV versus IM medetomidine-Zoletil in ostriches, future research should validate these findings across larger populations and additional ratite species, such as emus and rheas.

Conclusions

The medetomidine-Zoletil combination is effective and safe for ostrich immobilization, with IV administration demonstrating clear advantages over IM in terms of induction time, depth of sedation, and recovery profile. Adoption of this protocol can improve outcomes in routine veterinary procedures, transportation, and conservation management of ostriches.

Acknowledgments: The authors thank the veterinary, management and animal care staff at Vantara, Gujarat, for their invaluable assistance with the study.

Disclosure Statement: The author declares no conflicts of interest.

References

1. Fowler ME, Miller RE. *Fowler's Zoo & Wildlife Medicine.* 9th Ed. St. Louis, MO: Elsevier; 2020.

2. Mans C. Sedation and general anesthesia of pet birds. *Vet Clin North Am Exot Anim Pract.* 2013;16(1): 59-82.

3. Lierz M, Korbel R. Anesthesia and analgesia in birds. *J Exot Pet Med.* 2012;21(1):44–58.

4. Doneley B, Monks D, Johnson R, Carmel B. *Avian Medicine and Surgery in Practice.* 2nd Ed. Boca Raton, FL: CRC Press; 2018.

5. Hawkins MG, Pascoe PJ, Shofer FS. Clinical techniques for anesthetizing birds. *Vet Clin North Am Exot Anim Pract.* 2004;7(2):349-392.

6. Zehnder AM, Hawkins MG, Pascoe PJ, et al. Evaluation of indirect blood pressure monitoring in awake and anesthetized red-tailed hawks (*Buteo jamaicensis*): effects of cuff size, cuff placement, and monitoring equipment. *Vet Anaesth Analg.* 2009;36:464-479.

7. Ladds PW. *Pathology of Australian Native Wildlife.* Clayton, VIC: CSIRO Publishing; 2009.

8. Mans C, Sladky KK. Avian anesthesia and analgesia. *Semin Avian Exot Pet Med.* 2011;20(1):18-34.

Acupuncture and Holistic Treatment Modalities in Birds

Felicitas Tantiyatyanon, DVM, CVA

Session #159

Affiliation: From Leo Veterinary Care, Phoenix, AZ 85028, USA.

Abstract: Acupuncture is a major branch of the medical system known as Traditional Chinese Veterinary Medicine. Acupuncture has been used to treat animals in China for thousands of years; however, it is relatively new in the Western world. The traditional acupuncture texts mostly described acupuncture treatment in species such as cattle, pigs and horses. In Western society dogs, cats and birds are popular companion animals. Their popularity resulted in an increased understanding of acupuncture in these species during recent history. Acupuncture and holistic remedies have become increasingly popular in birds. There is limited research related to acupuncture in avian species; however, anecdotal evidence provides promising results encouraging consideration of these treatments as a helpful adjunct to Western remedies and to help improve the well-being of birds.

Introduction

Acupuncture has several effects that include:

- anti-inflammatory,
- endorphin releasing,
- calming, and
- circulating.

Documentation of acupuncture effects in avian species is currently limited to case reports. In rodents, "a number of acupuncture studies have shown anti-inflammatory effects of acupuncture treatment, mostly known at specific point ST36."[1] In dogs, one study utilized electroacupuncture to provide successful postoperative analgesia in dogs undergoing ovariohysterectomy. The dogs that received electroacupuncture were compared to dogs receiving only butorphanol prior to surgery. Plasma endorphin levels increased significantly after electroacupuncture. Prolonged analgesia of at least 24 hours was seen in all dogs that received the electroacupuncture.[2]

Acupuncture Technique

It can be challenging to treat birds with acupuncture. It is important to have an experienced handler and to keep a state of calm mind and manner around birds. With holistic medicine, it is important to consider our own energy and stress that could be projected onto the patient.

Psittacine birds will sometimes allow acupoints treated on their head and back while standing on a perch or hand; however, gentle restraint with a towel is often needed for the treatment of distal acupuncture points in the feet and legs. Chickens have the best response I have observed among avian species. I have often seen chickens relax or sway gently within minutes of starting the acupuncture treatment. Raptors can accept acupuncture treatment if they are used to handling. They can be treated while standing on the handler's gloved hand while gently holding jesses.

Most acupuncture in birds is performed with dry needling technique. This is where a sterile acupuncture needle is inserted into an acupuncture point and retained for up to 15 minutes or longer if the patient allows. If the avian patient is very mobile and stressed, a brief retention technique can be used where the needle is inserted and immediately removed. Another common technique is aqua-acupuncture. Where a liquid such as vitamin B12 or saline is injected into an acupuncture point to stimulate the point. This technique is very helpful in treating an especially stressed patient. I recommend limiting the B12 aqua-acupuncture injections to distal acupuncture points in the legs to avoid the risk of injecting into an air sac which could be potentially fatal.[1]

Electroacupuncture is a technique where a gentle electrical current is connected to the acupuncture needle to increase the stimulation of the acupuncture point. This technique is often avoided in birds due to the stress it can induce on the patient. I have seen this successfully used in anesthetized birds.

The veterinarian should obtain training, including avian acupuncture and Traditional Chinese Veterinary Medicine pattern diagnosis prior to providing treatment. A thorough description of avian acupuncture point locations and indications is available.[3] Conditions commonly treated with acupuncture in birds include anorexia, osteoarthritis, paresis, feather destructive behavior, and pododermatitis.[4,5]

Case Reports

The following is a collection of my case reports that describe the use of avian acupuncture in different species of birds. The + symbol is an indication of patient sensitivity at treated acupuncture points: + is mild sensitivity, ++ is moderate sensitivity, and +++ is severe sensitivity.

Chicken

A 6-year-old female chicken was presented for evaluation and treatment of anemia on May 26, 2021. Initial symptoms were quiet, slow to eat, and a pale and thin appearance. The owner was a veterinarian who ran bloodwork, which revealed anemia. The bird went into respiratory arrest and needed assisted breathing for 30 minutes. A blood transfusion was given, which stabilized the patient. The morning of the examination the bird vomited a small amount of bile, had slightly loose stool, and was reluctant to walk. There was no cough and no sneezing.

Physical examination revealed thin body condition, prone position and weakness, able to stand briefly with support. An intraosseous catheter was placed at the left tibiotarsus. Differential diagnoses for anemia included neoplasia, autoimmune, infectious, toxin, and adrenal fatigue.

The chicken was calm and accepted acupuncture well. The acupuncture points treated were: GV14, bilat ST36, lt SP6, bilat BL17, 18, Bai Hui (+) bilat Shen Shu.

Aqua-acupuncture was performed: 0.3 cc vitamin B12 + 0.7 cc saline injected subcutaneous at GV14 (++ and the patient started briefly grooming feathers then stopped), bilateral ST36, BL17, 18.

The owner reported that bird's appetite and mentation was better the day after acupuncture. It started to eat well the same night and did not require assisted feeding after a few days. The bird was also able to move back outside with her friends and did well for about 1 week. The bird passed away on June 10, 2021 and necropsy revealed an abdominal tumor.

Eclectus parrot

My 21-year-old female Eclectus parrot developed an episode of chest feather plucking in 2013. The parrot was diagnosed with bornavirus after a positive bornavirus PCR result on blood and cloacal swab. We suspected that the sudden feather plucking behavior was secondary to a neuritis from bornavirus. The last year of the bird's life, it struggled with complications secondary to bornavirus. In addition, a squamous cell carcinoma developed on the bird's left wing, which was diagnosed with histopathology.

A combination of acupuncture, physical therapy, aqua-acupuncture and both Eastern and Western therapeutics were often used to help improve quality of life. Some of the acupuncture points included GV20, GV14, BL11, 18, 23, 26, Bai Hui, Shen Shu, ST36, ST40, Liv3. The parrot enjoyed physical therapy and acupuncture treatments in the garden where it could hear birds, which seemed to bring comfort. I frequently saw great improvement in appetite, GI motility, energy and mentation after treatments. After the last treatment, my parrot continued to remain weak and cold. It was then that I knew there was not much time left so I stayed close that night and my bird passed away in my arms early in the morning in February 2022.

Scarlet macaw

A 23-year-old male scarlet macaw first presented for examination and acupuncture on October 23, 2024 for treatment of a right leg injury. It had a history of knuckling and weakened grip at the right foot affecting mostly the third digit for approximately a 12-week duration. Besides the weakness in the right foot, the bird has been doing well, with normal activity and appetite.

Upon examination the bird appeared bright, alert, responsive, sweet, a little nervous but allowed handling and was overall calm and social. The macaw was able to walk and climb. There was knuckling of the right foot during walking and trouble with gripping and placement of the right foot during climbing, with no apparent pain. Differential diagnoses included neuritis, impinged nerve, muscle or tendon injury, muscle weakness, hip injury/subluxation/neuropathy, sciatic nerve injury.

The bird was a bit nervous during acupuncture but allowed treatment including: GV20, GV14, BL13, BL17, rt ST36, rt ST37, rt Liv3, rt SP3, rt GB31, rt top of thigh: BL54 (+++); it immediately started flapping and trying to get away after this acupoint and there was improved grip strength of the right foot immediately afterwards.

Three more treatments of acupuncture were administered. During the second treatment we also performed aqua-acupuncture by injecting 0.1 cc vitamin B12 diluted with 0.7 cc saline injected into acupoints, which the bird allowed with towel restraint. The following acupoints were injected rt ST36, 37, BL54, GB29, 30.

The next two treatments were only acupuncture without vitamin B12 injections. The owner felt that the bird was weaker and had more trouble with the function of his right foot for 5 days after the vitamin B12 injections, so we continued with only acupuncture for the remaining treatments.

After a total of four acupuncture treatments the owner did not feel that there was improvement of the right foot function. The owner noticed that the macaw appeared happier and more social since acupuncture treatment.

I suspect that this is a case where despite the acupuncture not improving function and mobility of the right foot it could have provided a general improvement of stress reduction and contributed to overall mental wellbeing for the patient.

Saker falcon

A 6-year-old female saker falcon was seen for examination and acupuncture on February 13, 2025 to help improve appetite and encourage eating after a history of a traumatic attack from a bobcat 2 weeks prior. This attack resulted in a puncture wound over the left sinus and left oronasal fistula, weak grip at bilateral feet since trauma, able to stand with full weight bearing bilateral, but stronger on the right foot. The bird exhibited excessive scratching of the left sinus area, scratching with the left foot, and not scratching with the right foot. Although not eating on its own, the bird accepted syringe feedings well. There was also a puncture wound at the crown of the head, which penetrated the skull. Brain surgery was performed by a neurologist to remove bone fragments from the brain and the bird made a remarkable recovery.

During examination the falcon was bright, alert, and responsive. It had reduced grip strength at bilateral feet, was reluctant to bear full weight on the left foot, was able to bear full weight on the right foot, was able to stand, jump and fly briefly with slight lift. The puncture wound in the skull was closed with skin stitched over the skull defect post brain surgery to remove bone fragments and the surgical site appeared healed. A ruptured left eye was secondary to the trauma.

Differential diagnoses for reluctance to eat included pain, stress and recovery from recent trauma and surgery, infectious, inflammatory or behavioral.

The bird was nervous but considerably calm for a falcon and accepted acupuncture with gentle handling standing on a gloved hand while holding jesses. It did best with 0.5-inch size 32 Jing Mei copper needles. Acupuncture points treated included GV14, bilat BL11 (ht assoc), BL16 (Liv assoc), Shen Shu and Shen Shu alt, bilat ST36, brief lt BL62, lt SP3 (few drops blood post acupoint removal), LI4 left, LI11 left. The bird was a little more nervous during acupuncture at the right side (visual on the right). Most points were briefly treated, retention around 5 minutes, then replaced. The longest duration of retention was at left SP3, lt ST36 and GV14.

Immediately post acupuncture the falcon was standing stronger on its left foot and able to scratch its head with its right foot.

Vitamin B12 aqua-acupuncture was also performed: 0.2 cc vitamin B12 diluted with 0.4 cc saline injected 0.2 cc IM at bilat acupoint st 36, and 0.2 cc IM at left pectoral muscle. Full towel restraint was needed for the vitamin B12 injections; the bird was briefly stressed with full restraint then calmed immediately upon return to its cage. It started to eat on its own on February 22, 2025. There was only one acupuncture treatment so it is unclear how much acupuncture contributed to her recovery.

Conclusions

This presentation with case reports is intended to provide an overview and introduction of acupuncture and holistic treatment in birds. I encourage all practitioners to consider integration of these modalities into your veterinary practice.

References

1. Harrison TM, Churgin SM. Acupuncture and Traditional Chinese Veterinary Medicine in zoological and exotic animal medicine: a review and introduction of methods. *Vet Sci.* 2022;9(2):74. doi:10.3390/vetsci9020074.

2. Groppetti D, Pecile AM, Sacerdote P, et al. Effectiveness of electroacupuncture analgesia compared with opioid administration in a dog model: a pilot study. *Br J Anaesth.* 2011;107(4):612-618.

3. West C. TCVM for avian species: introduction, general overview, acupuncture point locations, indications and techniques. In: Xie HE, Lisa T, ed. *Application of Traditional Chinese Veterinary Medicine in Exotic Animals.* Reddick, FL: Jing Tang Publishing; 2011:53-71.

4. Eckermann-Ross C. An integrated approach to the treatment of regurgitation, anorexia, and pyuria in a 12-year old severe macaw. *Am J Trad Chin Vet Med.* 2009;4:58-61.

5. Oh JE, Kim SN. Anti-inflammatory effects of acupuncture at ST36 point: a literature review in animal studies. *Front Immunol.* 2022;12:813748. doi:10.3389/fimmu.2021.813748.

Successful Treatment of Wing Sarcoma Using Combined Surgical Excision and Strontium in a European Starling

Lauren Thielen, DVM, DABVP (Avian Practice), and
Josh Henry, DVM, ACVR

Session #208

Affiliation: From Cornell University Hospital for Animals, Ithaca, NY 13068, USA.

Abstract: A 3-year-old male starling (*Sturnus vulgaris*) presented for a 1-cm globoid, alopecic, erythematous dermal mass on the dorsocaudal aspect of the left carpus. The mass was present approximately 1.5 months prior to presentation. The mass was surgically excised under general anesthesia to determine diagnosis, prognosis and in hopes to preserve flight. Grossly incomplete margins deep to the mass were achieved due to invasion of surrounding wing soft tissue structures and aiming to preserve wing function prior to biopsy results. The skin was apposed and closed over the defect. Histologic examination revealed a mass consistent with lobules, bundles and streams of spindle cells. Based on the atypia of the cells and the degree of mitosis, a neoplastic process is favored with features that may be consistent with a soft tissue sarcoma with bone effacement. At the 3-week follow-up, the patient underwent a chemistry, CBC, and radiographs to screen for visible bony involvement of the wing as well as to screen for metastasis. At the 6-week follow-up examination, a CT with contrast found no evidence of retained tumor on the left wing, bony changes, or metastasis. Four months post-surgery the patient then underwent strontium radiation treatment over the area of the healed incision. Two simultaneous strontium treatments of 9 minutes each, one on the lateral aspect and another on the medial aspect, were performed under general anesthesia. At the follow-up examination 9 months later, there was no evidence of mass regrowth and the patient is still able to fly in the same fashion as before the mass presence. This case report describes the use of surgical excision and radiation therapy for the successful treatment of a sarcoma on the wing to preserve flight in a captive starling.

References

1. Nemetz L, Broome M. Strontium-90 therapy for uropygial neoplasia. *Proc Annu Conf Assoc Avian Vet.* 2004;5-19.

2. Pignon C, Azuma C, Mayer J. Radiation therapy of uropygial gland carcinoma in psittacine species. *Proc Annu Conf Assoc Avian Vet.* 2011;263.

3. Zehnder A, Swift LA, Sundaram A, et al. Clinical features, treatment, and outcomes of cutaneous and oral squamous cell carcinoma in avian species. *J Am Vet Med Assoc.* 2018;252(3);309-315.

Section 9

Species-Specific and Potpourri I Session

Zoë Selby, DVM
Session Chair

Essential Pigeon Medicine

Anthony A. Pilny, DVM, DABVP (Avian Practice)

Session #124

Affiliation: From Arizona Exotic Animal Hospital, Phoenix, AZ 85024, USA.

Abstract: Avian veterinarians might work with pigeons (*Columba livia*) as pets, as wildlife patients, or as birds maintained by breeders, fanciers, and keepers. Knowing how to keep them healthy and advising on preventive medical programs is essential for their well-being. Pigeon fanciers rarely seek veterinary advice with a sentiment that veterinarians do not have knowledge or understand pigeons so this summary will help make veterinarians more aware of health and disease. This lecture provides an overview of the most significant pigeon diseases, treatments, preventive medicine concepts, and general medical care.

Introduction

Archaeologic findings and records from Mesopotamia and ancient Egypt indicate pigeon domestication for food and other purposes started as early as 4,500 BC thus making them the oldest domesticated birds. They have been kept by fanciers for show, work, art projects, entertainment, and raised for food in places where squab is considered a delicacy. Pigeons also have served in wars and have been named as war heroes, as their efforts have helped save countless lives. They are also used as biomedical research models for many important diseases such as atherosclerosis and studied for their cognitive abilities and intelligence.[1-9]

Preventive Medicine: The Eight Basics

1. Buy healthy birds from quality breeders.
2. Practice quarantine and biosecurity.
3. Align with a knowledgeable veterinarian.
4. Examine birds for and treat any ectoparasites.
5. Perform fecal floatation, oral and crop cytology, and direct evaluation for endoparasites.
6. Establish a vaccination program that may include avian avulavirus-1 AAvV-1 (formerly known as pigeon paramyxovirus (PPMV-1), pigeon pox, and *Salmonella*.
7. Keep all pigeons in clean, dry, hygienic and well-ventilated lofts and avoid overcrowding.
8. Feed high-quality foods and necessary supplements, according to individual needs.

Common Medical Problems by Body System

Dermatologic issues

- Pigeon pox. Pox virus leads to skin lesions and secondary infections.
- Cutaneous mycobacteriosis. Caseous nodules in immunosuppressed birds.
- Ectoparasites. Lice, mites, and hippoboscid flies.
- Traumatic injuries. Predator attacks.
- Neoplasia. Fibrolipomas are common.
- Iatrogenic feather lesions. Treatment with benzimidazoles (use should be avoided), or potentiated sulfonamides.
- Dermatophytosis. *Trichophyton megnini* can be found.

Digestive and hepatic system

- Trichomoniasis or canker. Caused by *Trichomonas gallinae,* also referred to as canker when discussing oral lesions.
- Endoparasites. Coccidiosis, ascarids, *Capillaria.*
- Salmonellosis. Paratyphoid infection, most significant bacterial infection. One syndrome leads to boils over joints.
- *E. coli*. Also called colibacillosis.
- Candidiasis. Can lead to crop issues in young birds.
- Pigeon circovirus. Leads to immunosuppression in young birds; cause of young bird sickness.
- Adenovirus. Main clinical sign is hepatic necrosis.
- Pigeon herpesvirus. Can lead to inclusion body hepatitis.
- Iatrogenic hemorrhagic enteritis and intestinal crypt necrosis from benzimidazoles.

Respiratory system

- Chlamydiosis. Ornithosis complex, or eye cold, one-eye cold, or one-eye croup. High numbers of birds are carriers of serotype B (low pathogenicity compared to types in psittacine birds).
- Mycoplasmosis. Role is controversial.

- Aspergillosis. Ubiquitous fungal infections.
- Ammonia or other toxicosis or respiratory irritants from poor cleaning, lack of ventilation in lofts.

Neurologic system

- PMV-1/AAv-V1. Significant viral disease. Birds can survive. Vaccination is recommended.
- Poisoning or toxicity. Can be intentional for control of feral birds (i.e., mycotoxins, heavy metals, pesticides, rodenticides).
- Hypocalcemia. Occurs in breeding hens, typically after the first egg is laid.
- Thiamine deficiency. Can cause neurologic signs.
- Lead toxicosis. Seen in urban settings where pigeons commonly live.

Reproductive system

- Egg binding. Numerous factors can be the cause.
- Disease of reproductive tract. Neoplasia, prolapse, salpingitis, often cessation of laying is a sign of disease.

Congenital musculoskeletal abnormalities

- Splay leg. Common developmental issue, hobbling may work in young birds. Fluoroquinolone use, genetics, poor flooring or nests, nutritional deficiencies.
- Angel wing. Developmental wing anomaly often linked to nutritional causes.
- Metabolic bone disease. Birds denied vitamin D3 or sunlight.
- Pododermatitis or bumblefoot. Infection occurs with poor housing.
- Trauma. Predator attacks.

References

1. Tudor DC. *Pigeon Health and Disease.* Ames, IA: Iowa State University Press; 1991.
2. Marx DA. Veterinary approach to pigeon health. Lake Charles, LA: Racing Pigeon Digest Co.; 1997.
3. Hooimeijer J. Management of racing pigeons. In: Harrison GJ, Lightfoot TL, Flinchum GB, eds. Clinical Avian Medicine. Palm Beach, FL: Spix Publishing; 2006:849-860.
4. Rupiper DJ. Diseases that affect race performance of homing pigeons. Part I: husbandry, diagnostic strategies and viral diseases. *J Avian Med Surg.* 1998; 12(2):70-77.

5. Rupiper DJ. Diseases that affect race performance of homing pigeons. Part II: bacterial, fungal and parasitic diseases. *J Avian Med Surg.* 1998;12(3):138-148.

6. De Herdt P, Devriese L. Pigeons. In: Tully TN, Lawton MPC, Dorrestein GM, et al, eds. *Avian Medicine.* Oxford, England: Butterworth Heinemann; 2000:312-338.

7. Harlin R, Wade L. Bacterial and parasitic diseases of Columbiformes. *Vet Clin North Am Exot Anim Pract.* 2009;12:453-473.

8. Hooimeijer J, Dorrestein GM. Pigeons and doves. In: Altman RB, Clubb SL, Dorrestein GM, et al, eds. *Avian Medicine and Surgery.* Philadelphia, PA: WB Saunders; 1997:886-909.

9. Howard LL, Papendick R, Stalis IH, et al. Fenbendazole and albendazole toxicity in pigeons and doves. *J Avian Med Surg.* 2002;16(3):203-210.

Care and Diseases of Starlings and Mynas of the Family Sturnidae

Brynn McCleery, DVM, DABVP (Avian Practice)

Session #156

Affiliation: From Red Bank Veterinary Hospital, 100 Schulz Drive, Red Bank, NJ 07701, USA.

Abstract: This session briefly reviews the family Sturnidae, including important taxonomic relationships, conservation concerns, and social characteristics. There is a focus on species seen more commonly in clinical practice and zoological settings with a review of husbandry and diet, as well as discussion of common diseases.

Introduction

Starlings and mynas belong to the Old World family Sturnidae. Sturnids exhibit wide morphologic variation, including differences in color and body shape, with a high incidence of iridescent plumage in the African species. Overall, these passerine birds are beautiful, playful, and excellent mimickers, making them desirable pets and display animals in zoological collections.

Taxonomy

The family Sturnidae consists of 125 species (117 extant) within 36 genera.[1] The family includes starlings, mynas (also called mynahs), *Rhabdornis* (aka Philippine creepers), and allies, many of which are commonly found in zoological collections. Taxonomic relationships between sturnids and other bird families have changed dramatically throughout history. Phylogenetic studies based on DNA hybridization and physiologic traits have shown this family to be most closely related to Mimidae (mockingbirds, trashers, and allies). Genus *Buphagus*, consisting of two species of oxpeckers, was previously considered part of the Sturnidae family but has been reassigned as a basal lineage to both Sturnidae and Mimidae in a separate family, Buphagidae. The relationship between these three families has been supported by more recent phylogenetic studies based on mitochondrial and nuclear DNA sequences.[2]

Conservation and Range

A few starling species have gone extinct and several additional sturnids are considered critically endangered, endangered, vulnerable, and threatened, with *Leucopsar rothschildi* (Bali or Rothschild myna) the most well-known and the only species with a conservation program in place.[1,3] Causes of sturnid population decline include habitat loss and fragmentation, introduced predators, and capture of wild birds for the pet trade, aviculture, and zoos. Although some species, like the wattled starling, are valued for their consumption of locusts or other insects that are destructive to crops and pasturelands, other sturnid species are considered agricultural pests.

Species within the Sturnidae family are restricted to the Old World, except when introduced elsewhere by human intervention. Southeast Asia and Africa serve as the centers of diversity for these birds.[2] In the USA, European starlings are invasive and are therefore not protected by the Migratory Bird Treaty Act (MBTA). Unlike protected native avian species, wild European starlings can be kept as pets and may present to the veterinary clinic for care. Truly orphaned or found fledgling European starlings and injured adult starlings presented to wildlife rehabilitation centers may be refused or euthanized, depending on the rehabilitator's policy on treatment of invasive species.

Pet Trade

Sturnid species popular in the pet trade include the hill mynas (*Gracula* spp., including several subspecies of *Gracula religiosa*), common myna (*Acridotheres tristis*), superb starling (*Lamprotornis superbus*), and European starling (*Sturnus vulgaris*). An average lifespan of 7-12 years is most common in captivity with some birds living up to 20-25 years.[4] Although not robust singers, many starlings and mynas are prized as pets due to their excellent ability to mimic sounds, including human speech patterns.[5] In addition to habitat loss, Bali myna populations have suffered due to poaching for the pet trade since these birds are considered a status symbol in many parts of the world, and less than 100 adults are estimated to remain in the wild currently (with much lower numbers reported a few decades ago).[6]

Social Characteristics

Most sturnid species are very gregarious and social, living in flocks or larger family groups for the majority of the year.[3] Despite their social nature, these birds are usually kept alone as pets or in pairs in zoological institutions. Many members of Sturnidae act aggressively toward smaller avian species with harsh calls, displacement, or chasing, especially during the breeding season. Intraspecies squabbling is less common but may also occur during the breeding season, particularly associated with nest guarding. More aggressive fighting with locked feed and pecking has been observed in several species of captive sturnids.[3] Species differences in aggression should be considered when adding sturnids to a mixed flock.

Diet

Compared to other softbills, sturnids are quite omnivorous. Most starlings eat insects, especially species with increased terrestrial locomotion. Mynas are more frugivorous, particularly the more arboreal species. However, both will take advantage of readily available food sources including eggs of other birds and small lizards.[7] In captivity, a varied diet is recommended with a commercially produced softbill pellet as the base. Multiple pelleted diets specific to mynas were previously available and used for sturnids and other softbills.[7] However, myna diets often contained excessive iron (500 to > 1000 ppm).[8] Due to their susceptibility to hemochromatosis (i.e., iron storage disease), the iron content of the diet should be minimized. Most authors anecdotally recommend providing sturnids dietary iron concentrations similar to domesticated species.[7] Although variable depending on the study, iron recommendations for poultry range from 45-136 ppm.[9] Commercial diets should, therefore, contain a maximum of 50-100 ppm for starlings and mynas.[8] For bali mynas, the Species Survival Plan recommends approximately 66.7 ppm iron per dry weight of diet.[7] This suggested range was confirmed in European starlings in a study where the optimal dietary iron concentration for this species was reported to be between 34-125 ppm.[10]

Unfortunately, it can be difficult to achieve iron levels within the recommended ranges when utilizing traditional ingredients and ensuring the production of a nutritionally complete diet.[10,11] Although many low-iron commercial pellets and gel foods exist, not all of these diets have iron at or below the recommended levels due to specific ingredients and/or nutrient interactions (with other minerals, ascorbic acid, sugars, and polyphenols) in mixed diets.[8,11] Furthermore, the iron concentration does not always match that on the packaging, which is a guaranteed maximum level or, more commonly, a calculated level.[4,810,11] Discrepancies between reported (calculated or maximum guaranteed analysis) and actual analytic values can be caused by inadvertent contamination from metal equipment used in production of the pellet and iron contamination of dietary ingredients (including mineral mixes and supplements).[8] It is also important to note that iron levels can vary between batches of the same diet due to fluctuating soil conditions affecting iron levels in crops used as ingredients.[4,8] Examples of low iron commercial diets include Red Apple Jungle (Marion), Bird of Paradise diet (Zeigler), and Mazuri Softbill Diet.

A low-iron pelleted diet should be supplemented with a variety of fruits and vegetables. Ascorbic acid (vitamin C) may increase iron absorption when consumed with foods containing nonheme iron and should be minimized in the sturnid diet. Ascorbic acid levels of < 100 ppm have been recommended for iron-sensitive species; alternatively, vitamin C-containing foods can be fed separately.[8]

Live insects should be offered as < 5-10% of the total diet by weight and should be gut loaded on a commercial insect diet prior to feeding. The amount of insects offered should be increased for breeding birds. Live foods have been shown to be essential to chick rearing in multiple sturnid species.[7] Historic diets for breeding and nonbreeding sturnids have often included animal protein such as hard-boiled eggs, soaked cat or dog kibble, raw meat, or chopped pinkie mice.[7] Caution is recommended with offering too much animal protein as total iron absorption from heme-based iron sources such as meat is three times greater than that from nonheme foods with equivalent iron concentrations;[8] however, when European starlings were fed a low-iron formulated diet supplemented with meat-based dog food, no significant increase in iron absorption was observed compared to birds fed the low-iron diet only. The birds receiving the dog food also had subjectively better feather quality. The authors of the study suggested that meat supplementation at or less than 10% of the total diet may not promote iron absorption if the baseline diet contains adequate, but not excessive, iron.[10]

Enclosure

Enclosures should be based on the bird's natural history and mimic the natural habitat as much as possible, which is variable amongst sturnid species. Although breeding has been reported in cages as small as 3 square feet, the enclosure should be as large as possible to encourage flight. Various zoological institutions recommend a minimum enclosure size of 8 (L) × 3 (W) × 8 (H) feet [2.4 × 0.9 × 2.4 meters] for a single breeding pair of starlings. The species survival plan (SSP) recommendations for cage size for a pair of Bali myna is 4 × 7 × 15 ft [1.2 × 2.1 × 4.5 meters].[3]

Sturnids should be provided with ample perches (variable diameters with a preference for those between 0.5-1 inch) and other cage furniture due to their active nature. Plants (nontoxic live, dried vegetation, and/or artificial) can be used to provide visual barriers and areas for the birds to hide. However, some species, including superb and European starlings, enjoy spending time in the open to watch the activity of the surrounding environment or household.

Bathing and foraging

Large variation is noted in the frequency with which sturnids bathe in captivity and their preferred method. For the commonly kept pet species, bathing has been noted as frequently as daily when housed in zoological facilities. Bathing opportunities should, therefore, be provided frequently in the form of a shallow water dish and/or misting. An ideal dish for bathing should be at least 6 inches in diameter and 1-2 inches deep.

Sturnids lacking appropriate foraging and substrates are prone to elongation of the beak and nails. Providing a foraging box with firmer substrates (including leaf litter, wood chips, grass sod, pebbles, etc.) to sift through for food items can enhance self-grooming. Most starlings are active and require enrichment to keep them busy and mentally stimulated!

Sexing

DNA-sexing is more commonly performed than the more invasive option of surgical (endoscopic) sexing. A few starling species are sexually dimorphic, including the amethyst starling and wattled starling (during breeding season). Sexing by gape color has been reported for some starling species with males having a black or blue-tinged gape (extending to variable levels on the inside of the beak/oropharynx) and females having a pale, yellow, pink, or gray gape.[12] Anecdotally, this has been reported in European, East African (including superb, Hildebrandt's, and gold-breasted starlings), emerald, and some glossy and red-winged starlings.[3] Additional slight variations have been reported in various species in size or feather patterns and coloration; however, these differences are not reliable methods of sexing.

Reproduction

Clutch size is typically 2-4 eggs for most sturnids with an incubation time of 12-14 days from the date of the last laid egg.[3] Nestlings stay in the nest 19-22 days for starlings and up to 30 days for mynas.[3,13] Parental feeding of fledglings continues an additional 10-14 days.[3] Information is available on hand-rearing some sturnids, including Bali mynas and hill mynas. Chicks should be fed every 1.5-2 hours.[3] Self-feeding and weaning has been reported at 21-30 days for parent-reared or hand-reared sturnid species in captivity.[3,13]

Common Diseases

Systemic isosporiasis

Systemic isosporiasis (formerly atoxoplasmosis) is the most concerning protozoal infection of sturnids, especially the Bali myna.[6,14] The parasite has a prolonged life cycle with stages affecting both the intestinal submucosa and reticuloendothelial system.[6] Transmission occurs via the fecal-oral route. After ingesting oocysts, the sporozoites are released in the intestinal lumen and invade intestinal epithelial cells. Sporozoites also invade lymphocytes and macrophages, leading to dissemination throughout the body.[5,6] This disease causes high fledgling mortality and can affect adults less commonly. Clinical signs may include weight loss, diarrhea, lethargy, hyporexia, fluffed feathers, and death.[6] Diagnosis is through blood work and serial fecal analysis. Repeated fecal examination is recommended due to the possibility of intermittent shedding of oocysts.[5,6] Quantitative PCR can also be utilized with testing of 5 consecutive daily fecal samples providing an average diagnostic sensitivity of 86%.[15] Organisms may be identified on blood smears within mononuclear cells with increased incidence of identification on buffy coat smears due to concentration of the white blood cells.[6] Treatment and prevention can be attempted, but success rates vary. Drugs utilized include sulfachlorpyrazine, sulfachlorpyridazine, toltrazuril, and ponazuril. Preventive therapy during breeding to reduce fecal oocyst shedding is most effective and may reduce exposure of chicks to the parasite.

Other parasites

- Protozoa: *Toxoplasma gondii*,[5] *Plasmodium*, *Coccidia*[6]
- Vermicular endoparasites: nematodes (gapeworm/*Syngamus trachea*, *Diplotriaena* spp., tetramers), ascarids, cestodes, *Capillaria* spp.[3,5,13]
- Ectoparasites: Northern fowl mites, feather lice[3]

Hemochromatosis

Hemosiderosis is an increased intracellular accumulation of hemosiderin in tissues without pathology. When accumulation is severe enough to cause pathology and impaired function, the disease is referred to as hemochromatosis or iron storage disease (ISD).[11,16] Although multiple organs can be affected, the liver is commonly involved.[5] ISD has been reported in multiple sturnids, especially the more frugivorous species (Bali mynas, *Gracula* spp.). Clinical signs include difficulty flying and walking, coelomic distension secondary to hepatomegaly and/or ascites, depression, dyspnea, weight loss, cachexia, and acute death.[5,16] Definitive diagnosis requires a liver biopsy and histopathology.[5,8]

In humans, hemochromatosis can be a primary (inherited) or secondary condition (from some anemias, repeated transfusions, or excess iron in the diet).[8,11,17] The pathogenesis of iron-storage disease in birds is unknown and may be multifactorial.[5,6] Similar to humans, a genetic defect in iron absorption or metabolism has been theorized in sensitive species.[5,8] Many of the iron-sensitive avian species are frugivorous or, less commonly, insectivorous and are adapted to a low-iron diet in the wild. Due to the low iron concentration of the natural diet, these species may have evolved to absorb iron more efficiently, putting them at increased risk of ISD when dietary iron concentrations increase.[6,8] Mynas were found to have a higher rate of mucosal transport of iron compared to doves, a non-iron-sensitive species; similar results were found in European starlings.[8] Additional studies have shown overexpression of iron transporter proteins in the duodenum of iron-sensitive species.[18] In addition to primary inherited causes of ISD, dietary causes have been proposed in birds. High dietary iron has been shown to be a contributing factor in studies of both non-sensitive (pigeons, doves) and iron-sensitive (European starlings) species.[8]

Treatment of ISD can include phlebotomy, iron chelation therapy, and reducing iron in the diet.[8] In European starlings, both chelation with deferoxamine (100 mg/kg SC q24h) and phlebotomy (1% body weight q7d) were shown to decrease liver iron concentrations when used in combination with a low-iron diet.[17] An oral chelator, deferiprone, has also shown promise in reducing liver iron stores in birds, including chickens, pigeons, and hornbills.[18,19] In a study of ISD treatments in common mynahs, birds treated with deferoxamine (100 mg/kg SC q24h), deferiprone (75 mg/kg PO q24h), and each chelation drug with phlebotomy (1% body weight q7d) all had a significant reduction in iron after 1 month of therapy.[18] Some authors have recommended offering ISD birds an iron-deficient diet temporarily;[8,10] however, two studies in European starlings found that the rate of liver iron reduction in birds on an iron-deficient or low-iron diet only was too slow to be clinically helpful in birds with ISD.[10,17]

Prevention is focused on the diet including offering a low-iron diet and avoiding excess ascorbic acid. Diets should not contain meat or animal by-products, as these sources of iron have increased bioavailability.[8] In a study evaluating tea-derived tannin supplementation, European starlings in the non-treatment group on the iron-enriched diet without tea supplementation had increased hepatic iron concentrations after 6 months while those receiving tea supplementation did not.[20] European starlings fed an extremely high iron diet supplemented with

20 g/kg of phytate (inositol) and tannic acid did not have an increased iron concentration in the liver.[10] While tannins and other natural chelator (fiber, phytates) supplements may be an inexpensive alternative to purchasing or formulating a low-iron pelleted diet, the effectiveness of individual supplements on chelating iron is unknown, and chelators may affect the entire complement of minerals, not only iron.[8]

Molting

Although molting is not a disease, some sturnids undergo a very severe or heavy molt that may at first appear to be caused by underlying illness. Heavy feather loss, especially on the head and neck, has been observed in superb starlings, amethyst starlings, and Bali mynas. Other species less affected include golden-breasted, emerald, Asiatic glossy, purple glossy, and wattled starlings, as well as Indian hill mynas.[3] Most species molt annually, usually after the breeding season.

Other reported diseases

- Mycotic infection: aspergillosis (natural and experimental),[5,6,13,21] candidiasis,[5] avian gastric yeast[6]
- Viral infection: herpesvirus,[6] poxvirus,[22] influenza type A (H4N8), Newcastle disease virus, other paramyxoviruses (PMV-2, PMV-3)[5]
- Bacterial infection: *Pasteurella multocida* (avian cholera),[5] *Yersinia tuberculosis*,[5,6] *Salmonella* sp. (*S.typhimurium, S. saint paul, S. weltevreden*),[5] *Escherichia coli*,[5,6] *Klebsiella pneumoniae*,[5,6] *Campylobacter* sp., *Erysipelithrix insidiosa,* mycobacteriosis (*M. tuberculosis),* nocardiosis, *Coxiella burnetti* (causative agent of Q-fever), *Chlamydia psittaci* (carrier)[5]
- Trauma: interspecific aggression, intraspecific aggression, head trauma, predator attack, band construction secondary to hyperkeratosis of leg skin[6]
- Renal disease: nephrosis, renal insufficiency, tubular insufficiency, gout.[5,6,13]
- Cardiac disease congestive heart failure, atherosclerosis, endocardiosis[5,23]
- Ocular diseases: corneal ulceration, keratitis, chronic keratoconjunctivitis, hypermature cataract[5]
- Metabolic disease: obesity, hepatic lipidosis (associated with consumption of broiler diet or concurrent hemochromatosis), atherosclerosis, metabolic bone disease (unspecified)[23]
- Neurologic disease: encephalomalacia, seizures (suspected epilepsy)[5]
- Reproductive and cloacal disease: egg binding,[6] cloacal prolapse
- Neoplasia: lymphoid (chronic lymphocytic leukemia with paraneoplastic hyperviscosity syndrome, multicentric lymphoma with retroviral nucleic acid in neoplastic tissue, lymphosarcoma), hepatic (hepatocellular carcinoma and lymphosarcoma commonly reported in mynas with ISD),[5] other (hepatoma, pancreatic adenocarcinoma, ovarian adenocarcinoma, renal adenoma, intracoelomic lipoma, myelolipoma, squamous cell carcinoma)[5,6]
- Suspect behavioral: night frights, feather damaging syndrome[6]
- Suspected pancreatic insufficiency[5]

References

1. Winkler DW, Billerman SM, Lovette IJ. Sturnidae starlings. Available at: https://birdsoftheworld.org/bow/species/sturni1/cur/introduction#genusMino. Accessed April 1, 2025.

2. Lovette IJ, Rubenstein DR. A comprehensive molecular phylogeny of the starlings (Aves: Sturnidae) and mockingbirds (Aves: Mimidae): congruent mtDNA and nuclear trees for a cosmopolitan avian radiation. *Mol Phylogenet Evol.* 2007;44:1031-1056.

3. Bockheim G, Congdon S. *The Sturnidae Husbandry Manual and Resource Guide.* Lake Buena Vista, FL: Disney Animal Kingdom; 2001.

4. Boyer S. Are you feeding your myna to death? *AFA Watchbird.* 2000;27(4):58-59.

5. Panigahy B, Senne DA. Diseases of mynahs: review article. *J Am Vet Med Assoc.* 1991;199(3):378-381.

6. Norton TM. Mynah: medicine, natural history, and conservation: focus on the critically endangered Bali mynah (*Leucopsar rothschildi*). *Proc VMX.* 2019.

7. Dierenfeld ES. Nutrition. Husbandry guidelines for the bali mynah (*Leucopsar rothschildi*) species survival plan. Available at: https://nagonline.net/wp-content/uploads/2013/12/Bali-Mynah-Nutrition9.pdf. Accessed April 1, 2025.

8. Sheppard C, Dierenfeld E. Iron storage disease in birds: speculation in etiology and implications for captive husbandry. *J Avian Med Surg.* 2002;16(3):192-197.

9. Han M, Fu X, Xin X, et al. High dietary organic iron supplementation decreases growth performance and induces oxidative stress in broilers. *Animals.* 2022;12(13):1604.

10. Olsen GP, Russell KE, Dierenfeld E, et al. Impact of supplements on iron absorption from diets containing high and low iron concentrations in the European starling (*Sturnus vulgaris*). *J Avian Med Surg.* 2006;20(2):67-73.

11. Crissey SD, Ward AM, Block SE, et al. Hepatic iron accumulation over time in European starlings (*Sturnus vulgaris*) fed two levels of iron. *J Zoo Wildl Med.* 2000;31(4):491-496.

12. Lindholm JH. A non-surgical method for sexing starlings. *AFA Watchbird.* 1994;21(6):22-23.

13. Dorrestein GM. Veterinary problems in mynah birds. *Proc Annu Conf Assoc Avian Vet.* 1988;263-274.

14. Barta JR, Schrenzel MD, Carreno R. et al. The genus *Atoxoplasma* (Garnham 1950) as a junior objective synonym of the genus *Isospora* (Schneider 1881) species infecting birds and resurrection of *Cystoisospora* (Frenkel 1977) as the correct genus for *Isospora* species infecting mammals. *J Parasitol.* 2005;91:726-727.

15. Landolfi, JA, Adkesson MJ, Ahmed N, et al. A qPCR assay and testing guidelines for the molecular diagnosis of systemic isosporosis (formerly atoxplasmosis) in passerine birds. *J Zoo Wildl Med.* 2020;51(2):391-397.

16. de Oliveira AR, dos Santos DO, de Paula Martins Pereira M, et al. A retrospective study of hepatic he-

mosiderosis and iron storage disease in several captive and free-ranging avian species. *J Zoo Wildl Med.* 2022;53(2):455-460.

17. Olsen GP, Russell KE, Dierenfeld, et al. A comparison of four regimens for treatment of iron storage disease using the European starling as a model. *J Avian Med Surg.* 2006;20(2):74-79.

18. Jalili N, Mahdieh Z, Razmyar J, et al. Comparison of iron chelators used alone or in combination with phlebotomy in common mynahs (*Acridotheres tristis*) with experimental iron storage disease. *Avian Pathol.* 2021;50(4):350-356.

19. Whiteside DP, Barker IK, Mehren KG, et al. Clinical evaluation of the oral iron chelator deferiprone for the potential treatment of iron overload in bird species. *J Zoo Wildl Med.* 2004;35:40-49.

20. Seibels B, Lamberski N, Gregory CR, et al. Effective use of tea to limit dietary iron available to starlings. *J Zoo Wildl Med.* 2003;34(3):314-316.

21. Atasever A, Gümüssoy, KS. Pathological, clinical, and mycological findings in experimental aspergillosis infections of starlings. *J Vet Med Assoc.* 2004;51:19-22.

22. Landolt M, Kocan RM. Transmission of avian pox from starlings to Rothchild's mynahs. *J Wildl Dis.* 1976;12:353-356.

23. Norouzi A, Mirshahi A, Razmyar J, et al. Radiographic reference intervals for cardiac size in common mynahs (*Acridotheres tristis*). *J Avian Med Surg.* 2022;36(3):272-277.

Vulture Culture: The Current Status of Avian Scavengers Worldwide and Threats to Their Survival

Jesse A. Fallon, DVM, PhD

Session #211

Affiliation: From Avian Conservation Center of Appalachia and Cheat Lake Animal Hospital, Morgantown, WV 26508, USA.

Abstract: Vultures play an integral role in ecosystems around the world. These scavenging raptors remove carcasses from the environment, thereby reducing environmental pollution, preventing the spread of disease, and recycling nutrients quickly back into the environment. Therefore, vultures are integral for maintaining ecological balance. Unfortunately, many of these species are under threat from anthropogenic sources such as pesticides, improper use of veterinary pharmaceuticals, direct persecution, and habitat loss. Of the 23 extant species of vultures, 14 are threatened with extinction, while many of the remainder are in population decline. In this presentation, we review the current status of vulture populations worldwide and discuss the most important threats to their survival.

Introduction

From their bald-headed, beady-eyed appearance, to their preference for dead and sometimes decaying prey, vultures have a much-maligned reputation throughout the world. Even famed ecologist Charles Darwin, when observing a turkey vulture from the deck of the Beagle, described it as "a disgusting bird" whose bald head was "formed to wallow in putridity."[1] But vultures provide an invaluable service to the environment and are a pillar of healthy, functioning ecosystems. Often referred to as nature's clean-up crew, these scavengers fill a critical niche by removing carcasses from the environment, thereby reducing environmental pollution, preventing the spread of disease, and recycling nutrients quickly back into the environment.[2]

Unfortunately, many species of vultures are in decline throughout the world, resulting in negative, system-wide consequences. In Asia, for example, the decline of vultures has created a void in the food web, with uneaten carcasses serving instead as fodder for mammals, including feral domestic dogs, leading to an increase in prevalence of rabies virus circulating among mammalian scavengers. This has put humans at significant risk particularly in India, where the loss of healthy vulture populations may have led to an increase in human deaths from rabies virus by as much as 500,000 individuals.[3,4] In Africa, declining vulture populations contribute to increased burden of harmful pathogens such as anthrax, increasing water contamination, and spread of disease.[5] The loss of vulture populations has clear and well-established consequences. Therefore, an understanding of the current status of vulture populations is vitally important to ecosystems throughout the world

Current Conservation Status

Of the 23 species of vultures throughout the world, nine are critically endangered, two are endangered, two are vulnerable, and populations of almost all species are in decline (Table 1). Of 16 Old World vulture species, all but four are threatened or endangered. Clearly, efforts to understand the reasons for vulture declines worldwide are critically important; however, there are a few clear contributing factors, all of which have the potential for intervention and change.

Table 1. Conservation status and range of all vulture species.

Species	Range	Conservation status
Old World		
Cinereous vulture (*Aegypius monachus*)	Europe and Asia	Least Concern
Griffon vulture (*Gyps fulvu*)	Europe, Africa, Asia	Least Concern
White-rumped vulture (*Gyps bengalensis*)	Asia	Critically Endangered
Rüppell's vulture (*Gyps rueppell*)	Africa	Critically Endangered
Indian vulture (*Gyps indicu*)	Asia	Critically Endangered
Slender-billed vulture (*Gyps tenuirostri*)	Asia	Critically Endangered
Himalayan vulture (*Gyps himalayensi*)	Asia	Near Threatened
White-backed vulture (*Gyps africanu*)	Africa	Critically Endangered
Cape vulture (*Gyps coprothere*)	Africa	Vulnerable
Hooded vulture (*Necrosyrtes monachu*)	Africa	Critically Endangered
Red-headed vulture (*Sarcogyps calvu*)	Asia	Critically Endangered
Lappet-faced vulture (*Torgos trachelioto*)	Africa	Endangered
White-headed vulture (*Trigonoceps occipitali*)	Africa	Critically Endangered
Bearded vulture (*Gypaetus barbatu*)	Europe, Africa, Asia	Near Threatened
Egyptian vulture (*Neophron percnopteru*)	Europe, Africa, Asia	Endangered
Palm-nut vulture (*Gypohierax angolensis*)	Africa	Least Concern
New World		
Black vulture (*Coragyps atratus*)	North and South America	Least Concern
Turkey vulture (*Cathartes aura*)	North and South America	Least Concern
Lesser yellow-headed vulture (*Cathartes burrovianus*)	South America	Least Concern
Greater yellow-headed vulture (*Cathartes melambrotus*)	South America	Least Concern
California condor (*Gymnogyps californianus*)	North America	Critically Endangered
Andean condor (*Vultur gryphus*)	South America	Vulnerable
King vulture (*Sarcoramphus papa*)	South America	Least Concern

Reasons for Vulture Population Declines

Research has shown a few key anthropogenic causes of vulture population declines. These include but are not limited to habitat loss, climate change, toxins, targeted persecution, and unintentional trauma.

Habitat loss

As with many wildlife species, vultures are affected by habitat destruction and fragmentation. Habitat loss reduces nest site availability for disturbance-sensitive, tree-nesting vultures, including hooded, white-backed, lappet-faced, and white-headed vultures.[6] Urban expansion, deforestation, and the conversion of grasslands to agriculture also reduce the availability of nesting and foraging areas for vultures.[7] This can particularly affect large vulture species, such as the California condor (*Gymnogyps californianus*), which require vast distances to forage. Further, in areas where natural habitats are degraded, vultures are forced to move into human-dominated landscapes, increasing the likelihood of adverse events and disrupting breeding success for some species.

Climate change

Climate change continues to alter natural habitat and weather patterns, and these consequences can impact vulture populations. In central India, for example, a study predicted that at the current rate, climate change will lead to a decrease in suitable vulture habitats over the next several decades.[8] Similarly, in West Africa, habitat loss due to climate-induced changes in vegetation has been identified as a significant threat to vulture populations.[9] As habitats become less suitable, vultures may be forced to migrate to new areas, potentially leading to competition with other species and increased vulnerability to predation or human activities. For example, a study predicted that the habitat of the hooded vulture would shift significantly to the historic north of its current range.[10] Similarly, rising temperatures and extreme weather events can affect movement dynamics in Cape vultures, which are predicted to shift south along their range away from core colonies, resulting in a decrease in use of protected areas.

Climate change can also increase the risk of wildfires, which can directly impact vulture populations that nest in trees or depend on specific vegetation types for their habitat, as occurred in recent years in California condors. Further, because climate change affects the abundance and distribution of prey species, there is the potential for reduced food availability for vultures. This can lead to food shortages and increased competition among vultures, particularly during harsh weather conditions. In addition, climate-induced droughts can concentrate carcasses, increasing the risk of poisoning events for vultures.

Toxins

One of the most prominent drivers of vulture population declines in Asia is the use of veterinary pharmaceuticals, particularly diclofenac. This non-steroidal anti-inflammatory drug (NSAID), commonly used in livestock in South Asia, can be fatal to vultures when they consume the carcasses of treated animals.[11] In India, Pakistan, and Nepal, vulture populations—especially those of the white-rumped vulture (*Gyps bengalensis*)—plummeted by more than 95% since the 1990s due to diclofenac poisoning.[12] The drug can cause acute kidney failure in vultures, leading to death within days of ingestion. Although the governments of India and neighboring countries banned veterinary diclofenac in the mid-2000s, illegal use continues in some areas, undermining conservation efforts.[13,14]

Another well-documented toxin that is known to cause morbidity and mortality in vultures is the ingestion of spent lead ammunition. There is evidence that lead exposure remains a key factor in the loss of many endangered California and Andean condors, and that these losses have population effects in these species.[15,16] Clinical signs of affected birds include lethargy, ataxia, weakness, dull mentation, seizures, anorexia, and regurgitation. Treatment for lead toxicity typically includes decontamination, aggressive fluid therapy, supportive care, and chelation therapy with calcium edetate calcium disodium (CaEDTA) and dimercaptosuccinic acid (DMSA). Mortality rates are generally dose-dependent but can be as high as 100% even with small quantities of lead ingestion.[17]

Targeted persecution

Unfortunately, poisoning is a widespread and often intentional cause of vulture deaths. In Africa, vultures are frequently poisoned either directly or as non-target victims of poisoned bait used by poachers or pastoralists. Poachers sometimes poison carcasses to kill vultures, which might otherwise reveal the location of illegal kills to authorities through their circling behavior.[18] In southern and eastern Africa, hundreds of vultures have been killed in single poisoning events. For instance, in Botswana in 2019, over 500 vultures died after feeding on poisoned elephant carcasses left by poachers.[19]

In some cultures, vultures are killed for traditional medicine or belief-based practices, further exacerbating population declines. For example, in 2020 during the covid-19 pandemic, a large number of hooded vultures were poisoned and killed in Guinea-Bissau, with over 2000 deaths reported. The poisoning was due to the intentional use of methiocarb, a banned pesticide, to collect vulture heads which were believed to help prevent humans from contracting coronavirus.

Unintentional trauma

Infrastructure development also poses physical threats to vultures. Power lines and wind turbines can be deadly, especially for large, slow-flying species. Vultures often perch or roost on power pylons and risk electrocution. Studies have shown high mortality rates from power line collisions and electrocution in parts of Africa, Europe, and Central Asia.[20] Recreational activities can also result in vulture mortality, as seen during India's kite festivals. The use of manja, or wiry kite strings made with powerful and abrasive materials like powdered glass, can inflict severe injuries on birds, including vultures. Birds can become entangled, suffer wing lacerations, nerve damage, fractures, and death.

Infectious disease

Recent strains of avian influenza have the potential to affect vulture population. Identified in wild birds and poultry in North America in late 2021, the current H5N1 strain has a high mortality rate in black vultures and California condors.[21] Because of the risk posed to this endangered species, the U.S. Department of Agriculture's (USDA) Animal and Plant Health Inspection Service (APHIS) granted emergency approval for the use of an existing vaccine developed for domestic poultry and waterfowl and tested in black vultures for use in condors.[22] At this point, it is unclear how many California condors have succumbed to influenza virus since 2022, and the efficacy of the vaccination program remains uncertain.

Intervention Strategies

Recognizing the gravity of the worldwide vulture crisis, several conservation initiatives have been launched. In India, a major breakthrough was the government ban on veterinary diclofenac in 2006, followed by the promotion of a vulture-safe alternative anti-inflammatory, meloxicam. This policy change, combined with captive breeding programs and vulture "safe zones," has shown signs of stabilizing vulture populations. Organizations like the Vulture Conservation Foundation (VCF) in Europe and the Peregrine Fund in Africa and South America are actively involved in habitat protection, education, and rescue efforts. In Africa, the Multi-Species Action Plan to Conserve African-Eurasian Vultures, adopted under the Convention on Migratory Species (CMS), has provided a unified strategy to address threats across countries.

In North America, grassroots efforts to replace toxic lead ammunition with non-lead alternatives have produced modest, albeit slow, movement away from hunting with lead ammunition. California has banned lead ammunition for hunting in areas where California condors are found, and subsequently statewide, to mitigate the threat of lead poisoning. Despite this, wildlife veterinarians working in these areas continue to see lead toxicosis in condors and other species.

In addition to government regulations, community involvement is also crucial to reverse vulture population declines. In Nepal, for example, local communities participate in maintaining vulture restaurants—safe feeding stations that provide uncontaminated food for vultures. These programs have not only supported vulture recovery but also boosted eco-tourism and local economies.

Conclusions

Vultures may not be the most charismatic animals, but they are among the most essential. Their decline serves as a warning of the fragility of ecosystems and the unintended consequences of human actions. Conservation efforts must continue to focus on banning harmful substances, protecting habitats, and educating the public about the importance of these birds. With coordinated global action, it is possible to halt the decline of vulture populations and restore the ecological balance they help maintain.

References

1. Willson MF, Armesto JJ. The natural history of Chiloé: on Darwin's trail. *Revista Chilena de Historia Natural.* 1996;69:149-61.

2. Santangeli A, Lambertucci SA, Margalida A, et al. The global contribution of vultures towards ecosystem services and sustainability: an experts' perspective. *Iscience.* 2024;27(6).

3. Markandya A, Taylor T, Longo A, et al. Counting the cost of vulture decline—an appraisal of the human health and other benefits of vultures in India. *Ecol Econ.* 2008;67(2):194-204.

4. Frank E, Sudarshan A. The social costs of keystone species collapse: evidence from the decline of vultures in India. *Am Econ Rev.* 2024;114(10):3007-3040.

5. Ogada DL, Keesing F, Virani MZ. Dropping dead: causes and consequences of vulture population declines worldwide. *Ann N Y Acad Sci.* 2012;1249(1):57-71.

6. Bamford AJ, Monadjem A, Anderson MD, et al. Trade-offs between specificity and regional generality in habitat association models: a case study of two species of African vulture. *J Appl Ecol.* 2009;46(4):852-860.

7. Botha AJ, Ogada DL, Virani MZ. Multi-species action plan to conserve African-Eurasian vultures (Vulture MsAP). *Convention on the Conservation of Migratory Species of Wild.* 2017.

8. Jha KK, Jha R. Study of vulture habitat suitability and impact of climate change in Central India using MaxEnt. *J Resour Ecol.* 2021;12(1):30-42.

9. William G, Saqib Z, Naeem N, et al. Predicting climate driven habitat shifts for the Egyptian vulture in Punjab, Pakistan. *J Nat Conserv.* 2025;83:126774.

10. Mosisa GB, Tassie N, Yesuf F. Global change drives potential niche contraction and range shift of globally threatened African vulture. *Environ Chall.* 2024;17:101038.

11. Swan GE, Cuthbert R, Quevedo M, et al. Toxicity of diclofenac to *Gyps* vultures. *Biol Lett.* 2006;2(2): 279-282.

12. Green RE, Newton IAN, Shultz S, et al. Diclofenac poisoning as a cause of vulture population declines across the Indian subcontinent. *J Appl Ecol.* 2004;41(5):793-800. doi: 10.1111/j.0021-8901.2004.00954.x.

13. Cuthbert R, Parry-Jones R, Green RE, Pain DJ. NSAIDs and scavenging birds: Potential impacts beyond Asia's critically endangered vultures. *Biol Lett.* 2007;3(1):91-94. doi: 10.1098/rsbl.2006.0554.

14. Jimenez-Lopez O, Ponder J, Nault A, Bueno I. Non-steroidal anti-inflammatory drugs (NSAIDS) and their effect on Old World vultures: a scoping review. *J Raptor Res.* 2021;55(3):297-310.

15. Lambertucci SA, Donázar JA, Huertas AD, et al. Widening the problem of lead poisoning to a South-American top scavenger: lead concentrations in feathers of wild Andean condors. *Biol Conserv.* 2011;144(5): 1464-1471.

16. Cade TJ. Exposure of California condors to lead from spent ammunition. *J Wildl Manag.* 2007;71(7): 2125-2133.

17. Fallon JA, Redig P, Miller TA, et al. Guidelines for evaluation and treatment of lead poisoning of wild raptors. *Wildl Soc Bull.* 2017;41(2):205-211.

18. Ogada D, Botha A, Shaw P. Ivory poachers and poison: drivers of Africa's declining vulture populations. *Oryx.* 2016;50(4):593-596. doi: 10.1017/S0030605315001209.

19. Henriques M, Buij R, Monteiro H, et al. Deliberate poisoning of Africa's vultures. *Science.* 2020;370(6514):304.

20. Lehman RN, Kennedy PL, Savidge JA. The state of the art in raptor electrocution research: a global review. *Biol Conserv.* 2007;136(2):159-174. doi: 10.1016/j.biocon.2006.09.015.

21. Harvey JA, Mullinax JM, Runge MC, Prosser DJ. The changing dynamics of highly pathogenic avian influenza H5N1: next steps for management and science in North America. *Biol Conserv.* 2023;282:110041.

22. Katzner TE, Blackford AV, Donahue M, et al. Safety and immunogenicity of poultry vaccine for protecting critically endangered avian species against highly pathogenic avian influenza virus, United States. *Emerg Infect Dis.* 2025;31(6).

Pharmacokinetics of Amoxicillin in Domestic Chickens after Multiple-Day Oral Dosing and Comparing Three Formulations

Cheryl B. Greenacre, DVM, DABVP (Avian Practice), DABVP (Exotic Companion Mammal Practice),
Craig R. Reinemeyer, DVM, PhD, Dipl ACVM (Parasitology),
Nicole Szafranski, DVM, PhD,
Sherry Cox, PhD

Session #184

Affiliation: From University of Tennessee, College of Veterinary Medicine, Knoxville, TN 37996, USA (Greenacre, Cox), East Tennessee Clinical Research, Inc., Rockwood, TN 37854, USA (Reinemeyer, Szafranski).

Abstract: A previously published study in chickens found no measurable amoxicillin plasma concentrations at any time point after receiving a chewable tablet formulation of amoxicillin-clavulanic acid at 125 mg/kg orally every 12 hours for nine doses, although the clavulanic acid metabolites had measurable plasma concentrations for 8 hours.[1] Our random, cross-over study evaluated amoxicillin plasma concentrations in 11, 20- to 28-week-old white Leghorn hens after receiving either a Clavamox® chewable tablet crushed in water (CW, n = 7; 250 mg/kg) instilled into the crop, or whole tablet (TO, n = 6; 161-234 mg/kg) given orally, or Clavamox® oral suspension (OS, n = 7; 250 mg/kg) instilled into the crop, every 8 hours for 13 doses. Blood samples (~ 0.3 ml) were collected at ten time points (5, 10, 20, 30 minutes, and 1, 2, 4, 6, 8, and 12 hours) after delivery of the last dose. Mean maximum concentration (Cmax) for the CW, OS, and TO groups was similar (29.82 µg/ml, 33.96 µg/ml, and 24.71 µg/ml, respectively). The range of Cmax was similar for the CW and OS groups but was more variable for the TO group (24.46-63.57 µg/ml, 21.30-52.80 µg/ml, and 7.05-50.98 µg/ml, respectively). Time to maximum concentration (Tmax) for the CW, OS, and TO groups was 1.0, 1.1, and 1.75 hours, respectively. All birds in both the CW and OS groups had amoxycillin plasma concentrations above the minimum therapeutic plasma concentration for humans (0.5 µg/ml) for at least 6 hours, and all birds in the TO group for at least 4 hours.

Acknowledgments: This research was made possible by a grant from the Harrison Avian Research and Conservation Collaborative.

References

1. Shannon L, Cox SK, Bailey J, et al. Pharmacokinetics and drug residue in eggs after multiple-day oral dosing of amoxicillin-clavulanic acid in domestic chickens. *J Avian Med Surg*. 2020;34(1):3-8. doi:10.1647/1082-6742-34.1.3.

Welfare Concerns with Backyard Poultry

Anthony A. Pilny, DVM, DABVP (Avian Practice)

Session #123

Affiliation: From Arizona Exotic Animal Hospital, Phoenix, AZ 85024, USA.

Abstract: The rise in popularity of keeping backyard chickens as pets and the increased presentation of these birds to veterinarians have highlighted some serious welfare concerns. The poultry industry is important to the agricultural economy and is heavily regulated, unlike the keeping of chickens as pets. This lecture introduces several welfare issues faced by owners, veterinarians, and most importantly the birds themselves. It is also important to acknowledge the risk of zoonotic disease transmission with this increasingly popular pet hobby. The focus of this presentation is keeping pet chickens.

Introduction

Keeping backyard poultry, especially egg laying hens, has become a popular trend throughout the world. People who are interested in raising and consuming local foods with less environmental impact and improved welfare for the chickens are driving this trend. Despite this desire for better welfare, these birds still face serious challenges. From a purely research perspective backyard flocks are difficult to track and sample using random, unbiased methodology for scientific reporting purposes. Available research shows welfare concerns related to health and disease transmission, poor access or desire for veterinary care, and public health issues. The dissemination of reliable information to flock owners is very much needed.

Despite the popularity of backyard chickens in the United States, few studies have provided general information about these flocks. An online survey of chicken owners was conducted through websites and social platforms that consisted of fifty-six questions about flock history, husbandry, health care, and owner attitudes and demographics. The 1487 surveys were received almost equally from urban, suburban, and rural areas. According to the results, 71% of respondents owned fewer than 10 chickens and had kept chickens for less than 5 years. Of the respondents, 95% kept chickens as food for home use and 57% as pets. Owners thought that eggs and meat from their chickens were more nutritious (86%), safer to consume (84%), and tasted better (95%) than store-bought products, and that the health and welfare of their chickens was better (95%) than on commercial farms. Almost 60% reported no flock health problems in the last year; however, there was a lack of awareness about many health conditions. Specifically, most knew nothing about exotic Newcastle or Marek's disease, and 61% did not vaccinate against Marek's disease. As expected, the Internet was the main source of information used by backyard flock owners (87%), followed by books and magazines (62%) and feed stores (40%). Almost half of respondents cited the need to avoid predation as the biggest challenge followed by providing feed at low cost (28%), yard management (25%), and complying with zoning regulations (23%). The information from this survey will help to determine information and resources needed to maintain adequate biosecurity and improve the health and welfare of backyard flocks of pet chickens.[1]

Little is known about chicken's welfare status or the specifics of the human-chicken relationship. Another questionnaire survey study collected a comprehensive and holistic dataset on owners' care-taking practices, owners' attitudes towards chickens, the human-chicken relationship, and challenges regarding chicken welfare.[2] Significant results showed variable care-taking practices, largely positive attitudes towards chickens, mid-level closeness regarding the human-chicken relationship, and a desire to prevent chickens going to slaughter as a top motivator of chicken care. Most respondents had not heard of Suprelorin,[3] which is a hormonal implant used to prevent egg laying and modify rooster behavior that is not permitted in the United States. This study also showed a high prevalence of health conditions and high number of deaths like in the literature such as egg yolk peritonitis as a leading health problem and cause of death. Most chicken caretakers consumed chicken meat and eggs. Results of this study may help anyone interested in chicken welfare, including veterinarians, owners, and sanctuaries.

A study published from findings of a questionnaire in Finland[4] showed backyard flocks are small and mostly chickens. The results of the questionnaire indicate the health of flocks is good, mortality is low, and vaccinations are not commonly used. Standard biosecurity practices are not generally applied and contact with wild birds, pets and farm animals is frequent. This study also identified the postmortem necropsy findings on a total of 132 non-commercial laying hens with the most common postmortem findings being Marek's disease (27%) and colibacillosis (17%). Another study found, of 4,424 chicken veterinary consultations, that non-specific clinical signs were commonly recorded (17.6%) as were those indicative of advanced disease. Euthanasia comprises 29.8% of consultations and antimicrobials were commonly prescribed (33%), 43.8% of which included antimicrobials considered 'highest priority critically important' by the World Health Organisation.[5] A retrospective study published in the *Journal of the American Veterinary Medical Association* concluded backyard flock chickens were evaluated on an emergency basis most commonly because of trauma and reproductive tract disease. Although approximately half of the evaluated chickens were euthanized, the remainder were discharged from the hospital and required medical management.[6] A 2015 study from the *Journal of Exotic Pet Medicine* found Marek's disease and *Escherichia coli* infection were the most common disease conditions diagnosed. They also found ascaridiasis and *Mycoplasma gallisepticum* as well as neoplasia of the reproductive tract, ascites syndrome and fatty liver hemorrhagic syndrome. In their study, the two notifiable diseases of poultry, Newcastle disease and avian influenza, are considered rare in backyard poultry and neither of these diseases was diagnosed.[7]

Lastly, under the U.S. Department of Agriculture's Animal Welfare Act Standards for Birds, the Animal Welfare Regulations (AWR) define poultry as any species of chickens, turkeys, swans, partridges, guinea fowl and peafowl; ducks, geese, pigeons, and doves; grouse, pheasants, and quail. Under the AWR, poultry are further defined as either a farm animal or as a wild animal or exotic animal. All birds that are bred in captivity and used in research are exempt under the animal welfare regulations, including those species listed in the regulatory definition of poultry, regardless of further classification as a farm animal, wild animal, or exotic animal under the regulations.[8]

Welfare Issues

Exposure to predators and wildlife

Chickens have an extensive list of predators, both wild and domestic animals, with young birds being most vulnerable. Depending on the predator, some birds may survive an attack but often are in shock following the event and suffering some degree of trauma. Often owners do not know what type of predator harmed their chickens and most are just found injured. Birds that survive an attack may have physically endured anything from superficial skin damage to extensive mutilation. Animal bite wounds can cause fractures, spinal injuries, skin tearing, and damage to vital organs. Treatment for predator attacks depends on the type of attack, severity of the injury, and the overall health condition of the chicken. Chickens that survive an animal attack may initially appear unharmed, but several days later may get very sick. Backyard poultry are often injured, harmed, or harassed by free-roaming cats. In many cases birds suffer serious and painful injuries including degloving trauma and fractures. Many of these birds are not identified at the time of or right after injuries occur leading to needless suffering. Owners may try home remedies to attempt to doctor these injuries leading to further stress and pain. Many chickens are euthanized because of traumatic injuries and often days after they occur and often due to veterinary costs of treatment.

Lack of willingness to pay for veterinary care or lack of available veterinary services

Many veterinarians will not treat pet chickens and others will provide basic flock medicine, usually culling sick or injured birds. But clients presenting pet chickens may want a different level of care, and avian veterinarians have become the default when seeking care for pets. This does not always mean that owners are willing to pay for veterinary expenses and many will opt for humane euthanasia when presented with large estimates for care and what may be guarded prognoses. It is also strongly recommended that backyard poultry owners or those keeping pet chickens understand the need for necropsy and pathologic evaluation when birds in a flock die. These clients often present remaining healthy birds when other birds have died where the value is in the deceased pet and post-mortem evaluation. This may also be more than owners are willing to spend, and they may not understand the value of protecting the remaining flock. Veterinarians presented with sick or injured chickens and most owners are not willing to pay for veterinary care or attempting treatment at home is emotionally stressful and lowers morale.

Lack of preventive medicine programs

Chicken owners rarely have a preventive medical care plan or even understand what this entails. Birds are not seen for routine veterinary exams, may not have been vaccinated, and most owners rely on medicated feeds for parasite control. Chickens often have lice or mites that require repeated treatment and different nutritional requirements throughout life. Many owners look for cheap feed and few offer supplements. Chicks born at homes of friends and neighbors are often exchanged and these birds are not vaccinated and rarely do owners do any type of quarantine before introduction to the existing flock.[9]

Abandonment of chickens when no longer wanted

Chicken rescues and sanctuaries are often full and often these birds are hard to place when someone tires of or no longer wants to or can keep their chickens. Some animal shelters are faced with abandoned chickens and they are ill-equipped to deal with this situation. Convenience euthanasia is taxing on veterinary staff.

Risk of disease spread and zoonotic diseases

Perhaps the most important aspect of keeping chickens is the potential for zoonotic disease from a human welfare perspective and how a diagnosis could cause an owner to want to depopulate. Owners of backyard chickens and other poultry (e.g., ducks, turkeys) should be made aware of the risks these pets pose to humans and take basic biosecurity steps to protect against zoonotic disease transmission. Few owners are familiar with the terminology and tenants of biosecurity. Backyard poultry can appear healthy but can carry *Salmonella* spp., *Campylobacter* spp.[10], *E. coli*, and avian influenza. Since the 1990s, numerous widespread outbreaks of human *Salmonella* spp. infections linked to contact with backyard chickens have been documented in the United States.[11] Children younger than 5 years, those with weakened immune systems or 65 years of age or older, and pregnant women are at higher risk for serious illness from poultry-borne zoonotic diseases.[12,13] Veterinarians should advise owners of backyard chickens and other poultry about zoonotic risks and how to reduce the risk for disease transmission.

Hands should always be washed thoroughly with soap immediately after touching poultry or anything in their habitat. Hand sanitizer should be used if soap and water are unavailable. Poultry should not be allowed to enter homes, especially areas where food or drinks are prepared, served, or stored. Owners should designate a pair of shoes to wear while caring for poultry and avoid bringing those shoes into the home. Food and drink should not be consumed in areas where poultry live or roam. Birds and other poultry should never be kissed or snuggled, and touching of the face or mouth after handling birds should be avoided until hands can be washed. Equipment and materials (e.g., cages, feed or water containers) used to raise or care for live poultry should be cleaned outside the home.

Public nuisance complaints, pests, and vermin

Chickens can be considered a public nuisance if they violate local ordinances, such as zoning or noise ordinances. For example, some ordinances may prohibit keeping animals that cause offensive odors or require owners to remove animal waste regularly. Other issues reported when keeping chickens in residential areas are noise from roosters, attraction of flies and rodents to coops, poor waste management, violation of homeowner's association regulations and city ordinances, and safety code standard violations and lack of permitting or inspections. Chickens may venture into neighbor's yards or escape to the street. They have been known to graze in neighborhood gardens and plant beds digging up ornamental plants and damaging property.

Keeping chickens by their nature will attract pests or vermin. Rodent and roach control are imperative and often not taken seriously enough. The very nature of feeding, nesting, and maintaining a coop will attract rodents and undesired insects. Owners must remain vigilant in coop hygiene and safe pest control methods.

Inhumane DIY procedures and surgeries

There is a lot of information online, on social media, and in forums instructing lay people on performing medical procedures and surgeries on chickens and always without sedatives, anesthesia, or analgesia. Most notable is caponization or neutering of roosters. From a welfare perspective, the concern is this surgery (technically vivisection) is performed by lay people or non-veterinarians on fully conscious animals that have not been anesthetized nor provided any pain relief and with crude instruments and high risk of complications including death. Caponization is banned in the United Kingdom for humane reasons.[14] Other commonly seen topics include surgical management of pododermatitis, medical management of dystocia, spur removal, and a host of other procedures readily claimed to be doable by untrained lay persons.

Conclusions

Keeping backyard chickens as pets is increasingly popular. Despite their popularity, there are numerous welfare concerns raised when chickens are kept as pets. For additional information on backyard flock production, see Blatchford 2018.[15]

References

1. Elkhoraibi C, Blatchford RA, Elkhoraibi, C. et al. Backyard chickens in the United States: a survey of flock owners. *Poult Sci.* 2014;93:2920-2931.

2. Mace JL, Knight A. From the backyard to our beds: the spectrum of care, attitudes, relationship types, and welfare in non-commercial chicken care. *Animals.* 2024;14:288.

3. Virbac. Suprelorin Information Sheet. Available at: https://vet-us.virbac.com/home/products/in-clinic-use/suprelorin-f-deslorelin-acetate.html. Accessed July 12, 2025.

4. Pohjola L, Rossow L, Huovilainen A, et al. Questionnaire study and postmortem findings in backyard chicken flocks in Finland. *Acta Vet Scand.* 2015;57:3.

5. Singleton DA, Ball C, Rennie C, et al. Backyard poultry cases in UK small animal practices: demographics, health conditions and pharmaceutical prescriptions. *Vet Rec.* 2021;188.

6. Vaught ME, Gladden JN, Rozanski EA, Graham J. Reasons for evaluation on an emergency basis of and short-term outcomes for chickens from backyard flocks: 78 cases (2014-2017). *J Am Vet Med Assoc.* 2019;254(10):1196-1203.

7. Crespo R, Senties-Cue G. Postmortem survey of disease conditions in backyard poultry. *J Exot Pet Med.* 2015;24:156-163.

8. Animal and Plant Health Inspection Service. AWA Standards for Birds. Available at: https://www.aphis.usda.gov/awa/bird-standards. Accessed July 12, 2025.

9. Rasidi EK, Cornejo J. Managing the health of captive flocks of birds. *Vet Clin North Am Exot Anim Pract.* 2021;24(3):521-530.

10. Behravesh CB, Brinson D, Hopkins BA, Gomez TM. Backyard poultry flocks and salmonellosis: a recurring, yet preventable public health challenge. *Clin Infect Dis.* 2014;58(10):1432-1438.

11. Basler C, Nguyen T, Anderson TC, et al. Outbreaks of human *Salmonella* infections associated with live poultry, United States, 1990-2014. *Emerg Infect Dis.* 2016;22(10):1705-1711.

12. Iowa State University. The Center for Food Security and Public Health. Zoonotic Diseases of Poultry. Available at: https://www.cfsph.iastate.edu/Assets/zoonotic-diseases-of-poultry-table.pdf. Accessed July 12, 2025.

13. Centers for Disease Control and Prevention. Backyard Poultry. Available at: https://www.cdc.gov/healthy-pets/about/backyard-poultry.html. Accessed July 12, 2025.

14. Department for Environment, Food & Rural Affairs. 2002. Code of Practice for the Welfare of Meat Chickens and Meat Breeding Chickens. Available at: https://assets.publishing.service.gov.uk/media/5ab525afed915d78b9a45b0f/meat-chicken-code-march2018.pdf. Accessed July 12, 2025.

15. Blatchford RA. Backyard flock production. In: Mench J, ed. *Advances in Poultry Welfare.* Cambridge, MA: Woodhead Publishing; 2018: 339-350.

When Feather Picking is Not Feather Picking

Marli Lintner, DVM

Session #145

Affiliation: From Avian Medical Center, Lake Oswego, OR 97035, USA.

Abstract: The feather is a truly amazing structure—beautiful, strong and functional, yet fragile. A bird with perfect feathers is a joy to behold, while one with broken, missing, or damaged feathers presents a distressing sight. Owners often express frustration and concern, not only about the bird's appearance but also about the associated behaviors: picking, pulling, squawking and signs of discomfort. For veterinarians, feather-damaging behavior is one of the biggest challenges. This can be a result of physiologic, medical, psychological, or management issues. Sometimes it's not the bird's fault.

Molting

Feathers have a natural lifespan and are not designed to last longer than the next molt. Most captive parrots molt twice annually. In the Pacific Northwest, this typically occurs in late winter and late summer, but the timing may vary based on species. Keeping birds indoors can disrupt this cycle. Extended daylight from artificial indoor lighting, consistent indoor temperatures, and year-round high-protein diets all can interfere with natural hormonal cues. Hormonal changes to at least three systems are involved: the thyroid, gonadal steroid hormones and prolactin. More information on what is known about these hormones is available.[1]

These factors can lead to abnormal or skipped molts. A heavy molt can easily be mistaken for feather picking. All feathers come in through a pin or blood feather phase, which can cause a certain degree of discomfort and itching. A heavy molt may resemble the feeling of bark dust in your underwear. Birds normally preen 10-20% of the day. This preening increases during a molt, and with feathers dropping in piles it is understandable for owners to mistake preening for plucking. It is important to be able to differentiate between the two. Molting feathers are usually intact, while plucked feathers can be broken and damaged in the process. Even in a heavy molt, bare skin will rarely be observed. Molting involves the entire plumage to various degrees, while picking can occur in a more localized area and without a corresponding new crop of feathers coming in.

Poor Feather Grooming Routine

All feathers erupt wrapped in a dermal sheath also known as pin feathers. A normal healthy bird will work to preen these off, unraveling and smoothing out the new feathers. The exception being where they can't reach. Preening their own head is a problem. Without an accommodating mate, the owner should be instructed on how to properly preen the feathers open. Feathers left for extended periods of time in their sheaths will tend to degrade. When it finally opens, the color is muted and the structure is damaged. This is especially true for retained tail sheaths. Macaws that have either not learned proper feather grooming routine, which could be a by-product of decades of not being parent-raised perhaps or physical issues that inhibit reaching around to fully preen out their tails, will suffer from this issue. Veterinary staff may need to assist with preening, keeping in mind that skin around feathers is highly innervated and sensitive.

When a feather naturally molts, a new blood feather emerges from the follicle within 7-10 days. Depending on the size of the feather, another 7-10 days is needed to fully develop with the pulp regressing and drying out. At this point the follicle goes dormant. If the feather is damaged but not pulled from the follicle, this damage will remain until the next molt. This is an important concept when treating feather-damaging behavior. The patient may have stopped the behavior but show minimal improvement for months, until the next molt. Instruct owners to monitor feather debris on the cage floor. Once again, many owners mistake normal preening for abnormal behavior.

Handling

Feathers are not designed to be handled. Unlike fur, bird plumage can be damaged by excessive handling. Green feathers show damage most dramatically and the green reflective coating can wear off. Pressure points on the elbows, shoulder blades, vertebrae and pelvis can be worn black. Feather edges will eventually also wear black with time. Amazon parrots are notorious for skipping molts and while otherwise healthy, may have very worn dark plumage. The currently popular green cheek conure and its color crosses are another good example.

Oil and grease present a unique problem in feather care. The beak as a preening tool is unable to remove grease from plumage but can spread it around. A regular water bath is insufficient to remove oil and grease. Heavy grease will break down the structure of fine down and cause permanent damage that not even a good detergent bath will be able to clean and repair. Sources of oil can include:

- human hands with lotions and moisturizers,
- clothing that contains fabric softeners and deodorants, and
- cosmetics including makeup and lipstick.

These substances attract dirt and dust, which adheres to feathers and leads to dull, grey-brown, dingy plumage.

Unzipped Feathers

When the microscopic Velcro structure of the feather gets worn or packed with grease, dirt or powder down, the barbules lose their cohesion. The plumage looks unzipped. This can be a full body look but is most common on tail and torso plumage. Major improvements can often be obtained with a full detergent bath (Dawn is often used for grease) followed by a blow dry. Be sure the dryer is on medium-high heat and blown from a safe distance to avoid drying out or burning the skin. The heat is also important to bring the barbules up to function again. Back blow to fluff up the down.

Wing Trims

Severe wing trims often result in tail feather damage. Trimming wing feathers does not reduce the bird's natural instinct to fly. As the bird tries to fly and obtain lift, the tail moves under the body for support. When crashing down, feathers get bent and kinked. As the bird tries to preen out the damage the kink gets worked and splits even further until it eventually breaks. Owners may inadvertently mistake this for feather picking. Insidiously as each feather breaks off, the tail loses more lift, worsening the fall until eventually the whole sheath of tail feathers is lost. Continued falling may result in rachis being pushed into the tail head such that it ruptures through the follicle base. This can result in feather picking and tail mutilation from discomfort.

Baldness

Feather loss on the top of the head and back of neck is only rarely caused by the bird itself. There are examples of pulling cheek plumage with the feet, hormonal male baldness and skin issues, but most frequently this is the result of picking by a cage mate. A separation period of at least one month can be used to determine whether plumage regrows, while also monitoring for any stress-related behaviors that may develop during this time. Pin feathers have blood still in the shaft and blood tastes salty. When a bird craves salt, this could have reinforced the plucking by the mate to become a habit and thereby might be difficult to break.

Scarring

Feather follicles are tough structures but even these have their limits. When repeatedly pulled, this eventually can lead to scar tissue resulting in the inability of a feather to regrow. Chronic bare patches, often on the torso, may not be the result of current plucking, but rather reflect plucking in a more distant episode. Placing a collar for 4-6 weeks should make this apparent, with careful consideration of the need for this intervention prior to placing the collar. If no new growth occurs, continuing in a collar at this point is not productive and could be detrimental.

Chronic skin damage from self-mutilation or trauma can leave follicles macerated into the dermis. When spring or fall molting cycles occur this tissue will be activated. Unable to construct a full feather, this tissue will make subcutaneous pearls of feather keratin (i.e., feather cysts), which can be uncomfortable and can cause a new round of picking and mutilation. Even with a collar placed and the bird unable to reach the area, keratin will work its way to the surface and result in skin irritation.

Conclusions

Not all feather damage is self-induced. It is important to differentiate the cause before a treatment course is set. Birds are tougher than we give them credit for, both physically and mentally. I strongly believe most behavioral picking issues are the result of discomfort and irritation that cannot easily be diagnosed with a physical examination or extensive testing. Living a sedentary lifestyle, eating an unnatural diet, and aging can all lead to a variety of aches and pains, numbness and tingling. We should closely observe birds' feather-picking patterns and resulting damage. Our patients are trying to tell us something.

References

1. Dawson A. Avian molting. In: Scanes CG, ed. *Sturkie's Avian Physiology.* 6th Ed. Waltham, MA: Elsevier Academic Press; 2015:907-918.

What is the International Union for the Conservation of Nature's Species Survival Commission Wild Parrot Specialist Group and Why Should an Avian Veterinarian Care?

Patricia J. Latas, DVM, MS

Session #125

Affiliation: From IUCN SSC Wild Parrot Specialist Group, 7223 E. Camino Valle Verde, Tucson, AZ 85715, USA.

Abstract: This is a basic introduction to the International Union for the Conservation of Nature, Species Survival Commission and the brand-new Species Survival Commission Wild Parrot Specialist Group (WPaSG). Included are conservation functions of the WPaSG to improve the plight of parrots in the wild, the urgency and diversity of work that needs to be accomplished, and an introduction for avian veterinarians to the exciting, demanding and interesting possibilities for participation in a very worthy cause.

Introduction

The International Union for the Conservation of Nature (IUCN) Species Survival Commission (SSC) Wild Parrot Specialist Group (WPaSG) was formally hatched on World Parrot Day, 31 May 2024.

Parrots are in trouble. Overexploitation for the pet trade, loss of habitat, human incursion, disease, impact of invasive species, and changing climate are all contributing to the decline of most species of Psittaciformes across the globe. They need a strong voice in conservation. The IUCN SSC provides the umbrella under which the WPaSG is commissioned.

The WPaSG is structured to detect, monitor and manage, advise and inform decision-makers, funders, scientific communities and interested members of the public, communities, and other stakeholders about the perils impacting this important and adored taxonomic group world-wide. Global geographic subgroups assess the status of parrot populations within their regions, input data and research to the Redlisting Authority to evaluate status of each species (Critically Endangered, Endangered, Vulnerable, Near Threatened, Least Concern, Extinct), plan actions to initiate and improve conservation within and between regions, inform regional governmental policy, and develop multilateral collaborations among all stake holders. Thematic subgroups inform and support the regional assessment on broader umbrella topics, such as Trade and Use and Health and Welfare of wild parrots.

We have been working very hard this year to lay a strong and lasting foundation, planning basic functions and administration and amazing projects that will re-define parrot conservation. This presentation introduces the WPaSG and outlines its goals and dreams. Specifically, we discuss why a formal parrot conservation group is important to avian veterinarians, and how avian veterinarians can contribute to the work of WPaSG.

The Plight of Parrots in the World

Hundreds of thousands of parrots are caught for the illegal and unsustainable trade of wild birds. Human incursion on native habitat limits their food, water, and nesting ability. They are increasingly vulnerable to climate change. Disease from agriculture, humans and the pet trade threaten epidemics. Invasive species kill or squeeze out vulnerable populations.

Why do parrots need a specialist group?

From a tiny buff-faced pygmy parrot to a palm cockatoo, any member of the Order Psittaciformes is immediately recognizable as a parrot. Despite being the most recognized and sought-after group for pets, relatively little is known about many species in the wild. Parrots are an ancient group of birds, and their needs are complex and specialized. The IUCN SSC has recognized that they need their own specialist group for conservation.

Parrots are among the most threatened of all groups of birds with close to one in three classified as Threatened with extinction on the IUCN Red List. Parrots are inherently vulnerable to threats because:

- many species have naturally restricted ranges,
- there is large scale habitat loss,
- some are single-country/island endemics,
- many species of parrots are specialist feeders or nest-cavity dependent, and
- parrots have low reproductive rates and cannot replace the losses.

To add to the urgency, it is all happening at once.

Why wild parrots?

Parrots have a unique relationship with people, having been kept in captivity for millennia. Incorporating wild within the group name reflects the group's focus on the conservation of populations in the wild. Psittacine birds in human care have many organizations which speak for them; wild parrots have very few.

Who We Are

With a broad-based, diverse membership drawn from the global parrot research and conservation community, the WPaSG is inclusive and welcoming of a diversity of skills, experiences, and specializations, including both in situ and ex situ expertise. It provides a cohesive structure to address the many needs of parrots in peril as it is set to support conservation risks assessments for wild species and populations, prioritize planning needs, initiate, enable, and facilitate appropriate and inclusive conservation planning.

What are the IUCN, SSC, and Specialist Groups? How does all that work to save parrots?

The IUCN is an enormous and complex organization, which can be thought of as the equivalent of the United Nations for conservation. It includes several major divisions and commissions, including the Species Survival Commission. The SSC is a science-based network of more than 10,000 volunteer experts from almost every country of the world, representing taxa from slime molds to elephants, geological formations, oceans, and many thematic groups. There are more than 180 SSC Specialist Groups, and Red List Authorities, Conservation Committees and Task Forces.

What does the SSC Wild Parrot Specialist Group do?

The IUCN SSC WPaSG brings together expertise and experience from around the world to support a coordinated response to the most pressing and complex problems facing parrots in the wild. We employ conservation planning to prevent extinction and deterioration of conservation status of parrots around the world

Our goals

The goals of the WPaSG are as follows:

- Bring together diverse expertise to tackle the problems facing the parrots in the wild.
- Prevent the extinction and deterioration of conservation status of parrots around the world.
- Assess the needs and knowledge gaps in wild parrot conservation.
- Produce formal conservation action plans to address multifaceted and complex problems.
- Act to mitigate and halt the threat of extinction of this beautiful and interesting taxon.

Assess-Plan-Act (A-P-A)

We aim to explore all available conservation tools to encourage and support conservation action in alignment with the IUCN One Plan Approach: Assess-Plan-Act, Network, and Communicate (Fig 1).

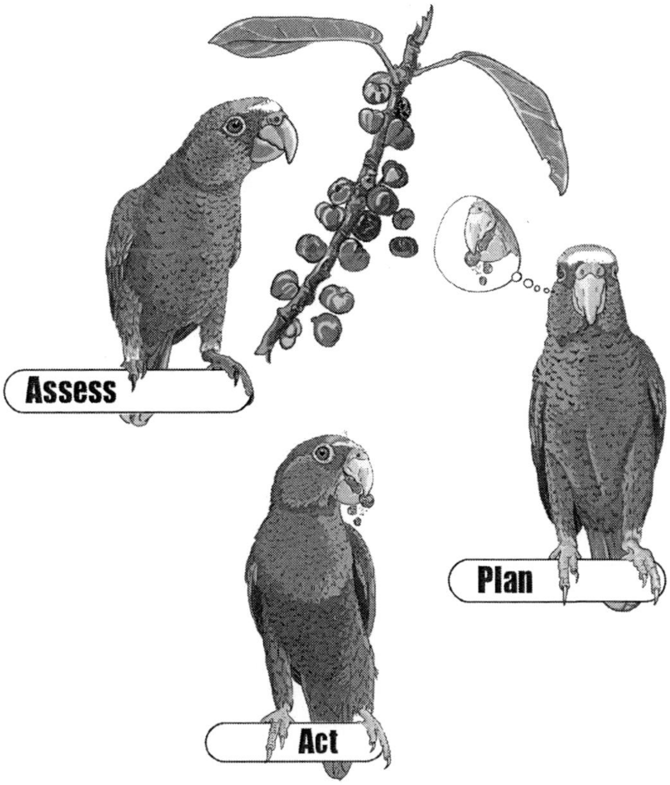

Figure 1. Species Survival Commission Specialist Groups contribute to conservation action in alignment with the IUCN One Plan Approach: Assess-Plan-Act, Network, and Communicate. Illustration of A-P-A using a thoughtful white-fronted parrot (*Amazona albifrons*) by Dr. Pat Latas.

The WPaSG is made up of geographic and thematic subgroups (Fig 2). One subgroup is Health and Welfare, and our experts there concentrate on the impact of health and disease on wild parrot species, populations, flocks and individuals and the effects on conservation planning. Proper evaluation is important to conservation planning for in-situ and ex-situ translocations, relocations, repatriation and rehabilitation and release to the wild. This subgroup is where avian veterinarians can have the greatest impact and influence. The threat of disease can move a species very rapidly from Least Concern to Critically Endangered, or even Extinct.

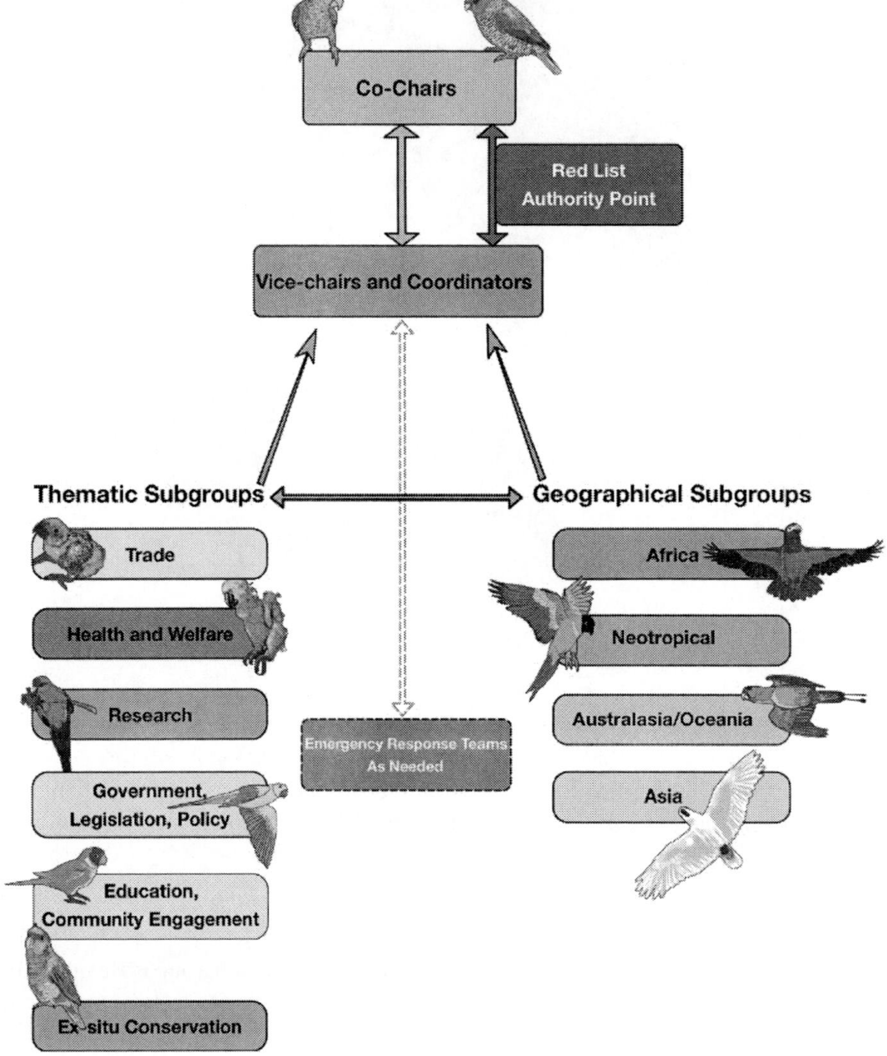

Figure 2. Because psittacine bird conservation is complex, the IUCN SSC Wild Parrot Specialist Group is organized into geographic and thematic subgroups. Veterinarians can contribute valuable work in any subgroup or level of leadership but are especially needed in Health and Welfare. Organizational chart illustrated by Dr. Patricia J. Latas.

Where We Are Going and Why Avian Veterinarians Should Care

Veterinarians with knowledge and passion for wild parrot conservation are welcome in the WPaSG. All people who work with parrots in human care should be concepted about their wild cousins. The extinction of 100 wild parrot species would greatly diminish the world.

Parrots are in trouble and your knowledge is valuable

The veterinary world and conservation world can deeply benefit from partnership. Avian veterinarians are not common, avian veterinarians with wildlife experience are rare, and avian wildlife veterinarians with wild parrot experience are nearly non-existent. Most veterinarians are non-conversant in conservation matters of in-situ wild parrots.

Conservation efforts can inform and improve the lives of parrots in human care

Knowing more about the requirements that wild parrots need to maintain health and well-being in the wild can improve understanding of the needs of parrots in human care; and knowing what wild parrots require can be an active field of veterinary interpretation and input.

Trade and use of wild parrots

Veterinary expertise is essential for evaluating, ameliorating and reversing the ongoing horrific impact of illegal wildlife trafficking on health, welfare, and conservation of wild populations.

Health and Welfare of Wild Parrots

Veterinarians are invaluable for proper evaluation and assessment of parrot health and welfare. For example: standards for short term care, assessment of welfare, guidelines for research, rehabilitation and field work, and assessment of impact of conservation action plans on health and welfare. Projects that are in the WPaSG plan for 2025 and the next quadrennium include the following:

- Development of a standardized welfare examination for wild parrots
- Guidelines and Standards for parrot welfare in fieldwork and field investigations
- Placement of non-releasable parrots from trafficking confiscations into conservation planning
- Recommendations for emergency preparation and planning for HPAI
- Impact of climate change on wild parrots' health and well-being
- Guidelines for including health and welfare considerations in conservation planning
- Long-term monitoring and database of disease in wild parrots

Collaborations

We have an exciting opportunity to collaborate with the Human-Wildlife Conflict and Coexistence Specialist group, to provide a chapter on parrots. The Health and Welfare subgroup is also in contact with the Wildlife Health Specialist Group and multiple other bird Specialist Groups.

Resources

Membership

If you have an interest in applying for membership:

The IUCN SSC Application for Membership is available at https://www.iucn-wildparrot-specialist-group.org/home/application-for-membership-to-iucn-ssc-wpasg

Websites

Wild Parrot Specialist Group: https://www.iucn-wildparrot-specialist-group.org/

WPaSG Subgroups: https://www.iucn-wildparrot-specialist-group.org/subgroups

WPaSG Mandates: https://www.iucn-wildparrot-specialist-group.org/home/iucn-ssc-wpasg-mandate

Social media

WPaSG on LinkedIn: https://www.linkedin.com/in/iucn-ssc-wild-parrot-specialist-group/

WPaSG on Bluesky @wildparrotsg.bsky.social:

https://www.iucn-wildparrot-specialist-group.org/home/iucn-ssc-wpasg-mandate

Acknowledgments: The WPaSG appreciates the persistence, inspiration and support of the IUCN SSC Chair, Dr Jon Paul Rodriguez. We are also indebted to Mr Sam Ivande, Bird Conservation Coordinator of the Global Center for Species Survival at Indianapolis Zoo, for his kind help and advice whenever needed. And of course, we acknowledge all the wild parrots and the people who work with, love and protect them.

A Photographic Image Says More Than a Thousand Words: Specific Instrumentation and Techniques in Exotic Animal Ophthalmic Photography

Rüdiger Korbel, Prof, Dr med vet, DipECZM (Avian)

Session #206

Affiliation: From Center for Clinical Veterinary Medicine, Clinic for Birds, Small Mammals, Reptiles and Ornamental Fish, University Ludwig Maximilian of Munich, Germany.

Abstract: Appropriate photographic imaging, including the anterior and posterior eye segment, is indispensable in ophthalmology. Thus, discussions on the best camera are a permanent, inevitable topic within ophthalmology. From a historical point of view, ophthalmologic photography and specific cameras for eye photography may be seen against the background of (1) invention of photography in 1839; (2) the invention of the ophthalmoscope by Helmholtz in 1851; (3) the first fundus photography in man by Howe, Jackman and Webster in 1886; (4) the first drawings of animal (avian) related physiologic fundic conditions by Woods in 1917; and (5) the first fundus photographs using a camera in animals in 1930 by Leonardi.[1-3] Besides the necessity for high image resolution and a broad exposure latitude range, which for a long time was superior in classical analogue film-based techniques (e.g., using the legendary Kodakchrome® material) over digital sensors, the general technical challenge for designing an appropriate camera for ophthalmologic use becomes more important (especially in fundus photography) the smaller the eye response and the pupillary diameter is because of the necessity for an almost coaxial illumination and viewing beam. Therefore, especially in exotic animal eye photography, finding an appropriate camera is not easy. Based on a historical view with milestones in general and specialized camera development for the ophthalmologist, a review from classical analogue equipment and procedures to most recent developments in digitally based camera equipment and practical photographic recording techniques to avoid common mistakes in archiving appropriate image results is provided.

References

1. Woods CA. *The Fundus Oculi in Birds Especially as Viewed by the Ophthalmoscope.* Chicago, IL: The Lakeside Press; 1917.

2. Korbel RT. Avian ophthalmology. Air sac perfusion anaesthesia, ocular disorders and ophthalmological photography, Vet Habil Thesis. Munich, Germany: 1996.

3. Korbel RT. Specific techniques for ophthalmic photography in exotic animals. *Proc Annu Conf Eur Coll Vet Ophthalmol.* 2024.

Identifying the Normal Nasochoanal Microbiome of Cockatiels (*Nymphicus hollandicus*) Utilizing Next-Generation Genomic Sequencing

Nicole Sands, DVM,
Shachar Malka, DVM, DABVP (Avian Practice),
Gina Vecere, DVM,
Margaret Lee, DVM,
Kaylie Zapanta, PhD,
Janina A. Krumbeck, PhD

Session #105

Affiliation: From Long Island Bird and Exotics Veterinary Clinic, 333 Great Neck Road, Great Neck, NY 11021, USA (Sands, Malka, Vecere, Lee), MiDOG LLC, 17062 Murphy Ave., Irvine, CA 92614, USA (Zapanta, Krumbeck).

Abstract: Nasochoanal flush samples from 27 client-owned cockatiels (*Nymphicus hollandicus*) (5 females, 21 males, 1 unknown; juveniles: n = 7, 8 weeks to 18 months, mean 1.38 ± 1.1 years; adults: n = 17, 18 months to 14 years, mean 4.91 ± 3.9 years; geriatrics: n = 3, > 14 years, mean 23.67 ± 2.1 years) were analyzed using next-generation DNA sequencing (NGS) with metagenomic analysis. Health status was confirmed by physical exam and absence of clinical signs for 2 weeks prior to collection. Flushes were performed with 3 ml sterile saline per nostril using a standardized protocol and collected with a MiDOG Animal Diagnostics LLC kit containing sterile DNA preservative and swabs. Rigorous quality control was ensured by processing multiple positive and negative controls alongside samples, including extraction and library preparation negatives, no-template controls, and commercially available mock microbial communities.[1-5] Automated liquid handling minimized human error. Dominant bacterial genera included *Corynebacterium*, *Staphylococcus*, and *Pasteurellaceae*; common fungi were *Cladosporium*, *Alternaria*, and *Aspergillus-Eurotium*. Females had increased *Streptococcus suis*, *Massilia putida*, *Alcaligenaceae* spp., and *Ureaplasma* spp., while males had more *Sphingobacterium* spp., *Lactobacillus*, and *Comamonas terrigena*. Microbial diversity varied by age using Shannon, Simpson, and Observed Species indices, which assess richness and evenness; increased diversity with age may reflect uneven communities dominated by select taxa rather than increased species richness. These findings highlight age- and sex-related variation in the cockatiel nasochoanal microbiome, providing a foundation for future avian respiratory microbiome research and clinical applications. Elaboration of how these results compare to each other is further elaborated in the presentation and future publication.

References

1. Tang S, Prem A, Tjokrosurjo J, et al. The canine skin and ear microbiome: a comprehensive survey of pathogens implicated in canine skin and ear infections using a novel next-generation-sequencing-based assay. *Vet Microbiol.* 2020:247:108764.

2. Krumbeck JA, Turner DD, Diesel A, et al. Skin microbiota of Quaker parrots (*Myiopsitta monachus*) with normal feathering or feather loss via next-generation sequencing technology. *J Exot Pet Med.* 2022;42: 26-34.

3. Vecere G, Malka S, Holden N, et al. Comparison of ear canal microbiome in rabbits with and without otitis externa using next generation DNA sequencing. *J Exot Pet Med.* 2022;42:35-41.

4. Melgarejo T, Oakley BB, Krumbeck JA, et al. Assessment of bacterial and fungal populations in urine from clinically healthy dogs using next-generation sequencing. *J Vet Intern Med.* 2021;35(3):1416-1426.

5. Damerum, A, Malka S, Lofgren N, et al. Next-generation DNA sequencing offers diagnostic advantages over traditional culture testing. *Am J Vet Res.* 2023;84(8):ajvr.23.03.0054.

Section 10

Management, Business, and Potpourri II Session

Richie Burdeaux, Jr, DVM, DABVP (Avian Practice)
Session Chair

Recession Proof Your Small Animal or Avian Practice: For Owners and Associates

Len Donato, VMD, DABVP (Avian Practice)

Session #147

Affiliation: From Radnor Veterinary Hospital Wayne, PA 19087, USA, the Department of Clinical Sciences and Advanced Medicine, The University of Pennsylvania School of Veterinary Medicine, Philadelphia, PA 19104, USA.

Abstract: During the COVID-19 pandemic, clinical veterinary medicine did very well financially. As time moves on from the pandemic, veterinary medicine is seeing that client visits are beginning to drop to pre-pandemic levels. When the number of visits decreases, a decrease in gross revenue is typically observed. Veterinary medicine has not experienced that yet because inflation caused fees to increase significantly, thereby increasing the average client transaction (ACT). Raising fees alone will not sustain meaningful growth, and in fact contributes to practice stagnation and regression. Understanding a variety of strategies to promote more visits and increase ACT will benefit the financial health of the hospital overall. Exotic animal practitioners must be savvier than dog and cat counterparts as a majority of the patients do not get vaccines, flea and tick medication, and heartworm prevention. Those products and services help gross revenue by increasing the number of transactions and ACT. Strategies affecting the number of transactions and the ACT at the same time will benefit the practice owner, the non-owner veterinarians, and staff. Increasing the number of transactions involves veterinarians maximizing time management, properly following up and providing patient reminders, along with a comprehensive internal and external marketing campaign. ACT is addressed by appropriate fee increases, managing lost fees, and expanding services and skills. Knowing these strategies and how they affect gross revenue, veterinarian salary, and the health of the clinical environment will minimize the effects of a recession and will likely result in growth financially and professionally.

Introduction

Recessions are defined by the National Bureau of Economic Research (NBER) as a significant decline in economic activity that is spread across the economy and last for more than a few months.[1] Historical data demonstrates the most recent prolonged recessions were The Gulf War Recession (7/90-3/91), The Dot Com Recession (3/01-11/01), The Great Recession (12/07-6/09), and The COVID Recession, which is on-going.[2] When recessions are prolonged they are defined as depressions, but there is debate as to definition.[1,2]

It is widely thought that veterinary medicine is recession-proof, particularly with the most recent experience. In reality there have been industry-wide effects from the beginning of the COVID-19 pandemic. Food-animal supply chains were dramatically affected,[3] zoo income was reduced by lack of visitors and donations,[4] and a large number of veterinarians retiring from the profession or going part time has even made the everyday news.[5] The Great Recession in 2008 resulted in decreased gross revenue for veterinary practices and a significant decrease in net that was previously 35-45% and fell to 10% or below for some practices. The effects of that are still felt today as net revenue for a non-specialty practice reached 15%.[6] Non-owner veterinarians also feel recessions as numbers of clients typically decrease as well as their number of visits (N) and how much they spend at each visit (i.e., average client transaction, ACT). Pro-sal is a common method to pay associates which is based on revenue generated by the associate. Since N and ACT decrease, that affects additional pay over base salary. This makes

Pro-sal a bit riskier for the non-owner veterinarian during a recession. Also, if a practice has a decrease in net revenue, there are fewer resources that could go to additional staff, equipment, or capital improvements for the hospital building, all of which make for a better-quality hospital and work environment.

Recently our profession felt good financially because of the COVID and post COVID bump, but data show the number of visits decreased by 4% in 2022[7] and that trend is continuing. As seen from the recessions listed above, it is not if a recession occurs. The questions are: how many recessions will be experienced in a career, and how severe will they be? What can be done to maintain income during those times? There are strategies to make sure a similar drop-off is not seen as has been seen with other recessions. It is better to implement and practice the strategies now, rather than when the economic environment is not healthy.

Understanding Profitability

Veterinary practices should be managed to be profitable. This is a function of gross revenue and expenses to run the practice. While this simplifies what a practice needs to achieve, executing is a big challenge. Focusing on gross revenue, but not ignoring the expenses, will make for a profitable practice. Expenses need to be monitored as most will be within a percentage range of gross revenue. As an example, as the number of staff increases, so does payroll. The goal behind the increased staff is to accommodate more patients, which increases gross revenue. In this example, as the expense of staff increases, so does revenue. There are two exceptions to expenses where a difference can be made; these are discussed later. The focus should be on revenue generation.

Managing gross revenue has the greatest impact on profitability. To understand income there is a simple formula:

$$\text{\# of transactions (N)} \times \text{average client transaction (ACT)} = \text{gross revenue}$$ [8]

It is important to understand the components that make up both factors in the equation, and the best ways they can be affected to produce higher income or gross revenue. Increasing ACT has a greater impact on gross revenue than N. Both are important to manage. This is why specialists typically generate more than general practitioners as one transaction could be in the thousands regularly.

One of the easiest ways to increase revenue through ACT is to raise prices. This has occurred in the last few years because of the extreme increase in N during the pandemic. It was meant to handle overall increases in payroll, increases due to inflation, and to become more profitable as a practice. The continued price increases during a recession eventually catch up and surpass what owners are willing to pay.[7] A precipitous drop in N occurs because the high prices keep clients from scheduling. This was observed in the Great Recession as the years before were quite prosperous. This isn't to say prices should not be raised. It should be a thoughtful process understanding the value of services, inventory costs, lab fees, and other expenses. When fees are raised to a proper level, there will be resistance from just a few clients.[9] This will serve as a good barometer for you. If there are complaints about fees multiple times per day, they are genuinely too high. Also, raising fees too high will weed out a lot of clients during a recession.

Comprehensive Raising of ACT

Yes, raise fees

As outlined above, make this a thoughtful process with good intentions.

- Exams and surgery, generally value based.
- Lab fees, generally 2 × cost ± a technician fee. Monitor monthly as fees from reference lab or materials for in-house equipment may increase suddenly.
- Inventory, generally 2 × cost ± a pharmacy fee. Monitor invoices as they arrive for fee increases.

Practice quality medicine

This recommendation sounds obvious. Practicing quality medicine is the greatest part of our careers, as not only does it satisfy our desire to be an avian or exotics veterinarian, but it also increases ACT. It allows us to offer a variety of services. Consider adding a new service this coming year.

- Preventive care.
- In-hospital diagnostics.
- Advanced imaging, radiology, ultrasound, CT, and endoscopy.
- Soft tissue surgery and orthopedics.
- Behavior consultations.
- Senior wellness programs.
- Many others; you know what you like!

Address missed fees

Missed fees refers to fees not charged by the veterinary practice for services that were performed. This is a long-standing problem that can lose a practice 5-10% of gross revenue a year.[10] A practice that generates $2,000,000 in gross revenue can be missing up to $200,000/year.

- Identify areas where fees may be missed and address.
- Institute checks-and-balances systems that work for the doctors and staff.

Time management

Managing time sounds like it is obvious and intuitive. It is not. It must be practiced like sports or playing an instrument. Having more time allows you to see more cases or at least offer more services, which can affect ACT too. So how should time be managed?

Delegate

Delegation is an art and a skill. The art is you must be able to trust others.[8] The skill is that requires follow up to be sure the task was completed according to hospital procedures. Even though you delegate a task, it does not mean you are not to check on said task and provide feedback. It is the most forgotten step of delegating. As a veterinarian one should perform exams and surgery and follow up with the clients on patient care. No one else can do those tasks. Everything else can be delegated like phlebotomy, taking radiographs, marketing, scheduling appointments, and collections.[9] Delegation builds trust. To get the best from your team and make them understand that they themselves are a valuable member of the team of your team, trust is key to being a successful medical director or leader.[11]

Have the right people

Have proper staffing type, numbers, and skills in the practice. Your external team is just as important. These are the payroll company, accountant, attorneys, and financial planner.[8] Performing these financial and legal tasks on your own will end just as badly as if you tried to be the only one handling all the patients and their needs.

Evaluate scheduling

Scheduling becomes a topic of debate. There are many scheduling strategies ranging from straight booking of timed appointments to high density schedule with catch-up slots. There are even schedules with overlapping appointments considering support staff contact and doctor contact for each patient. Scheduling considers the veterinarians and staff as a whole entity. Evaluate your schedule to determine if it is efficient and fits the practice philosophy.

Be paperless or paper light

Time savings looking for files alone is worth every penny spent on an electronic medical record system. This also aids in addressing missed fees.

Client reminders

These reminders are not for just annual examinations and vaccines. Set reminders for blood work when monitoring chronic patients, behavior visits for young birds to make sure clients get into good habits early, examination and husbandry reviews for reptiles before winter, and anything important to the specific patient.

Recall system

Call clients far more than if they were in for a sick visit. Call clients or have staff call for weight checks, diet conversions, recheck reminders, blood work monitoring, setting spay/neuter, post-surgery check in, and whatever you think is remotely important. It helps with owner compliance, checks on the pet, increases visits, and builds trust, which results in return visits.

Marketing

Managing a successful veterinary practice involves strategic planning within and between several functional areas of the business; just a few of those areas include finance, human resources, inventory, operations, and marketing. The hospital's marketing strategy should provide a clear depiction of the specific elements that set your practice apart from the crowd.[12] The marketing strategy's key elements ideally include defined internal and external marketing components, brand messaging, print and digital-based marketing tools, word-of-mouth communications, and social media.[12] All these marketing mediums work together to reach and engage different types of consumers that would benefit from your services. Having a general understanding of the different types of marketing tools and their costs will help a practice owner determine specific marketing mediums to focus on, relative to their overall return on investment.

Expenses

As previously mentioned, expenses are not the focus when evaluating profitability and solvency.[9] They cannot be ignored and must be managed, but your profitability will not dramatically change by continually switching companies for electricity. An office manager or hospital administrator will make sure those expenses are kept in check, but not a lot of resources should be placed on expenses with a couple of exceptions.

Payroll

The cost of labor should be calculated as a percentage of gross revenue. It will be close to 20% for veterinary salaries and 20% for non-veterinary salaries, totaling 40% of gross revenue.[13,14] The office manager or hospital administrator should monitor overtime. That is the one area payroll can get out of hand. If overtime is managed and payroll is above 40%, review the amount and type of staff to find where costs are going.

Inventory management

Inventory or Cost of Goods Sold (COGS) is an expense that can get way out of control. There are a few ways this can happen. The first is that there no controls over how much is ordered or fear buying. There can be a lot of anxiety about running out of a product. Use ordering history to determine how much is needed over a given period. While taking longer than it has in the past, products still arrive in a fairly timely manner. Learn the patterns of delivery from your distributors. The second is that an owner of a practice may just want what they want. This can be challenging to overcome as everything and anything gets ordered. Cost of goods should be about 17.5% of revenue.[14] If it is above that, look for excess inventory, duplication of medications, lack of use of an online pharmacy, and even theft.

Conclusions

These outlined concepts are not challenging to understand. Implementation is the most challenging part. The strategies and tactics above will prevent having a challenging time during a recession. They require practice while growth and gross revenue is high so when it is not, you are prepared.

References

1. National Bureau of Economic Research. Frequently Asked Questions. Available at: https://www-nber-org. proxy.library.upenn.edu/research/business-cycle-dating/business-cycle-dating-procedure-frequently-askedquestions. Accessed March 18 ,2023.

2. Rodeck D. What is a recession? *Forbes.* 2022;7.

3. Hayes DJ, Schultz LL, Hart CE, KL Jacobs. A descriptive analysis of the COVID-19 impacts on U.S. pork, turkey, and egg markets. *Agribusiness.* 2021;37(1):122-141.

4. Fine L, Barnes C, Niedbalski A, Deem S. Staff perceptions of COVID-19 impacts on wildlife conservation at a zoological institution. *Zoo Biol.* 2022;41(3):234-243.

5. Columbia Broadcasting System (CBS). Available at: https://www.news9.com/story/62900baa45556307273f185c/future-of-veterinarian-care-uncertain-as-many-begin-to-retire. Accessed March 18, 2023.

6. Felsted KE. What is your veterinary clinic worth? *DVM360.* 2021;53(3):54.

7. Zirkle K. Reading the signs of healthy veterinary practices. *Trends.* 2022;15(1):26-28.

8. Donato L. Avian practice management: the art of delegating. *Proc Annu Conf Assoc Avian Vet.* 2008.

9. Miller E. How to salvage your business when you are in a cash flow crunch. Available at: https://econologicsfinancialadvisors.com/2020/03/11/. Accessed May 22, 2025.

10. Fuhrman J. Don't misfire on missed charges. *Today's Vet Bus.* 2021;5(1):46-47.

11. Roark A. What the best medical directors have in common. *Today's Vet Bus.* 2025;9(1):76-80.

12. Dewilde C. Outmarketing the big guys. *Today's Vet Bus.* 2025;9(2):86-87.

13. O'Neil T. 5 Ways to better manage veterinary hospital labor. *Today's Vet Bus. 2019;2(6):12-16.*

14. American Veterinary Medical Association. The Economic State of the Veterinary Profession 2019. Available at: https://ebusiness-avma-org.proxy.library.upenn.edu/files/productdownloads/HR-Secured-ECOStateofProfessionReport-20190927.pdf. Accessed March 18, 2023.

Management of Inadequate Molt in Two Grey-Backed Thrushes (*Turdus hortulorum*) with Photoperiod Modification

Jocelyn Choo, BVM&S (Hon), MRCVS, PGCert (EXAP)

Session #198

Affiliation: From Mandai Wildlife Group, 80 Mandai Lake Road, Singapore 729826.

Abstract: Abnormal molt and poor feathering are commonly seen in birds kept outside their normal distribution rate, particularly when birds with a 12:12-hour photoperiod are kept in temperate climates with a variable photoperiod, or vice versa, as the molting process tends to be regulated by the photoperiod. Two grey-backed thrushes (*Turdus hortulorum*) kept in an open aviary were presented for poor plumage due to a chronic inability to molt for the past 2 years. Levothyroxine treatment (levothyroxine 0.02 mg/kg PO q24h for 14 days) was given in an attempt to induce molting but showed limited success. The individuals were moved to a different aviary with more natural light access. After 9 months, the environmental conditions improved contour feather quality, except for the remiges and rectrices in both birds. Hormone therapy was attempted again with the use of melatonin (0.5 mg/kg PO q24h for the first 14 days), GnRH implant (4.7 mg SC on day 15) and thyroxine (20 µg/kg PO q24h from day 15 to 28) but limited improvements were seen. Two months after the last dose of thyroxine, conservative management through photoperiod manipulation using a smart timer in an enclosed environment was tested over a period of 6 months. Multivitamins and a higher calorie diet were given throughout the period. The molting of the remiges and rectrices in both individuals were achieved. This report demonstrates that photoperiod modification can be performed in a clinical set up.[1,2]

References

1. Meijer T. Photoperiodic control of reproduction and molt in the kestrel, *Falco tinnunculus*. *J Biol Rhythm.* 1989;4(3):351-364.

2. Chilgren JD. Effects of photoperiod and temperature on postnuptial molt in captive white-crowned sparrows. *Condor.* 1978;80(2):222-229.

Biomarkers of Inflammation and Novel Antibody Testing to Aid in the Diagnosis of Aspergillosis

Carolyn Cray, PhD,
Adriana Callico, MS,
Jeny Soto, MS

Session #193

Affiliation: From Division of Comparative Pathology, Department of Pathology & Laboratory Medicine, University of Miami Miller School of Medicine, Miami, FL 33136, USA.

Abstract: Antemortem diagnosis of avian aspergillosis is challenging despite advances in serodiagnostics over the past 20 years.[1] In the present study, samples from penguins, a species with known predilection to aspergillosis, have been examined to define the utility in diagnosis and prognosis using the following tests: Afmp1p antibody reactivity, galactomannan, gliotoxin, protein electrophoresis, and acute phase protein quantitation. A target of the humoral immune response, Afmp1p, was examined via ELISA using a recombinant antigen.[2] Using samples from penguins with confirmed (n = 18) and probable infection (n = 13) versus normal penguins (n = 33), the sensitivity was 94.4% and the specificity was 57.6%. Significantly higher reactivity was observed in both groups versus clinically normal penguins ($P < 0.0001$). Interestingly, increased reactivity also was noted in normal adult versus juvenile penguins ($P < 0.0001$). Testing included capillary zone electrophoresis, which is a recent technological improvement over agarose electrophoresis offering a resolution of seven protein fractions in this species.[3] Versus clinically abnormal but non-aspergillosis cases, penguins with aspergillosis had significant decreased A/G ratio ($P < 0.0001$) related to increased alpha 2 and gamma globulins as well as an increase in beta 1 globulins which was not observed in the other clinical group. Whereas galactomannan testing had poor sensitivity in penguin cases, gliotoxin was detected in the majority of confirmed cases.[4] Lastly, assays for serum amyloid A and haptoglobin were examined for cross reactivity to penguin acute phase proteins. While SAA reagents do not appear to be cross reactive with penguin species, HP may have utility as a diagnostic or prognostic marker.

Acknowledgments: This project was supported by the African Penguin Species Survival Plan as part of the Association of Zoos & Aquariums' Saving Animals From Extinction conservation initiative. We thank all the participating institutions.

References

1. Vieu S, Guillot J, Beaudeau F. Antemortem diagnostic tests for the detection of *Aspergillus* infection in birds: a systematic review. *Med Mycol.* 2024;62.

2. Wernery U, Tsang C-C, Hebel C, et al. Serodiagnosis of aspergillosis in falcons (*Falco* spp.) by an Afmp1p-based enzyme-linked immunosorbent assay. *Mycoses.* 2018;61:600-609.

3. Cray C. Protein electrophoresis of non-traditional species: a review. *Vet Clin Pathol.* 2021;50:478-494.

4. Reidy L, Desoubeaux G, Cardenas J, et al. Detection of gliotoxin but not bis(methyl)gliotoxin in plasma from birds with confirmed and probable aspergillosis. *J Zoo Wildl Med.* 2022;53:60-69.

Investigating Acute Phase Protein Seasonal Variance in Backyard Chickens

Anna Watson, DVM,
Carolyn Cray, PhD,
Jennifer Graham, DVM, DABVP (Avian Practice), DABVP (Exotic Companion Mammal Practice), Dipl ACZM,
Angela Lennox, DVM, DABVP (Avian Practice), DABVP (Exotic Companion Mammal Practice), DipECZM (Small Mammal)

Session #133

Affiliation: From Avian and Exotic Animal Clinic of Indianapolis, 9330 Waldemar Road, Indianapolis, IN 46832, USA (Watson, Lennox), Huntsville Veterinary Specialists, 800 Dr. Joseph E. Lowery Blvd, Huntsville, AL 35801, USA (Graham), Department of Comparative Pathology, University of Miami, Miller School of Medicine, Miami, FL 33136, USA (Cray).

Abstract: Markers of inflammation have been examined in chickens using experimental models of infection with various agents,[1,2] but reference intervals for these markers have not been established. In the present study, 42 hens of various breeds with a median age of 2 years (range 1.3-3.4 years) were examined. During winter, routine hematology and biochemistry testing, as well as measurement of protein electrophoresis and novel inflammation biomarkers including serum amyloid A (SAA), were conducted. Physical exams revealed healthy birds with a median body condition score of 4/9 (interquartile range 4-5), mild pododermatitis was observed in one bird and mild coelomic distention was noted in 14/42 birds. There was no significant correlation between SAA and total WBC or A/G ratio. Reference intervals (RI) were calculated for all measurands using ASVCP guidelines using the robust method on n = 27 birds without any clinical abnormality. The RI for packed cell volume was 26-40%. For the CBC, the RI for WBC, heterophils, and lymphocytes were 5.3-23.1, 1.6-8.0, and 2.0-15.2 × 10^3/µl, respectively. The reference interval for calcium, A/G ratio, and beta globulins were 4.7-32.1 mg/dl, 0.25-1.06, and 0.82-2.72 g/dl, respectively. For SAA, the interval was 0.1-24.8 ng/ml. In total, these RI may reflect some reproductive changes despite exclusion of chickens with coelomic distension. An additional assessment during summer months will be conducted to examine the effects of active egg laying on these measures. In addition, chickens with common presentations including pododermatitis, ectoparasites, and trauma will be assessed to gauge the clinical applicability of SAA quantitation in this species.

References

1. Horvatic A, Guillemin N, Kaab H, et al. Quantitative proteomics using tandem mass tags in relation to the acute phase protein response in chicken challenged with *Escherichia coli* lipopolysaccharide endotoxin. *J Proteomics*. 2019;192:64-77.

2. Riva F, Eckersall PD, Chadwick CC, et al. Identification of novel biomarkers of acute phase response in chickens challenged with *Escherichia coli* lipopolysaccharide endotoxin. *BMC Vet Res*. 2024;20:236.

Cerebellar Hypoplasia in a Port Lincoln Parrot (*Barnardius zonarius zonarius*)

Emmanuelle Asrow, DVM,
Stephanie Lamb, DVM, DABVP (Avian Practice),
Ethan Biswell, DVM, MS, DACVP (Anatomic)

Session #210

Affiliation: From Arizona Exotic Animal Hospital, Phoenix, AZ 85012, USA.

Abstract: An approximately 15-year-old adult female, Port Lincoln parrot (*Barnardius zonarius zonarius*) was presented after being rescued in 2016 with neurologic signs. Initial exam findings included ataxia, intermittent reduced grip strength, star-gazing, head tremors, and hock/tail-sitting. Infectious disease testing performed was negative for avian bornavirus (ABV) by PCR, lead and zinc toxicity, and chlamydia. Bloodwork showed a transient leukocytosis. Magnetic resonance imaging was performed and showed abnormal cerebellar anatomy. The parrot's neurologic signs were managed and did not progress despite treatment for multiple systemic diseases including neoplasia. In 2024 it became anorexic, lethargic, and was straining to defecate. Bloodwork showed a transient hyperuricemia and elevated bile acids. Radiographs showed a metal density in the ventriculus, but lead and zinc testing were negative. A barium study and ultrasound showed filling defects in the caudal coelom and cloaca. Repeat ABV PCR and serology testing were negative; however, an avian-ganglioside antibody test was positive. The patient passed away despite treatment and a necropsy was performed. The bird was diagnosed with adenocarcinoma of the distal colon, secondary bacterial proximal salpingitis, and cerebellar hypoplasia. Cerebellar hypoplasia is an uncommon condition diagnosed in parrots and was not the cause of death in this patient. This is the first report of a parrot diagnosed with cerebellar hypoplasia with magnetic resonance imaging (MRI). This case supports that parrots diagnosed with congenital cerebellar hypoplasia can live full lives.

Clinical History

An approximately 15-year-old adult female Port Lincoln parrot presented to Arizona Exotic Animal Hospital in Mesa, Arizona in 2016 to establish care and evaluation of neurologic deficits. The parrot was obtained from a parrot rescue and had previously lived in an outdoor aviary. The bird was evaluated initially by avian specialists in California through a rescue group. Initial exam findings including appropriate mentation, abnormal posture with extended pelvic limbs and bearing weight on its tail base, ataxia, slightly decreased body condition, increased head tremors with stimulation, mild bumblefoot, and fecal staining with undigested food in its feces.

Initial diagnostics included a normal complete blood count (CBC) and comprehensive chemistry, radiographs showing historic healed bilateral distal radius and ulnar fractures, mild cardiomegaly versus widened cardiohepatic silhouette, and a possibly altered nasal-frontal hinge joint and proximal maxilla. Based on these findings a cardiac ultrasound was performed that was normal. The parrot was started on an isoxsuprine trial during diet transition. One month later due to persistent neurologic signs, further testing was performed. Blood lead (0.03 ppm) and zinc (1.5 ppm) were normal, a chlamydia PCR was negative, an avian bornavirus (ABV) PCR was negative, and normal fecal and choanal gram stains. A repeat CBC showed a mild leukocytosis (12 k/µl) that later resolved. Treatment with meloxicam, gabapentin, and apple cider vinegar were implemented. Isoxsuprine was not continued.

On initial presentation in Arizona, examination findings were stable. Repeat radiographs still showed a widened cardio-hepatic silhouette, but it was improved. Since there was no response to previous medications along with concern for possible sarcocystis infection due to a number of reported cases at that time in Northern California where the bird had been housed outdoors, a short course of trimethoprim/ sulfamethoxazole was started.[1] A barium study was performed to determine the cause of the widened cardio-hepatic silhouette. The proventriculus was normal in size, but the intestines were being pushed cranially due to a filling defect in the cloaca. This was likely due to fecal retention secondary to neurologic issues and postural changes. The patient was transitioned to Celebrex from meloxicam, tramadol was added for additional pain control, and gabapentin was continued.

The patient was referred for magnetic resonance imaging (MRI). Findings included that the cerebellum lost its form, there was a homogenous distortion to 75% of the cerebellum especially caudally and ventrally, and the caudal left cerebrum had increased CSF compared to the right side. At this point, the presumptive diagnosis for the cause of this patient's neurologic deficits was cerebellar hypoplasia.

From 2017-2024 the patient lived in a modified enclosure and maintained a good quality of life. Over the years it had sinus infections and gastrointestinal infections that resulted from abnormal posture and difficulty voiding. The parrot was treated as needed for secondary infections and given pain control (celecoxib) for problems due to decreased mobility and historic injuries. In 2023 it had squamous cell carcinoma that was completely excised from the right carpus.

In August of 2024 the parrot became lethargic and anorexic. Bloodwork abnormalities included elevated uric acid (26 mg/dl), elevated bile acids (159.2 µmol/L), leukocytosis (12,184 /µl), and heterophilia (11478/µl). Chemistry changes resolved in a few days with supportive care. There was a mineral density in the proventriculus on radiographs, but lead (< 0.02 ppm) and zinc (1.5 ppm) testing were normal. A fecal gram stain was positive for clostridial enteritis that resolved with treatment, but *Candida* was diagnosed at recheck. To rule it out as a cause of chronic gastrointestinal issues, the ABV testing was repeated. A fecal quantitative PCR and ELISA serology submitted through the Schubot Center were negative. An anti-ganglioside antibody (AGA) test submitted through Veterinary Molecular Diagnostics was positive. On fluoroscopy with barium, a filling defect was identified in the cloaca. An ultrasound was performed that showed heterogenous material within in severely dilated cloaca and mass effect. A cloacal flush was performed under sedation and a large amount of fecal material was removed. A repeat ultrasound showed the mass effect was resolved afterwards, but there was concern for a cloacal mass, stricture, or motility disorder causing the delayed cloacal emptying.

The patient passed away 2 days later prior to further testing. The final diagnosis at necropsy and the cause of decline was cloacal/distal colon adenocarcinoma and secondary proximal bacterial salpingitis. Cerebellar hypoplasia and hydrocephalus were confirmed with histopathology. The cerebellum was decreased in size with decreased folia formation, decreased density of the granular layer, thinning of the molecular layer, poorly organized and differentiated Purkinje fibers, and vacuolar degeneration within the white matter. No other significant histologic abnormalities were seen in the cerebrum, brainstem, cardiovascular system, gastrointestinal system, liver, or kidneys. Electron microscopy was performed on brain tissue through the Michigan State University Veterinary Diagnostic Laboratory and no viral, bacterial, fungal, or protozoal organisms were identified.

Discussion

This report presents a case of cerebellar hypoplasia in a female Port Lincoln parrot that displayed chronic, non-progressive, neurologic deficits. Common differential diagnosis for these clinical signs include cerebellar abiotrophy, toxin exposure, historic trauma, or infectious disease (e.g., ABV, avian chlamydiosis, West Nile virus, fungal infections). Neurologic signs, diagnostic testing, advanced imaging, and necropsy findings were consistent with those found in multiple reported cases of cerebellar hypoplasia.[2-6]

Typical neurologic examination findings with cerebellar hypoplasia can include appropriate mentation, ataxia (characterized by intention tremors, nystagmus, and dysmetria), and delayed proprioception, a wide-based stance, and normal spinal reflexes. Most importantly, neurologic abnormalities should not worsen over time.[1,2,6] On initial examination this patient had appropriate mentation, ataxia, a wide-based stance with difficulty balancing, and tremors especially when excited. These changes persisted without progressing for the following 8 years.

Histologic changes with cerebellar hypoplasia, all of which were found in this patient, commonly include reduced cerebellar size, decreased folia, disorganized and decreased numbers of Purkinje cells, thinning of the molecular and granular layers, and spongiosis.[2,3,7,8] Cerebellar abiotrophy is associated with progressive neurologic signs and not typically associated with gross reduction in the size of the cerebellum, and in this case no signs of progression were observed.[4,5] A toxin (metal versus other) is unlikely in this case since there were no histologic changes to the rest of the brain as would be expected with toxin exposure and blood lead and zinc tests were negative.[2,6]

The parrot had a positive anti-ganglioside antibody (AGA) PCR in the month prior to passing. This test has been associated with ABV infections; however, this patient had a negative ABV PCR in 2016, negative PCR and serology in 2024, and did not have lesions consistent with ABV on histopathology. It is believed that other conditions can also lead to the development of these antibodies, but there is little available published evidence to support this theory. Electron microscopy did not identify any infectious organisms which further rules out potential infectious causes of secondary cerebellar hypoplasia.[2,3] The definitive cause of this patient developing antiganglioside antibodies is unknown, but these tests support that not all birds with positive AGA tests have ABV.

This is the first case report of a parrot diagnosed with cerebellar hypoplasia via MRI. There are very few reports on the use of MRI in parrots, but one publication using MRI for diagnosis of hydrocephalus, which was also found in this case.[9] The MRI findings were consistent with those expected in dogs and cats with cerebellar hypoplasia rather than cerebellar abiotrophy including marked reduction in cerebellar volume and increased volume of surrounding CSF.[10] Both hypoplasia and hydrocephalus were confirmed on necropsy. In most publications on cerebellar hypoplasia in parrots, diagnosis was made through necropsy, but this supports that MRI poses an excellent option for diagnostic confirmation antemortem in parrots as in other species.

This patient had a normal expected lifespan for its species and ultimately passed away secondary to cancer, not complications of cerebellar hypoplasia. To date, there are no other published case reports on the natural outcome of captive parrots diagnosed with this condition. In other available publications on parrots, Galliformes, and raptors, the birds were euthanized despite being apparently healthy on necropsy.[2,7,8] There is a report of pullets with neurologic signs (i.e., ataxia and fine tremors) being otherwise healthy on exam and able to lay eggs normally.[8] While cerebellar hypoplasia is uncommon in birds, this case supports that if a parrot is diagnosed with this condition but is otherwise healthy, they can have a good quality of life and normal lifespan with appropriate care.

References

1. Penderis J. The wobbly cat diagnostic and therapeutic approach to generalized ataxia. *J Feline Med Surg.* 2009;11:349-359.

2. Taggers A, Kelly-Bosma M, Mastakov A, et al. Hypoplasia in two juvenile African grey parrots (*Psittacus errithacus*). *J Avian Med Surg.* 2022;36(3):308-315.

3. Toyoda T, Ochiai K, Hatai H, et al. Cerebellar hypoplasia associated with an avian leukosis virus inducing fowl glioma. *Vet Pathol.* 2006;43:294-301.

4. Marusak RA, Guy JS, Abdul-Aziz TA, et al. Parvovirus-associated cerebellar hypoplasia and hydrocephalus in day old broiler chickens. *Avian Dis.* 2010;54(1):156-160.

5. Scott EY, Woolard KD, Finno CJ, Murray JD. Cerebellar abiotrophy across domestic species. *Cerebellum.* 2018;17(3):372-379.

6. Jones MP, Orosz S. Overview of avian neurology and neurological diseases. *Semin Avian Exot Pet Med.* 1996;5(3):150-164.

7. Armienn AG, McRuer DL, Ruder MG, Wunschmann A. Purkinje cell heterotopy with cerebellar hypoplasia in two free-living American kestrels (*Falco sparverius*). *Vet Pathol.* 2012;50(1):182-187.

8. Markson LM, Carnaghan RBA, Young GB. Familial cerebellar degeneration and atrophy—a sex-linked disease affecting light Sussex pullets. *J Comp Pathol.* 1959;69(2):223-229.

9. Fleming GJ, Lester NV, Stevenson R, Silver XS. Highfield strength (4.7t) magnetic resonance imaging of hydrocephalus in an African grey parrot (*Psittacus erithacus*). *Vet Radiol Ultrasound.* 2003;44(5):542-545.

10. MacKillop E. Magnetic resonance imaging of intracranial malformations in dogs and cats. *Vet Radiol & Ultrasound.* 2011;52(1):42-51.

Diagnosis and Treatment of Hemochromatosis Associated with Elevated Iron Levels in Drinking Water in a Toco Toucan (*Rhamphastos toco*)

Wm. Kirk Suedmeyer, DVM, Dipl ACZM,
Brian G. Stockinger, DVM,
Michael M. Garner DVM, DACVP

Session #163

Affiliation: From the Kansas City Zoo & Aquarium, 6800 Zoo Drive, Kansas City, MO 64132, USA (Suedmeyer, Stockinger), Northwest ZooPath 654 West Main Street, Monroe, WA 98272, USA (Garner).

Abstract: Toco toucans (*Rhamphastos toco*) are members of the Rhamphastidae family. They are exquisitely colored, popular exhibit animals, which are often privately owned.[1] Hemochromatosis is the pathologic deposition of iron in tissues, especially the liver, resulting in cirrhosis. Hemosiderosis is the non-pathologic general deposition of iron in the reticuloendothelial tissues of the spleen, liver and bone marrow.[2] Both entities are common in Rhamphastidae.[3] Historically these syndromes are associated with elevated levels of iron and a lack of natural chelators in captive diets. No reports of hemochromatosis or hemosiderosis from water sources have been documented for birds. A deteriorated plumbing source of drinking water for two toco toucans and 15 Chilean flamingos (*Phoenicopterus chilensis*) had high levels of iron compared to safe levels for humans.[4] Iron panels were evaluated for all flamingos and both toucans. One of two toucans had substantially increased iron values. The toucans were maintained on a low-iron diet. Hepatic biopsies and dry weight values for iron were compatible with hemochromatosis. The bird was treated for 6 months with daily chelation. Additional hepatic biopsies and iron panels identified a steady decline in iron values to acceptable levels. Subsequent serum iron panels demonstrated slight increases in iron values but within acceptable levels. Thirty months after cessation of treatment the bird developed non-responsive ascites, hepatomegaly, inspiratory compromise, inappetence, and died. Postmortem findings included cirrhosis associated with hemochromatosis. Iron deposition in renal and pancreatic tissue was not found indicating treatment was initially successful. Water sources should be included in the differential etiology of iron storage disease in susceptible animals.

Acknowledgments: The authors thank the Kansas City Zoo's Animal Care and Veterinary Health teams for their care, expertise and efforts caring for this toucan to ensure its welfare on exhibit.

References

1. Sick H, Barruel P. *Birds in Brazil: A Natural History*. Princeton, NJ: Princeton University Press; 1993.

2. Cornelissen H, Ducatelle R, Roles S. Successful treatment of a channel-billed toucan (*Ramphastos vitellinus*) with iron storage disease by chelation therapy: sequential monitoring of the iron content of the liver during the treatment period by quantitative chemical and image analyses. *J Avian Med Surg*. 1995;9:131-137.

3. Lowenstine LJ, Stasiak IM. Update on iron overload in zoologic species. In: Fowler ME, Miller RE, eds. *Zoo and Wild Animal Medicine.* Vol 8. St. Louis, MO: Elsevier; 2015:674-680.

4. World Health Organization. Iron in drinking-water: background document for development of WHO Guidelines for drinking-water quality guidelines for drinking-water quality. In: *Guidelines for Drinking-Water Quality.* 2nd Ed. Vol. 2. Geneva, CH: WHO; 1996.

Histologic Improvement of Iron Storage Disease in the Liver of a 32-Year-Old Congo African Grey Parrot (*Psittacus erithacus*)

Sydney Jones, DVM,
Stephanie Lamb, DVM, DABVP (Avian Practice),
Ethan Biswell, DVM, MS, DACVP (Anatomic Pathology)

Session #132

Affiliation: From Texas Avian and Exotic Hospital, Grapevine, TX 76051, USA (Jones), Arizona Exotic Animal Hospital, Phoenix, AZ 85012, USA (Lamb), Biosphere Lab, Louisville KY 40299, USA (Biswell).

Abstract: Iron overload disease or iron storage disease in birds is characterized by the accumulation of excess iron within organs, most commonly the liver. Lesions commonly seen include hepatocellular degeneration and necrosis, hepatic cord disruption, periportal fibrosis, bile duct hyperplasia, nodular hyperplasia, and cirrhosis in chronic cases. A case of iron storage disease diagnosed via hepatic biopsy in a 25-year-old male Congo African grey parrot (*Psittacus erithacus*) was presented in the Journal of the American Veterinary Medical Association's Pathology in Practice in 2022.[1] The histopathologic findings demonstrated extensive accumulations of golden-brown cytoplasmic pigments of Kupffer cells, and an area of mild fibroplasia associated with a portal triad. Prussian blue staining confirmed that the pigment granules were iron granules confirming a diagnosis of iron overload disease. Quantitative iron analysis via flame atomic absorption spectroscopy revealed an iron content of 988 ppm. Treatments initiated included feeding a low-iron, low-vitamin C diet, adding a black tea supplement to the bird's water, biweekly phlebotomies for four total treatments, and starting deferoxamine. Improvement in the histologic appearance of the liver was seen after 14 months on repeat hepatic biopsy; however, the hepatic iron content at that time was 8,930 ppm. The discrepancy between the histologic improvement of the liver on follow up biopsy with an increased iron content was suspected to be secondary to the small sample size provided for the first biopsy, suggesting the first iron content value may not have been accurate. The low-iron, low-vitamin C diet was continued long term. Approximately 7 years later, the bird presented to another specialty clinic for weight loss, hyporexia and a stroke-like episode. The patient was thin with a grade III heart murmur and a poorly regenerative anemia reflected on blood work. The patient was treated with isoxsuprine, pentoxifylline, and omega 3 fatty acids with the commercial liver supplement Hepasil due to the concern for atherosclerosis and historical liver disease. The patient also received two doses of iron dextran 2 weeks apart. An echocardiogram was scheduled but unfortunately the anemia worsened significantly to 14% and the patient declined clinically. Euthanasia was elected by the owner. Gross necropsy revealed severe aortic atherosclerosis, a proventricular adenocarcinoma, and marked improvement in the histologic changes to the liver from 7 years prior. Histopathology of the liver revealed few scattered small aggregates of macrophages that contained brown granular pigment consistent with hemosiderin, with none noted in hepatocytes. The hepatic iron content was improved from the second biopsy at 1,352 ppm. This presentation serves as an update to a previously published case demonstrating efficacy of various treatment modalities including long term diet modification in improving disease caused by excess iron accumulation in the liver.

References

1. Lamb S, Reavill D, Biswell E. Hepatic iron overload in a Congo African grey parrot (*Psittacus erithacus*). Pathology in Practice. *J Am Vet Med Assoc.* 2022; 259(S2): doi: org/10.2460/javma.21.04.0209.

Avian Stem Cell Therapy Program and Case Report at the Kansas City Zoo & Aquarium

Wm. Kirk Suedmeyer, DVM, Dipl ACZM,
Larry Snyder, DVM,
Brian G. Stockinger, DVM,
Jamie Gerlach, DVM

Session #164

Affiliation: From the Kansas City Zoo & Aquarium, 6800 Zoo Drive, Kansas City, MO 64132, USA (Suedmeyer, Stockinger), Ardent Animal Health 220 Bellaire Drive, Nicholasville, KY 40356, USA (Snyder).

Abstract: Stem cell therapy is a growing modality in exotic animals. In 2016 we initiated a prophylactic stem cell collection program for zoo animals including mammals, reptiles and birds. There are different definitions of stem cell therapy, but all are based on utilizing the patient's inherent healing properties. For the purposes of this presentation, stem cell therapy at the Kansas City Zoo & Aquarium involves utilizing mesenchymal stem cells (MSC) collected during quarantine, from ill animals or routine examinations and these cells are processed and stored. These processed cells can be stored for up to 25 years. Collection technique in avian species mimics mammalian surgery although specific sites differ. In selected cases, these cells are combined with platelet-rich plasma (PRP) collected at the time of mesenchymal therapy. Mammalian platelets and avian thrombocytes contain bioactive proteins including growth factors, chemokines, cytokines and other molecules that play a critical role in maintaining homeostasis and tissue remodeling.[1-3] The combination is activated by specific wavelengths of polychromatic light and the resultant product is utilized for mesenchymal therapy by intraarticular, intravenous intraperitoneal or intracoelomic injection. Upon activation, these cells release sustained continual growth factors, increased anti-inflammatory factors and reduced proinflammatory interleukin 1-a.[1] The PRP-MSC combination differentiates into chondrocytes, adipocytes or osteocytes resulting in regenerative therapy. This has been successful in reducing discomfort and improving mobility in numerous cases of degenerative osteoarthritis in mammals, two reptiles and a Pekin duck (*Anas platyrhynchos*). Utilization of stem cell therapy minimizes drug therapy, provides long-term relief of discomfort and improves welfare of the animal.

Acknowledgments: The authors thank the Kansas City Zoo's Animal Care and Veterinary Health teams for their care, expertise and efforts caring for the animals treated with stem cell therapy.

References

1. Karina K, Rosaliana I, Rosadi I, et al. Safety of technique and procedure of stromal vascular fraction therapy: from liposuction to cell administration. *Scientifica (Cairo).* 2020;2020:2863624.

2. Ferdous F, Scott T. The immunological capacity of thrombocytes. *Int J Mol Sci.* 2023;24:12950. doi: 10.3390. ijms24612950.

3. Pochini AC, Antonioli E, Bucci DZ, et al. Analysis of cytokine profile and growth factors in platelet-rich plasma obtained by open systems and commercial columns. *Einstein (Sao Paulo).* 2016;14(3):391-397.

Ack! Asper Attacks! Diagnosis and Treatment of Tracheal Aspergillosis in a Juvenile Bald Eagle (*Haliaeetus leucocephalus*)

Marcie Logsdon, DVM, CWR

Session #216

Affiliation: From Washington State University Veterinary Teaching Hospital, Exotics and Wildlife Department, 205 Ott Rd Pullman WA 99164, USA.

Abstract: A captive juvenile bald eagle (*Haliaeetus leucocephalus*) presented for acute voice change and increased respiratory effort. Tracheal endoscopy revealed a fungal plaque at the level of the syrinx consistent with aspergillosis. Initial endoscopic debulking of the lesion followed by nebulization and systemic anti-fungal treatment resulted in complete resolution of signs and lesions within three weeks. A follow-up site assessment identified likely predisposing factors at the facility and led to husbandry recommendations to prevent future recurrence.

Introduction

Diagnosis of Aspergillosis

Aspergillus spp. is a genus of ubiquitous environmental saprophytic fungi that can cause clinical disease in avian and mammalian species. These fungi are found across the globe with *Aspergillus fumigatus* being the most common cause of clinical disease in animals.[1]

Clinical signs associated with aspergillosis in avian species depend on the site of infection.[2,3] Tracheal or syringeal fungal granulomas are often acute in nature and can lead to an obstructive breathing pattern, dyspnea, or voice change. Air sacculitis and generalized respiratory infection can cause dyspnea, decreased athletic performance, and weakness. Invasive systemic infection can cause weight loss, weakness, anorexia and, in severe cases, neurologic signs.[2,3] None of these signs are pathognomonic for aspergillosis and differentials for aspergillosis of the respiratory tract include neoplasia, inhaled foreign body, other fungal or bacterial infections, and tracheal stricture.

Compatible history plays an important role in diagnosing aspergillosis. Susceptible patients that are immunocompromised or undergoing a stressful event are at a higher likelihood of developing aspergillosis. Some species appear more susceptible to aspergillosis and young individuals appear to be more susceptible as well.[2] Radiographs can often show signs of air sacculitis and occasionally are able to detect tracheal or syringeal granulomas; however, early lesions may be missed and radiographic changes are not specific to aspergillosis. CT imaging offers a more precise view and can be useful in identifying and characterizing the extent of intratracheal or bronchial lesions that may be obscured during traditional radiography; however, again, lesions detected are nonspecific to aspergillosis.[2]

Endoscopy of the trachea and/or air sacs allows for direct visualization of lesions and sampling. This procedure requires general anesthesia and is moderately invasive. Endoscopy is the only diagnostic tool that also offers an opportunity for treatment as granulomas can be debulked and medications applied directly to lesions via the working channel of the scope.[2]

Hematology changes such as a leukocytosis and monocytosis may be seen in some species of birds with chronic aspergillosis; however, acute disease isolated to the respiratory tract may not induce a leukocytosis and changes are not specific to aspergillosis.[2] Leukograms in conjunction with *Aspergillus* antibody levels and galactomannan (a component of the fungal cell wall) levels have increased sensitivity for detecting aspergillosis; however, some species such as raptors develop high levels of aspergillosis antibody from daily environmental exposure and may not show detectable levels of galactomannan in the blood, even with clinical infection.[2]

Culture and sensitivity of *Aspergillus* can be unrewarding in clinical settings due to the slow growth rate of fungi when compared to the need for rapid intervention, and potential for false negatives.[2] Characteristic branching hyphae on histology or cytology of lesions can provide a working clinical diagnosis of *Aspergillus* however culture or PCR are required for definitive diagnosis.

Treatment, Management, and Prognosis

Prevention of aspergillosis focuses primarily on husbandry and overall systemic health of the bird. Environmental management to decrease the number of fungal spores present in an enclosure can help prevent development of disease in susceptible animals.[4] These measures include increasing ventilation, ensuring enclosure materials are not harboring fungal growth, and avoiding bedding with high concentrations of fungal spores, such as hay or straw. Any conditions that can lead to immune suppression such as poor nutrition, stress, or long-term antibiotic or steroid administration should also be addressed.

Treatment of aspergillosis depends on location and severity of infection. For respiratory infections, treatment with nebulized medications such as amphotericin B or terbinafine offers the benefit of high medication concentrations at the site of infection with minimal to no systemic absorption.[4] For systemic medications, voriconazole is currently the anti-fungal of choice for disseminated or severe aspergillosis;[4] however, there have been some reports of voriconazole-resistant *Aspergillus fumigatus* strains isolated from domestic and wild birds.[4] While itraconazole has been used in treatment and prevention of aspergillosis in birds, studies have shown poor lung concentration levels when administered orally and a high level of resistance in *Aspergillus fumigatus* strains found in avian patients.[4] Nebulized medications can be used in conjunction with systemic oral medications to treat a patient with respiratory aspergillosis.

Case Report

Presentation

A 3-year-old male bald eagle (*Haliaeetus leucocephalus*) presented on referral for evaluation of a sudden voice change and yellow oropharyngeal plaques diagnosed as candidiasis on impression smear by the referring veterinarian.

The eagle was housed in a large outdoor aviary and was used for educational programs. The primary trainer noted an acute voice change during a training session and presented the bird to their local veterinarian the next day. On physical examination, the referring veterinarian noted numerous small yellow plaques in the caudal oropharynx. Swab and cytology of the plaques was consistent with candidiasis and the patient was prescribed nystatin, 150,000 IU/kg, PO, q12h (Nystatin Oral Suspension USP, Pharmaceutical Associates Inc., Greenville SC). Whole body radiographs were performed, and a CBC and blood chemistry panel were submitted. The patient was then referred for further workup of the voice change.

On presentation the patient was bright, alert, and responsive. A healed partial (mid-metacarpal) left-wing amputation was noted. The patient's body condition was good with a weight of 3.4 kg. The eyes, nares and choanal slit were clean, clear, and free of discharge. The previously described oropharyngeal plaques were observed. The glottis was clear and free of inflammation or discharge. The crop was empty on palpation.

When at rest, a very mild increased respiratory effort with an intermittent, faint inspiratory wheeze was observed. With any amount of exertion or restraint, the patient rapidly progressed to open-mouth breathing with an audible inspiratory wheeze. This wheeze remained when the beak was opened and was loudest on auscultation over the trachea at the level of the thoracic inlet. Air sac and lung auscultation were consistent with a referred upper airway wheeze or stridor.

Case management and outcome

The radiographs, CBC, and chemistry panel provided by the referring veterinarian were reviewed. The radiographs showed no abnormalities to explain the dyspnea or voice change. There was no sign of air sacculitis or pneumonia. The chemistry panel was within normal limits and the CBC showed a normal total white blood cell count with a mild heterophilia and lymphopenia. The WBC count and differential were performed by the lab from a slide estimate with many smudge cells noted on the report, which may have affected the differential.

Based on clinical signs, a bronchial or incomplete tracheal occlusion was suspected. A CT scan or tracheal endoscopy were offered to the owner as the next diagnostic step. The owner elected tracheal endoscopy due to the ability to quickly move to endoscopic-guided removal if an obstruction was found.

A clean, non-sterile 4.1-mm flexible endoscope without a working channel was used for initial tracheal endoscopy (Visera Tracheal Intubation Videoscope LF Type V, Olympus Surgical Technologies America, Brooklyn Park, MN). This scope is routinely used for initial diagnostic tracheal and esophageal endoscopy because it is readily available and inexpensive. Once a problem is identified and endoscope needs are determined, the appropriate specialty scope is selected for treatment.

The patient was premedicated with 1 mg/kg midazolam IM (Midazolam Injection USP, Almaject Inc., Morristown, NJ). Anesthesia was induced and maintained with isoflurane (Isospire, Dechra Veterinary Products, Overland Park, KS) anesthesia. Prior to intubation, the glottis was visually examined and the cranial ~6" of the trachea was briefly scoped and found to be clear of obstruction or inflammation. The patient was then intubated with a 4.5-mm non-cuffed endotracheal (ET) tube with an adapter to allow passage of the endoscope; however, the added length of the adapter and ET tube connector prevented the scope from reaching the tracheal bifurcation. Once an adequate plane of anesthesia was achieved, the ET tube was removed and the scope was passed. The trachea was clean and clear until the level of the tracheal bifurcation where a large white to tan mass was found to be occluding the entrance to both mainstem bronchi. (Fig 1).

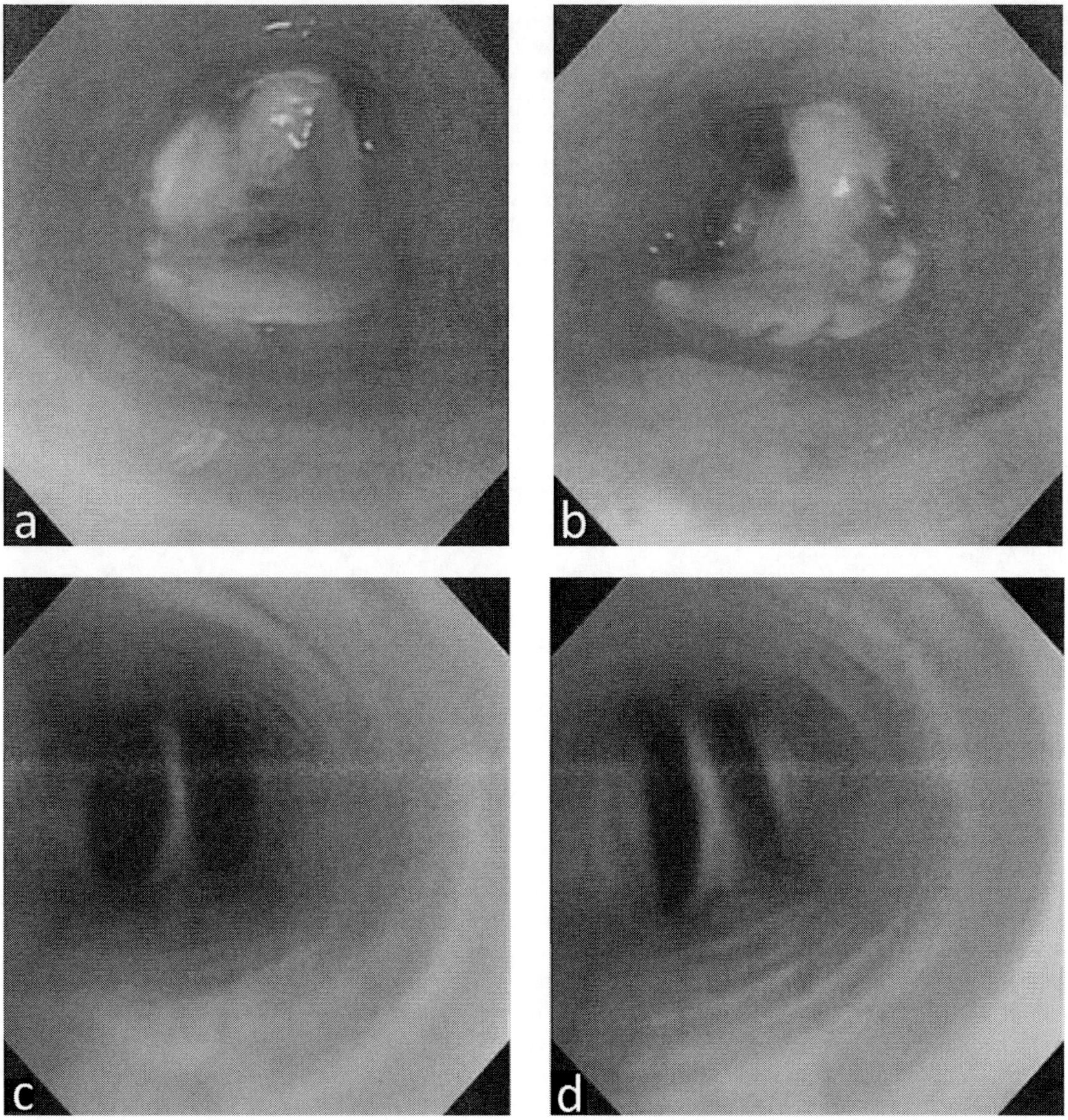

Figure 1. Initial endoscopic view of the tracheobronchial obstruction (a), tracheal bifurcation after initial debulking (b), and view of tracheal bifurcation/syrinx 3 weeks after initiation of treatment (c, d) showing complete resolution of lesions.

The scope was removed and the patient was reintubated. When the patient was repositioned, it ceased spontaneous respirations. Manual positive pressure ventilation was able to adequately inflate the patient despite the occlusion. The patient's heart rate remained steady.

After consultation with the owner, the decision was made to pursue removal of the obstruction. The patient was placed in left lateral recumbency. A 4-mm diameter non-cuffed endotracheal tube was used for placement of an air sac cannula into the right caudal thoracic air sac using standard technique. Positive pressure ventilation and isoflurane gas anesthesia were delivered via this cannula for the duration of the procedure.

A 2.9-mm flexible ureter-reno video-endoscope with a working channel (URF-V EndoEYE Flexible Uretero-Reno Videoscope 2.9mm, Olympus Surgical Technologies America, Brooklyn Park, MN) was used to sample and debulk the mass. Slide cytology of a sample collected during the procedure showed fungal hyphae consistent with *Aspergillus*. (Fig 2) The mass was debulked using a combination of suction and manual debridement with ratigator forceps. A sample was collected for histopathology. Once finished, a flat plaque remained along the dorsal aspect of the trachea. (Fig 1) A dose of 0.1 ml of amphotericin B 5 mg/ml (Amphotericin B for Injection, USP, XGen Pharmaceuticals DJB Inc., Big Flats, NY) was applied topically to the fungal plaque before withdrawing the scope.

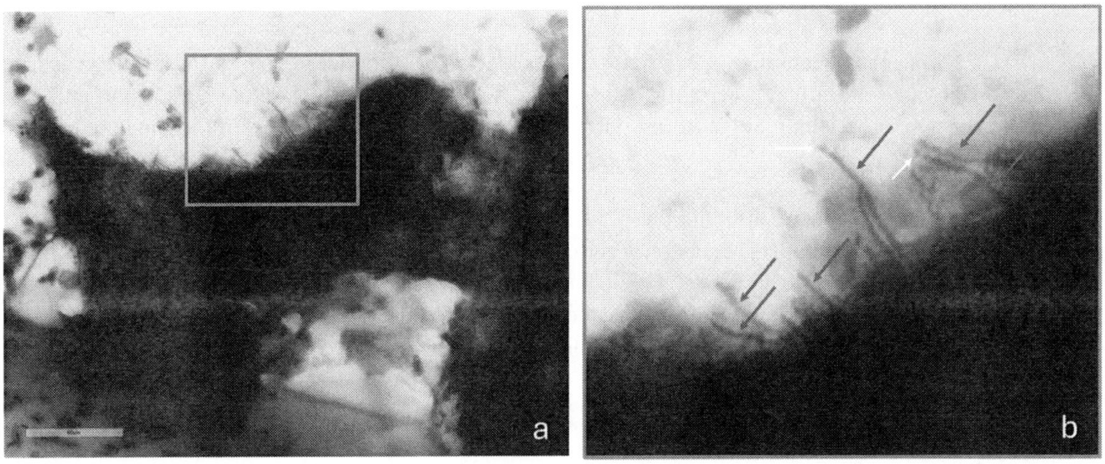

Figure 2. Cytology of the tracheal obstruction showing branching, fungal hyphae (red arrows) and conidia (yellow arrows) consistent with *Aspergillus*.

The patient recovered from anesthesia uneventfully with the air sac cannula in place and patent. A small flexible e-collar was placed to prevent the patient from removing the cannula.

The patient was started on voriconazole 15 mg/kg, PO, q12h × 21 days (Voriconazole 50 mg tablets, Ajanta Pharma, Bridgewater NJ); marbofloxacin 7.5 mg/kg, PO, q24h × 10 days (Marbofloxacin 25 mg tablets, Zenequin, Zoetis Inc., Kalamazoo, MI); and meloxicam 1 mg/kg, PO, q12h × 14 days (Meloxicam Tablets 7.5 mg, Cipla USA Inc, Warren, NJ). Nystatin was discontinued as *Candida* is also sensitive to systemic voriconazole treatment. Nebulization therapy with amphotericin B 1 mg/ml was performed for 15 minutes every 24 hours for 3 days starting the day after the procedure. (Fig 3) During each nebulization therapy, the air sac cannula was temporarily occluded to prevent nebulized amphotericin B from leaking into the room air and to test the patency of the patient's trachea. On the first day the patient's respiratory rate and effort increased moderately when the air sac cannula was occluded, but he did not become distressed, nor did he require any intervention or supplemental oxygen. On the second day there was no appreciable change to his respiratory effort when the air sac cannula was occluded. On the third day the air sac cannula was removed post-nebulization therapy and the patient was discharged. No sutures were required to close the cannula site.

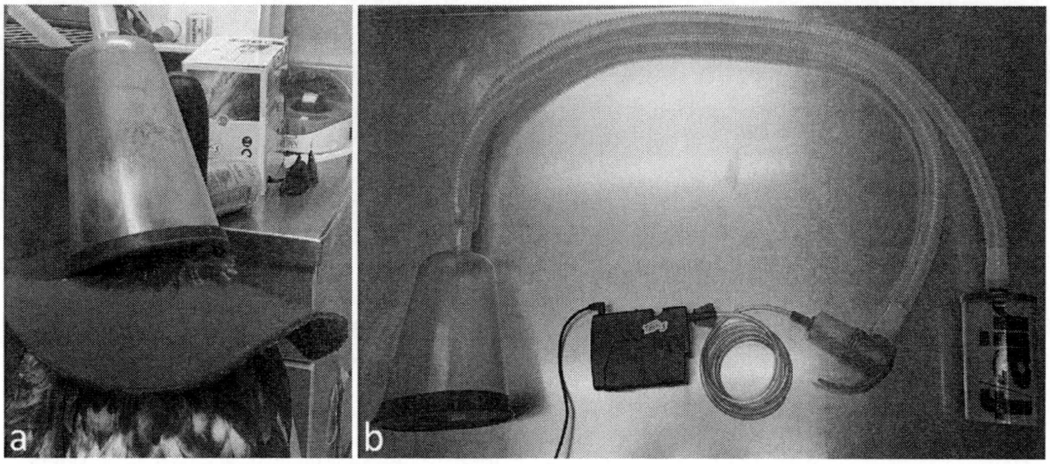

Figure 3. The patient undergoing nebulization therapy (a). The nebulizer (Trek S Portable Aerosol System, PARI Respiratory Equipment, Inc, Midlothian, VA) and F-air cannister are attached to either end of an anesthesia breathing circuit to minimize environmental and staff exposure to the nebulized medication (b).

Histopathology confirmed a heterophilic fungal plaque containing intralesional fungal hyphae consistent with *Aspergillus*. After 3 weeks the patient returned for repeat tracheal endoscopy. The handler reported no concerns and that the patient's voice and activity level had both returned to normal. The patient was premedicated with 1 mg/kg midazolam IM followed by induction and anesthesia with isoflurane. Tracheal endoscopy showed complete resolution of the lesion. (Fig 1) The oropharyngeal plaques had resolved as well, and the patient was otherwise apparently healthy on physical exam.

This was the second eagle to succumb to clinical aspergillosis at this facility within 2 years. An onsite evaluation of the aviary and surrounding environment was performed to identify a potential source of the fungal infection. A large number of bison were housed adjacent to the aviary and, during the winter months, were fed large quantities of hay directly upwind from the outdoor avian enclosures. A recommendation was made to shift the hay storage and bison feeding areas downwind of the aviary.

References

1. Tell LA, Burco JD, Woods L, et al. Aspergillosis in birds and mammals: considerations for veterinary medicine. In: Gupta A, Singh NP, eds. *Recent Developments in Fungal Diseases of Laboratory Animals.* 1st Ed. New York, NY: Springer Nature; 2019:49-72.

2. Fischer D, Lierz M. Diagnostic procedures and available techniques for the diagnosis of aspergillosis in birds. *J Exot Pet Med.* 2015;24(3):283-295.

3. Savelieff MG, Pappalardo L, Azmanis P. The current status of avian aspergillosis diagnoses: veterinary practice to novel research avenues. *Vet Clin Pathol.* 2018;47(3):342-363.

4. Krautwald-Junghanns ME, Vorbrüggen S, Böhme J. Aspergillosis in birds: an overview of treatment options and regimens. *J Exot Pet Med.* 2015;23(4):296-307.

Section 11

Wildlife and Then Some Session

Salina Locke, DVM, DABVP (Avian Practice), Cert AqV;
Evanna Lai, DVM
Session Chairs

Keeping an Eye on Rehabilitating Raptors

John Chitty, BVetMed, CertZooMed, MRCVS,
Michael Jones, DVM, DABVP (Avian Practice),
Scott Ford, DVM, DABVP (Avian Practice)

Session #128

Affiliation: From School House, Allington, Salisbury, Wilts, UK (Chitty), Jones Veterinary Services, Knoxville, TN 37932, USA (Jones), Avian Specialty Veterinary Services, Milwaukee, WI, USA (Ford).

Abstract: Hunting raptors are binocular species and require optimal eyesight to hunt and survive post-release. Unfortunately, ocular injuries are common and may be difficult to detect as ocular examination is not always straightforward. This class covers the anatomy and physiology of the raptor eye as well as practical evaluation of sight and examination of the eye. Common injuries and diseases are discussed along with their medical and, where appropriate, surgical management. The care of raptors with reduced visual capacity in rehabilitation is discussed and the release and post-release management of these cases is covered.

Avian Vision Form and Function: Understanding How and What Raptors See (Adapted from Jones, et al 2007.[1])

Nearly all vertebrates depend heavily on their visual ability to navigate their environment, identify conspecifics, capture prey, escape from predation, and find suitable mates. This is true of avian species, and in particular, birds of prey. This section will focus on key anatomic aspects of the eyes of raptors in relation to color vision, the fovea, visual fields and visual acuity, image stabilization, vestibular reflexes, accommodation, and flicker-fusion frequency. A more in-depth discussion of the anatomy and physiology of the avian and raptor eye is available in Jones, et al (2007) and other resources.[1-13]

Color vision

The ability to perceive color to varying degrees enhances visual activity by adding contrast, aiding animals in discriminating between objects, locating food, assessing threats, and attracting mates. Color detection requires two phenomena: (1) the eye's ability to separate wavelengths and relay information to the brain (color vision), and (2) the brain's capacity to process this information for behavior and survival (color perception). Both mechanisms involve light-receptive photoreceptors (rods and cones) in the retina. Rods respond to lower luminance and wavelengths around 500 nm, while cones respond to brighter light and specific wavelengths. For example, iodopsin-containing cones in humans peak at 560 nm (red spectrum).

Birds have more complex retinas compared to humans and many other species. Birds are typically tetrachromic, with cones sensitive to red, green, blue, and ultraviolet wavelengths while human retina is trichromic, containing cones sensitive to long wavelength (red cones) of 560 nm, medium wavelength (green cones) of 530 nm, and short wavelength (blue cones) of 420 nm, enabling the perception of millions of colors. Colored oil droplets in avian cones further refine light filtration, increasing their ability to discern wavelengths. These droplets act as filters, enhancing spectral sensitivity beyond photopigment capabilities.

Avian retinas enable greater wavelength discrimination compared to mammals. Experiments have primarily focused on passerines and waterfowl, leaving raptor cone classes, with the exception of tawny owls (*Strix aluco*), less studied compared to other avian species. Tawny owls possess trichromatic retinas (463 nm, 503 nm, and 555 nm) and lack UV sensitive cone opsin and red oil droplets but may see UV light at night. American kestrels (*Falco sparverius*) demonstrate ultraviolet or near ultraviolet sensitivity, using this capability to locate vole scent marks visible in UV light.

Polychromism provides subtle color discrimination and biological advantages, such as differentiating surfaces of varying chemical compositions. However, visual acuity—defined as resolving distinct points—decreases with increasing cone diversity due to lower cone density. For instance, human fovea optimizes visual acuity by limiting cones to red and green sensitivity. Raptors balance color discrimination and acuity according to ecological needs, highlighting the evolutionary adaptations of avian visual systems.

Fovea

Specialized regions within the retina, termed visual streaks, *areae centralis*, or *areae*, allow for enhancement of visual acuity significantly greater than in other areas of the retina. These specialized regions of the retina possess higher photoreceptor and ganglion cell densities. Fovea (small pit) are subspecialized portions within these areas designed to even further maximize visual acuity. Its features include increased cone photoreceptors, reduced convergence onto ganglion cells, and a parting of retinal layers internal to the photoreception layers, creating a characteristic depression. Cone photoreceptors confer greater acuity than rods because they have less convergence onto bipolar cells, and subsequently ganglion cells and cells in the visual cortex, which results in more cones in the visual cortex. In contrast, rod photoreceptors have a high degree of convergence, which enhances sensitivity but diminishes acuity. The foveal depression, or parting as mentioned before, enhances acuity by allowing photons to reach the photoreceptors with less attenuation although there is no published evidence for this assertion. In avian retinas, the foveal depression is nourished entirely from the choriocapillaris beneath the retina. Thus, the absence of superficial vasculature also optimizes image clarity in foveate animals. An additional benefit of the fovea in avian species is as a navigational aid—the light reflected off one wall of the depression and onto the other stimulates photoreceptor responses that differ depending on the plane of polarized light.

Many animals are afoveate (i.e., lack true fovea), and in fact, primates are the only foveate mammals. All raptors studied to date possess fovea, and nearly all avian species possess fovea, reflecting the importance of acute vision. Raptors often have two foveae: a central fovea for lateral monocular vision and a temporal fovea for binocular vision. Temporal fovea, which are placed closer to the central axis in raptors with more frontally placed eyes versus those with laterally placed eyes, aid in targeting prey within the frontal visual field, whereas the central fovea provides superior distance acuity. Bifoveality allows raptors, and other birds, to see clearly along more than one visual axis, which significantly aids in the perception of distance and the speed of their prey. Species such as owls, Andean condors (*Vultur gryphus*), and American black vultures (*Coragyps atratus*) deviate, with single fovea adapted to their ecological needs. Owls possess only a temporal fovea, probably owing to their more frontally placed eyes and increased binocular vision. The benefit for owls is that this arrangement helps to summate light from both eyes during glow light conditions. Furthermore, the fovea in owls is unique as it primarily contains rods, an adaptation to their nocturnal lifestyle. Condors and black vultures have only a nasal fovea.

Raptors, including peregrine falcons (*Falco peregrinus*), exhibit functional coordination between fovea. For example, peregrines follow a logarithmic spiral flight path, rather than direct approach, when sighting distant prey, optimizing the central fovea's acuity while reducing drag. This arrangement allows the falcon to maintain its head straight relative to the body but at a slight angle relative to its intended prey. The position of the head and central fovea allows for reduced drag and increased air speed. As the falcon approaches its prey, the temporal fovea supports binocular vision for precise targeting of their prey.

Central foveae are typically deeper (i.e., convexiclivate) and denser in photoreceptors than shallower (i.e., conaviclivate) temporal fovea. Deep fovea may magnify images via a telephoto effect, although studies suggest acuity aligns with anatomic limits. Both foveae are encompassed by a pronounced horizontal visual streak in raptors. The region has the highest concentration of ganglion cells and allows raptors to combine panoramic vision with excellent stereoscopic vision which is crucial for capturing fast-moving prey in flight. In diurnal birds that live in open country, the horizontal streak aligns with the horizon, aiding navigation.

Visual fields and visual acuity

Eye position is a critical determinant of visual fields in avian species. In birds, extraocular muscles are reduced to thin bands and torsional movement of the glove in many species is limited to between 2° and 5°. As a result, most birds rely on head and neck movements to scan their environment due to limited ocular mobility. Remember that birds are bifoveate, with two distinct visual fields enabling high acuity both frontally and laterally without substantial head movement.

Predatory birds, such as Falconiformes, have large, frontally placed eyes supporting binocular vision and depth perception (60-70° binocular field). In contrast, prey species like pigeons and sparrows have flattened, laterally placed eyes that provide a wide visual field (300°) at the cost of reduced acuity. Prey species often use monocular vision to detect predators without revealing their position. Raptors may also utilize monocular vision for distant objects, turning their heads sideways to achieve optimal acuity using the central fovea.

Notable adaptations include the American woodcock's (*Scolopax minor*) (360° horizontal and 180° vertical visual field, made possible by eyes positioned oppositely. Similarly, the mallard duck has a 360° horizontal field with a 20° binocular range.

Visual acuity, the ability to resolve distinct points, varies significantly among species. Behavioral acuity tests measure cycles/degree using pattern recognition. Raptors like the American kestrel (*Falco sparverius*) (160 cycles/degree) and wedge-tailed eagle (*Aquila audax* spp) (132-143 cycles/degree) exhibit acuities exceeding humans (33-73 cycles/degree). Kestrels' anatomic resolution (46-51 cycles/degree) raises questions about these findings.

In raptors, acuity is influenced by light (luminance) levels. Diurnal raptors, such as wedge-tailed eagles and kestrels, exhibit high but luminance-sensitive acuities. Crepuscular and nocturnal raptors, such as owls, demonstrate moderate acuities that are luminance insensitive.

Image stabilization and visual acuity

Maintaining stable visual images is crucial for avian species, particularly raptors capturing prey in flight or prey species evading predators. This sophisticated system of reflexes allows birds to maintain sharp visual focus and adapt to dynamic environments. Head and neck movements are primary challenges to ocular stability, risking retinal slip, or visual blurring caused by retinal displacement relative to the object's optical projection. Birds mitigate this using combined vestibular and visual reflexes: the vestibulo-ocular reflex (VOR), vestibulo-collic reflex (VCR), optokinetic nystagmus reflex (OKN), and optocollic reflex (OCR).

Vestibular reflexes

The VOR compensates for head movements by stabilizing gaze through vestibular inputs that relay head motion velocity directly to the oculomotor system. This facilitates coordinated eye, head, and neck movement as well as posture and balance. The VCR activates neck muscles in response to horizontal head motion sensed by vestibular organs in the inner ear.

Visual reflexes

Reflexive eye movements include OKN and OCR. The OKN employs retinal image velocity to induce eye movements matching external field motion, alternating between slow tracking and fast resetting phases. The OKN represents an alternating pattern of slow-phase eye movements in the same direction as the object(s) or stimulus motion and fast phases against the direction of the stimulus motion. OCR, which is a behavior compensatory head reflex, complements the OKN by stabilizing the head during changes in visual field, alternating between thrust (i.e., rapid forward movement) and hold (i.e., locked positioning) phases. Birds exhibit OCR when perched on moving branches, keeping their heads motionless despite limb motion.

Accommodation

Accommodation enables birds to rapidly adjust refractive power for sharp vision across varying distances and is vital for high visual resolution. Accommodation occurs primarily through two mechanisms: (1) corneal accommodation and (2) lenticular accommodation. Additional mechanisms used by birds to produce accommodation include corneal curvature changes and a softer lens; both facilitate rapid accommodation and production of larger retinal images to enhance vision. As in most species, accommodation declines with age (presbyopia). Presbyopia is believed to have multifactorial causes, with changes in eye geometry being a primary factor.

Corneal accommodation

Birds are emmetropic on land, relying on corneal accommodation for approximately two-thirds of refractive power. Corneal curvature changes increase refractive power, with distortion facilitated by the anterior sclerocorneal muscle in nocturnal species and hawks. Under water, the cornea loses refractive power due to similar refractive indices of aqueous humor and water, shifting reliance to the lens. Penguins and albatrosses possess flattened corneas to retain refractive power underwater, while pursuit divers exhibit strongly curved corneas and enhanced lenticular accommodation to compensate for corneal limitations.

Lenticular accommodation

Lenticular accommodation involves the direct action of the posterior sclerocorneal muscle on the lens, increasing curvature, particularly in diurnal species. Diving birds exhibit additional compression of the lens by iris sphincter muscles, enabling forward bulging through the pupil. Some birds, such as chickens, demonstrate independent accommodation in each eye. Relaxation mechanisms include intraocular pressure and ligament action to restore unaccommodated corneal curvature.

Accommodative range

Accommodative ranges vary across species, with 0.6-10 diopters (D) reported in owls. Aquatic birds likely rely more heavily on lenticular accommodation as the refractive power of the cornea decreases due to similar refractive indices of aqueous humor and water. Water birds may also exhibit extreme accommodative accommodation to mitigate underwater refractive loss. Accommodative ranges up to 70-80 D have been reported in hooded mergansers and goldeneyes. Corneal refractive power ranges in cormorants are reported between 52.1 D and 63.2 D.

Static mechanisms

Certain birds maintain simultaneous focus on objects at different distances, utilizing asymmetric corneal and lenticular designs for emmetropia in some visual fields and myopia in others. Prey species, like pigeons, benefit from this ability by scanning horizons for predators while foraging. Raptors lack lower field myopia, optimizing their vision for capturing mobile prey.

Flicker-fusion frequency

Flicker-fusion frequency (FFF) is defined as the threshold at which rapid motion becomes indistinguishable and is vital for birds' survival. Temporal resolution depends on photoreceptor density and neural complexity, with birds far surpassing humans in this ability. While human FFF averages 50-60 Hz, birds can resolve movements above 100 Hz. Budgerigars exhibit an FFF of 115 Hz, chickens 105-120 Hz, and pigeons 116-146 Hz. Raptors, like Cooper's hawks (*Accipiter cooperii*), use this high FFF to detect prey movement during high-speed chases, avoiding obstacles such as branches that appear blurred to humans. In addition, birds can detect slow movements, such as the sun's arc or celestial rotations, aiding navigation during migration.

Medical and Surgical Management

Medical management

Table 1 lists common conditions of the eyes of birds of prey. Potential causes, prognosis, and recommendations for management also are described (Table 1).

Normal ocular parameters including intraocular pressure (IOP), Schirmer tear test (STT), and phenol red thread test (PRTT) as reported in various species of birds of prey are provided in Table 2.

Table 1. Common conditions of the eyes of birds of prey.

Condition	Potential causes	Prognosis for release	Management
Conjunctivitis	Relatively unusual in rehabilitation setting but can be significant. Foreign bodies or irritation may be seen (e.g., in birds fallen down chimneys). Nematode infections have been reported in North America. Viruses including pox virus (often associated with eyelid lesions); highly pathogenic avian influenza (HPAI) reported usually associated with neurologic signs. Herpesvirus associated with conjunctivitis along with systemic signs and death. Bacterial (chlamydial or mycoplasma infection often associated with sinusitis) and yeast infections also reported.	Good to poor depending on underlying cause.	Cases require full examination and evaluation and barrier care and quarantine. Examination under anesthesia will enable location or foreign bodies and possible flushing of the conjunctival sacs. Bacterial sampling is mandatory and viral swabbing for isolation and PCR where such infections are suspected. Chlamydial/ mycoplasma swab PCR may be indicated. Concurrent cytology indicated in all cases. Treatment depends on cause, although general supportive care while awaiting results will include topical eye drops and lubricant along with systemic antibiosis and NSAIDs.
Blepharoconjunctivitis	May be associated with trauma. In the UK often associated with tick attachment and reaction. Usually the tick(s) is(are) still attached but may have already detached by the time of examination. This may cause extreme reaction, debility and is life-threatening.	Good if responsive to therapy.	Systemic short-acting corticosteroid with antibiosis (e.g., penicillins). Supportive care including fluids and feeding is essential.

Table 1 Continued.

Condition	Potential causes	Prognosis for release	Management
Corneal ulceration or abrasion	The relatively large eye size of raptors renders them vulnerable to corneal injury. The cornea is relatively thin compared to mammals making it harder to approach surgically and more likely to sustain full penetration injury. Corneal lesions may be seen as part of other ocular infectious disease (e.g., herpesvirus infection) and full ocular and systemic evaluation is essential.	Good if the injury is not too deep and the cornea can be treated or repaired.	Full ocular and systemic examination. Bacterial and viral swabbing and cytology in refractive cases. Fluorescein staining is effective in assessing ulceration as in other species. Tear film assessment may be required in refractory cases especially where there are obvious cranial neurologic defects. Medical management (e.g., topical antibiosis) in shallow ulcers. Deeper or refractory ulcers may require surgical management as per dogs and cats. Penetration injuries will require corneal repair as per other species and management of associated uveitis (see below). Third eyelid flaps (or tarsorrhaphy) may be used to protect post-surgery. Topical NSAIDs should be used with care and while monitoring ulcer size and depth. Corticosteroids are contraindicated.
Uveitis or hypopyon	Uveitis is relatively uncommon in rehabilitation settings unless associated with trauma (especially corneal penetration) or infection (hypopyon). If bilateral, investigation of systemic infections should be undertaken.	Guarded unless the underlying cause can be found and treated. Hypopyon carries a poor prognosis for release.	Euthanasia if rehabilitation chances are deemed poor. Requires full systemic and ocular examination; the latter should be performed under general anaesthesia or using topical muscle paralytics to enable pupil dilation and posterior segment examination. Where anterior flare or hemorrhage and pus preclude such examination, ocular ultrasound is useful. Where treatment is deemed necessary, cases will generally require topical and systemic NSAID and antibiotics. Pupillary dilation is required as this can be hard to achieve practically.
Trauma or ocular hemorrhage	This is common and ocular damage must be assessed in any case of trauma, especially known head trauma. This will involve full posterior segment assessment looking for hemorrhage and retinal detachments.	Anterior segment hemorrhage carries a poor prognosis. Posterior segment hemorrhage may be managed provided it is not too extensive and retinal detachment is minimal.	Euthanasia if recovery prognosis is poor. Otherwise, topical corticosteroids (ensure no corneal damage first) and systemic NSAID may be effective in reducing hemorrhage and adhesion formation to the retina. Eyes should be examined at least weekly and hemorrhage resolution monitored. If deemed successful, sight should be evaluated as thoroughly as possible pre-release.

Table 1 Continued.

Condition	Potential causes	Prognosis for release	Management
Glaucoma	This is unusual but may be seen as a sequel to uveitis or hemorrhage with iris distortion or debris causing blockage of the irido-corneal drainage.	Guarded outlook.	Rebound tonometry is recommended in such cases especially where globes appear of different sizes. Ocular ultrasound is useful to differentiate from retro-bulbar disease. Full ocular examination should include examination of the irido-corneal drainage angle. If treatment is attempted, this should include assessment and management of underlying causes as well as use of drugs such as dorzolamide (topically) or dichlorphenamide (orally). Eyesight assessment is mandatory following treatment and before release even if pressures are normal.
Lenticular damage	Unusual in rehabilitation settings. Young owls often have opaque lenses and these clear as they mature; these do not require treatment	Poor when seen.	Surgery is possible but unlikely to be recommended if release is the goal of therapy.
Neural damage	Accompanying neural damage is not rare in ocular trauma cases. Horner's syndrome has been noted in such cases.	Depends on the nature of damage.	Full cranial nerve evaluation is required in head trauma cases. Management depends on the type of injury and extent of damage. Some will self-resolve; anti-inflammatories may be indicated. Nonresolution carries a poor prognosis.

Table 2. Normal ocular parameters (intraocular pressure [IOP], Schirmer tear test [STT], and phenol red thread test [PRTT]) reported in various species of birds of prey. IOP values were obtained by applanation tonometry.[5]

Species	IOP Mean ± SD (median)	STT Mean ± SD mm/min (median)[5]	PRTT Mean ± SD
Bald eagle (*Haliaeetus leucocephalus*)	21.5 ± 1.7 mm Hg; range, 15-26 mm Hg[a] 20.6 ± 2.0[b]	14 ± 2; range, 8-19 mm/min[a]	
Golden eagle (*Aquila chrysaetos*)[b]	21.5 ± 3.0		
Red-tailed hawk *Buteo jamaicensis*)[b]	20.6 ± 3.4		
Swainson's hawk (*Buteo swainsoni*)[b]	20.8 ± 2.3		
Harpy eagle (*Harpia harpyja*)[c]	15.00 (median) Interquartile range 6.0	17.48 ± 5.48; range, 3-28 mm/min	
Black kite (*Milvus migrans*)[d]	17.1 ± 7.2 (14.0)	7.4 ± 5.7 (6.0)	12.6 ± 9 (9.4)
Common buzzard (*Buteo buteo*)	19.4 ± 3.9 (18.5)[d] 17.2 ± 3.53[e]	13.7 ± 4.4 (13.0)[d] 12.47 ± 2.6[e]	16.0 ± 7.7 16.7)[d]
European honey buzzard (*Pernis apivorus*)[d]	14.4 ± 2.8 (14)	7.5 ± 2.2 (8.0)	25.6 ± 8.6 (23.5)
Common kestrel (*Falco tunnunculus*)	11.9 ± 3.3 (11.6)[d] 8.53 ± 1.59[e]	5.8 ± 4 (5.6)[d] 6.20 ± 3.67[e]	29.6 ± 4.7 (28.5)[d]
Great horned owl (*Bubo virginanus*)[b]	10.8 ± 3.6		
Barn owl (*Tyto alba*)	18.0 ± 6.6 (16.3)	3.6 ± 2.2 (3.0)	19.5 ± 7.2 (19)
Eurasian scops owl (*Otus scops*)[d]	14.5 ± 3.9 (14.8)	1.0 ± 0.5 (0.8)	11.8 ± 5 (11.0)
Little owl (*Athena noctua*)[e]	9.83 ± 3.41	3.50 ± 1.96	
Tawny owl (*Strix aluco*)[e]	11.21 ± 3.1	3.12 ± 1.9	

Source: [a]Kuhn et al., [b]Stiles et al., [c]Grego et al., [d]Beckwith-Cohen et al., and [e]Giovanni et al.[2-6]

Surgery

As a clinician in support of captive birds of prey, the most common ocular surgery you may need to perform is removal of a diseased eye. Enucleation was initially the most recommended method, borrowed largely from mammalian techniques; however, scleral ossicles, lack of bony orbital enclosure, short optic nerves, and attendant risk of damage to the optic chiasm from tension on the optic nerve, combine to make a less invasive technique, such as evisceration, desirable. Enucleation still has application, particularly in cases where neoplasia, isolated to the globe, is suspected. But the most common scenario of eye trauma which has rendered an eye phthisic, dysfunctional, uncomfortable, or unattractive, is probably best addressed with evisceration.

Anesthesia and analgesia

Standard avian general anesthetic techniques apply and can be found elsewhere in the literature. No special modifications are generally required for ocular surgery. A subcutaneous block of diluted lidocaine or bupivacaine, 2 mg/kg total dose, is suggested around the surgery site, particularly the lateral canthotomy. Endotracheal intubation is required, even in cases of rapid surgery, because of the potential for blood or lavage fluids to access the oral cavity through the lacrimal ducts or a perforated infraorbital sinus. Pre-operative general analgesics are recommended such as buprenorphine (0.25 mg/kg SC) or butorphanol (2-4 mg/kg IM). Midazolam may also be helpful (0.5-2 mg/kg IM) prior to surgery to decrease anxiety during restraint and induction. NSAIDs such as carprofen or meloxicam are also described as beneficial.

Surgical preparation

Preparation consists of cleansing the skin and small feathers adjacent to the eye. Although described in the literature, complete plucking of the skin is not required and can cause significant trauma to the delicate skin of small species, particularly of the lids, as well as unnecessarily prolonging preparation time. If there are long feathers present that may interfere, they can be trimmed short. Dilute chlorhexidine or povidone-iodine is applied to the skin for aseptic preparation, taking care to not allow it to enter the opposite eye. A sterile transparent drape is applied to allow visualization of the head position and mouth.

Equipment and supplies

Small, lightweight eyelid retractors are helpful. Otherwise, in large patients, a plastic frame retractor (e.g., Lone Star Retractor, Cooper Surgical, CT, USA) can be utilized with small hooks or modified to accept stay sutures. The author has also utilized small vascular clamps, attached to the surgical drape, as lightweight anchoring points for stay sutures. Magnification loupes are highly recommended for this procedure in combination with a forehead-mounted strong light source. Microsurgical instruments or tenotomy scissors are also helpful, though standard small iris scissors can suffice in large eyes. A small (e.g., #11 or #15) scalpel blade is recommended for the canthotomy and initial limbal stab incision.

Specialized instruments such as lens loupes or cyclodialysis spatula are described for removing eye contents, but in the author's experience, sterilized cotton-tipped applicators (CTAs) are sufficient for blunt tissue dissection. Fine Adson or similar atraumatic thumb forceps are recommended for manipulation of tissue edges and closure.

Evisceration

The procedure begins with a lateral canthotomy to create room for retracting the eyelids and accessing the anterior half of the glove. The cornea is then incised in a full circle along the limbus using either a #11 blade or tenotomy scissors. The contents are then gently dissected and removed from the scleral shell using sterile CTAs. Bleeding in most cases is minimal and can be controlled with pressure. All content should be removed to reveal the white scleral layer of the globe. In cases of chronic trauma, scar tissue may interfere with complete removal of some structures by blunt dissection and so some sharp dissection may be required. As long as the scar tissue in question is free of glandular tissues, it is likely safe to leave in place and disruption of the scleral ossicles or scleral shell is not recommended.

After content is removed, the interior of the globe is flushed with warm sterile saline, blotted dry with sterile gauze or CTAs, and lightly packed with absorbable gelatin sponge (Gelfoam, Pfizer Inc, New York, NY, USA). Insertion of a silicon prosthesis is described in the literature but unnecessary and rarely utilized. In the author's opinion, the only benefit is slightly fuller, natural facial appearance but it comes at the risk of unnecessary weight and increased, albeit low, risk of complication. The lid margins are trimmed away taking care to not disrupt the superior or inferior lacrimal puncta present near the medial canthus. Some techniques describe removal of the third eyelid and some leave it in place. The author recommends leaving it in place unless it is compromised and presents a risk of blocking the puncta or creating a nidus of irritation or infection. Conjunctiva is left intact. Closure of the lids is accomplished in one layer, sometimes two if lids and conjunctiva are thick enough to support it, using an absorbable monofilament suture (e.g., 5-0 polydioxanone [PDS II] or polyglactin 910 [Ethicon Inc, Piscataway, NJ, USA]) in a simple continuous pattern.

Enucleation

The eyelids are apposed using fine monofilament suture (e.g., 5-0 PDS II) in a simple continuous pattern, leaving extra loops or tags of suture at each end to facilitate retraction. The skin of the eyelids is carefully incised 2-3 millimeters from the margins to expose the underlying palpebral conjunctiva. A fibrous connection is present at the medial canthus that requires careful dissection. The superior and inferior puncta must be preserved. The conjunctiva is bluntly dissected away from the eyelids and the globe elevated with gentle traction on the aforementioned tags or loops on the eyelid sutures. Excessive traction may pull on the optic nerve and damage the optic chiasm, which could result in contralateral blindness. Once the space behind the eye is accessible, sharp dissection of the muscles and fat behind the eye, combined with fluid deformation of the eye itself, provides more space for visual access to the optic neurovascular bundle (ONB). Alternately, vitreous can be partially drawn from the eye at any point during surgery, using a large-bore needle and syringe, to shrink the glove and facilitate access. The ONB is clamped off with small hemoclips or utilizing bipolar radiocautery. Monopolar cautery is not recommended, especially in very small patients, due to the risk of collateral heat damage to the nearby optic chiasm and brain. The ONB is then transected distal to the clip or seal and the globe removed. Though exceptional, sometimes small patient size (body mass < 100 g) or chronic scarring of structures does not permit visual access to the ONB. In these cases, it may not be possible to confidently clip or seal it prior to cutting and removal of the eye, even with partial evacuation of the eye. Alternatively, the eye can be freed as much as possible, the bundle cut blindly with curved microscissors, and the eye quickly removed, followed by immediate pressure with CTAs to the optic neurovascular bundle. It is not as optimal as pre-emptively ligating the structure but rarely results in critical blood loss if actions are swift. Following removal of the eye, the socket is inspected for remaining tissue and lightly packed with absorbable gelatin sponge. The eyelids are closed with the aforementioned suture in a simple continuous pattern.

Post-operative care

Opiate pain coverage using butorphanol (1-2 mg/kg q6h IM) or tramadol is recommended for the first 2-4 days post-operation. There are a few species-specific dosages in the literature for tramadol use in birds of prey, including 5 mg/kg PO q12h for bald eagles (*Haliaeetus leucocephalus*) and kestrels (*Falco sparverius*), 8-11 mg/kg PO q12h for red-tailed hawks (*Buteo jamaicensis*), and 15 mg/kg PO (frequency unspecified, presumed q12h) for tropical screech owls. Consult a current exotic animal formulary and select the dose closest to the species with which you are working. In addition, a nonsteroidal anti-inflammatory is recommended such as meloxicam at 1-1.5 mg/kg PO q12-24h, depending on patient response. Sutures need not be removed. Rechecks at 24 hours, 7 days, and 14 days after surgery are recommended to monitor for swelling, drainage, and discomfort.

Other considerations

Pathologic lesions should be submitted to a laboratory for histopathologic evaluation. If infection is suspected, swabs should also be submitted for culture and sensitivity.

References

1. Jones MP, Pierce K, Ward D. Avian vision: a review of form and function with special consideration to birds of prey. *J Exot Pet Med.* 2007;16(2):69-87.

2. Kuhn SE, Jones MP, Hendrix DVH, et al. Normal ocular parameters and characterization of ophthalmic lesions in a group of captive bald eagles (*Haliaeetus leucocephalus*). *J Avian Med Surg.* 2013;27(2):90-98.

3. Grego AL, Wanderlei de Moraes, Moore BA. The eye of the largest forest raptor, the harpy eagle (*Harpia harpyja*): morphologic observations and reference values for selected ophthalmic tests. *Vet Ophthalmol.* 2025;1-13.

4. Beckwith-Cohen B, Horowitz I, Tali B, et al. Differences in ocular parameters between diurnal and nocturnal raptors. *Vet Ophthalmol.* 2015;18(Suppl 1):98-105.

5. Giovanni B, Briganti A, Spratte JR, et al. Schirmer tear test Type I reading and intraocular pressure values assessed by applanation tonometry (Tonopen® XL) in normal eyes of four European species of birds of prey. *Vet Ophthalmol.* 2013;16(5):365-369.

6. Stiles J, Buyukmihci NC, Farver TB. Tonometry of normal eyes in raptors. *Am J Vet Res.* 1994;55(4):477-479.

7. Montiani-Ferreira F, Moore BA, Ben-Shlomo G. *Wild and Exotic Animal Ophthalmology.* Vol 1. Cham, CH: Springer Nature; 2022.

8. Williams DL. Raptors: ophthalmology. In: Chitty J, Lierz M, eds. *BSAVA Manual of Raptors, Pigeon and Passerine Birds.*Gloucester, UK: British Small Animal Veterinary Association; 2008:278- 283.

9. Murray M, Pizzirani S, Tseng F. A technique for evisceration as an alternative to enucleation in birds of prey: 19 cases. *J Avian Med Surg.* 2013;27(2):120-127.

10. Pigatto J, Vargas E, Torikachvili M, et al. Modified eye evisceration in a tropical screech owl (*Megascops choliba*). *Acta Scient Vet.* 2021;49(Suppl 1): 664.

11. Dees D, Knollinger AM, Maclaren NE. Modified evisceration technique in a golden eagle (*Aquila chrysaetos*). *Vet Ophthalmol.* 2011;14(5):341-344.

12. Christen C, Richter M, Fischer I, et al. Unilateral evisceration of an eye following cornea and lens perforation in a sulfur-crested cockatoo (*Cacatua galerita*). *Schweiz Arch Tierheilkd.* 2006;148(11):615-619.

13. Altman R. Soft tissue surgical procedures. In: Altmann RB, Clubb SL, Dorrestein GM, Quesenberry K, eds. *Avian Medicine and Surgery.* Philadelphia, PA: WB Saunders; 1997:704-732.

Selecting an Appropriate Raptor for Permanent Placement

Dana Franzen-Klein, DVM, MS

Session #142

Affiliation: From The Raptor Center at The University of Minnesota College of Veterinary Medicine, St. Paul, MN 55108, USA.

Abstract: Raptors are highly desired for public education due to their ability to relay important environmental education messages while enthralling the audience with the opportunity to see these powerful, majestic, elusive birds up close. Most raptors that enter captivity from the wildlife rehabilitation sector are placed due to an injury or disability that prohibits their release to the wild. Veterinarians have an opportunity to make a large impact on the individual animal's long-term welfare by ensuring that the bird has been appropriately assessed before signing the letter approving the bird for placement in captivity. A thorough medical and behavioral assessment should be performed to select a raptor that will succeed in its new role as an ambassador animal. Variables such as age, species, natural history, permanent injury, individual temperament, the bird's response to the initial training efforts, and the desired job for the bird at the receiving institution must all be considered when determining if placement is appropriate for the individual bird. This 1-hour lecture uses case-based discussions to highlight The Raptor Center's recommendations for raptor placement based on 50 years of experience in the wildlife rehabilitation and raptor ambassador sectors.

Laws and Regulations

Federal regulations regarding the placement of wild birds from rehabilitation into permanent captivity are outlined in the Code of Federal Regulations (CFR) Title 50, Chapter 1, Subchapter B, Part 21, Subpart C, §21.76, number 4 (https://www.ecfr.gov/current/title-50/chapter-I/subchapter-B/part-21/subpart-C/section-21.76). The key points from the regulations to take note of as a veterinarian assessing the bird for placement are as follows:

- "You must euthanize any bird that cannot feed itself, perch upright, or ambulate without inflicting additional injuries to itself... You must euthanize any bird that is completely blind, and any bird that has sustained injuries that would require amputation of a leg, a foot, or a wing at the elbow or above."
- The sending and/or receiving institution must complete the U.S. Fish and Wildlife Service (USFWS) Migratory Bird and Eagle Acquisition and Transfer Request form (3-202-12).
  - A component of this paperwork is a letter from a veterinarian outlining why the bird is non-releasable and why it is appropriate for placement.
    - The USFWS provides guidance in a document titled "Guidance for Veterinarian Certification of Condition That Renders Migratory Bird/Eagle as Non-releasable." This document outlines all the variables that must be assessed and information to be included in the letter from the veterinarian. This document can be found at this website: https://www.fws.gov/sites/default/files/documents/Template_Vet_Letter.pdf.
  - If you will not be providing long-term medical care for this raptor at its new facility, you must provide the name of the veterinarian you consulted with at the receiving institution who will be providing long-term care for this bird.

The role of the veterinarian in raptor placement

A letter of support from a veterinarian is required for a raptor to be placed in captivity as an educational ambassador. This includes performing a physical examination and talking to the rehabilitator about the bird's physical and behavioral suitability for placement. This is a prime opportunity for the veterinarian to provide an objective assessment of the bird's medical and behavioral welfare, and to provide input and feedback on the suitability of the bird for long-term captive placement. If you are not certain if the bird is appropriate for placement, consider consulting with another veterinarian, or a facility such as The Raptor Center that does both rehabilitation and has a collection of captive ambassador raptors, before you sign the letter of support.

Assessing a Raptor for Permanent Placement

Signalment

Signalment, including species age and sex, is the first component of assessment when determining if a raptor is appropriate for placement. In general, hatch-year birds (i.e., raptors in their first calendar year of life) more readily adapt to a life under human care than adults, and experience better long-term welfare.[1] For the purpose of this document, an adult raptor is defined as any raptor in its second year of life or older. A raptor is considered to be in its second year starting January 1st the year after the hatch year. Raptors can be aged using a combination of plumage and physical characteristics.

Males of many raptor species tend to be more energetic and high strung compared to females.[1] This can be explained by their natural history; during breeding season males are the primary food provider and territory defender, so they spend more time in flight.

Some species of raptors are high stress and do not adapt well to captivity after spending the first portion of their life in the wild. Examples include osprey (*Pandion haliaetus*), sharp-shinned hawks (*Accipiter striatus*), and snowy owls (*Bubo scandiacus*).[1] Conversely, other species such as American kestrels (*Falco sparverius*) and red-tailed hawks (*Buteo jamaicensis*) tend to adapt well to captivity.[1] The skill level of the staff at the receiving institution should also be considered when evaluating if an individual bird is appropriate for that institution. Some species are more challenging to train and manage, and require more experienced handlers for the welfare of the bird and for human safety.[1]

Physical assessment

Raptors are placed from rehabilitation facilities into captivity due to a physical or mental condition that has rendered them non-releasable. These permanent, physical disabilities can significantly impede the bird's welfare, either in the present, in the future, or both. The veterinarian plays a critical role in this component of pre-placement assessment, and is uniquely qualified to provide insight into the long-term physical welfare of the bird.

Along with the physical placement criteria outlined in the federal regulations, The Raptor Center has the following additions for birds that we feel should be euthanized instead of placed into captivity due to concerns for the animal's long-term welfare. These recommendations are based on experiences from The Raptor Center, where we have managed a collection of captive raptors for nearly 50 years, and based on experiences of colleagues. Additions are listed below.

- Birds with significant wing droops
- Birds with significant neurologic deficits
- Birds with balance issues
- Wing amputations at the level of the carpal joint or wrist, or above
- Any medical condition that has the strong potential to deteriorate, require chronic medications, or cause chronic discomfort; examples include but are not limited to:
    - Degenerative joint disease or arthritis present at the time of placement
    - Active uveitis that requires medical management at the time of placement

Behavioral assessment

While there are general guidelines on which species of raptors tend to do well in captivity, each bird is an individual and may not follow the general trends. Some individuals more readily adapt to captivity, while others remain highly stressed despite changes in management to try to decrease the source of stress. A bird that is developing captivity related injuries in rehabilitation prior to placement, either due to a physical injury or stress related behaviors, should be critically evaluated and may not be appropriate for placement.

Behavioral assessment begins by examining how the bird behaves in the captive rehabilitation setting. Does it eat voluntarily? Is it maintaining its feathers well, meaning it is not damaging feathers by exhibiting escape behaviors or due to a severe wing droop? How close can people get before the bird exhibits escape behaviors? Does the bird show its stress outwardly through behavioral changes or is it a bird that tends to internalize its stress, which makes it more challenging to train.

The primary behavioral reason for placement is human imprinting or habituation. Imprints can be great ambassador birds, especially when they are captive reared by a professional raptor breeder. Hard human imprints, or sexual imprints, may become aggressive towards humans when the bird reaches sexual maturity. Imprints can also be unpredictable and their behavior can be hard to read. Soft imprints, dual imprints, or birds that are strongly habituated towards humans may be safer placement options. An example of a habituated bird may be a bird that was parent reared until it reached fledging age, and then entered human care due to an injury or kidnapping by a well-meaning member of the public.

In an ideal situation, the training process would begin as soon as a raptor is identified to be non-releasable. This includes changing the bird's handling and management to create more positive associations with humans. For example, by using positive reinforcement you can train the bird to step onto a scale to obtain voluntary weights, or train the bird to step up and ride on the handling glove so it can be moved from place to place in a positive way. This is a great way to perform a thorough behavioral assessment and see how the bird responds and adapts to this new type of handling and relationship with humans. This can remove the need to physically restrain the bird for routine care tasks, and decrease the amount of negative experiences with people in the bird's memory bank. The longer the bird is being handled like a wild rehabilitation patient, the more negative experiences it will have with people and the harder it will be to gain that bird's trust. This ideal scenario is not possible in the majority of cases. Rehabilitators often do not have the experience or time to start this training. In addition, permitting processes can be prolonged, delaying the bird's transfer to the receiving institution where this training can begin.

Desired role

The final component of assessing a raptor for placement is evaluating the needs of the receiving institution. What role would they like the bird to be able to perform at their facility? Birds that are used for on-the-glove programs and travel to multiple different locations for programming have different behavioral criteria for placement compared to a bird that is for display only at a zoo and is not directly handled by people on a daily basis. The experience level of the staff at the receiving institution, and their ability to access mentorship, should also be considered. Taking the time to match the bird with the needs and capacity of the receiving institution ensures that both the people and the raptor are set up for success.

References

1. Arent L. *Raptors in Captivity: Guidelines for Care and Management.* Blaine, WA: Hancock House Publishers; 2018.

Falconry in the United States: An Introduction to the Sport for the Avian Practitioner

Dana Franzen-Klein, DVM, MS

Session #143

Affiliation: From The Raptor Center at The University of Minnesota College of Veterinary Medicine, St. Paul, MN 55108, USA.

Abstract: Falconry is the ancient sport of using a trained bird of prey to hunt wild quarry. The art of falconry is still practiced today throughout the world, and throughout the United States. Falconers often struggle to find access to veterinary care, as many practitioners do not feel comfortable providing care to these birds. This 30-minute lecture is an orientation to falconry, providing a high-level overview of the sport and how these birds are trained and used so that the avian practitioner can feel more confident understanding falconry clients. This presentation does not focus on medical diagnoses or treatments. The presenter has been providing medical care for falconry clients for 9 years at The Raptor Center at the University of Minnesota and is also a practicing falconer.

The Art and Sport of Falconry

Falconry is the art and sport of taking wild quarry (i.e., game animals) with trained birds of prey such as falcons, hawks, owls and eagles. Falconry is embedded in many cultures throughout the world, with over 4,000 years of history as a method of hunting and sport. In 2010 the United Nations Educational Scientific and Cultural Organization (UNESCO) added falconry to its list of the Intangible Cultural Heritages of Humanity.[1]

It is often said that falconry is more than a sport; it is a lifestyle. The relationship between the falconer and the bird is a priority, and takes a large amount of time, skill and patience to develop and maintain. The art of falconry is in the relationship between the falconer and their bird. A seasoned falconer is able to read subtle changes in the bird's behavior and tailor the training regime to each individual. A relationship built on trust and respect is required to successfully fly a raptor free and be able to recover that bird after the hunting expedition. Falconry birds are not pets in the modern sense of the word. They are wild animals that the falconer has developed a working relationship with. Falconers care for and respect their birds and want to ensure their health and welfare.

Most falconers are also conservationists. They are invested in the protection and sustainability of wild raptor populations, the wild populations of their intended quarry, and the conservation of the associated natural habitats. For example, falconers played a vital role in the recovery of the previously endangered peregrine falcon (*Falco peregrinus*). It is not uncommon to come across an accomplished falconer in the field of raptor biology and conservation.

Falconers in the United States hunt a wide variety of game species, including squirrels, rabbits, quail, grouse, pheasant and ducks. The intended quarry is matched with the falconry bird, using the natural hunting styles and instincts of the species to hunt in a cooperative team with the falconer.

Laws and Regulations in the United States

All falconers must hold a state permit for routine falconry activities; a duplicate federal permit is not needed. A separate U.S. Fish and Wildlife Service (USFWS) federal permit is required for abatement (permit 3-200-79) and propagation (permit 3-200-12). State falconry regulations are quite variable. State regulations can be more restrictive, but not less restrictive, than the federal regulations. Federal falconry regulations are located in the Code of Federal Regulations (CFR) Title 50, chapter 1, subchapter B, part 21, subpart C, §21.82 (https://www.ecfr.gov/current/title-50/chapter-I/subchapter-B/part-21/subpart-C/section-21.82).

There are three main classes of falconry permits: the apprentice, general and master falconer levels. An apprentice class falconer is the first level of permit a falconer can acquire. An apprentice must have a sponsor, pass a written examination, may only have one bird in their possession at a time, and they are restricted in the species and origin of the birds they are allowed to have in their possession (which varies by state). An apprentice falconer can apply for a general class permit after a minimum of two years. General class falconers are allowed to have up to three birds in their possession at once and are allowed to have a wider variety of species, including both captive bred and wild birds of various age classes. A general class falconer may apply for their master class permit after 5 years. Master class falconers may have up to five birds in their possession at once and have access to all species permitted for use in falconry including eagles. Facilities inspections are a component of the application and renewal process for all types of falconry permits.

Falconers must report the acquisition and disposition of each bird in their possession using the USFWS form 3-186A within 5 days of the transaction. This form is submitted at both the state and federal level. Falconers must also complete an annual report at both the state and federal level.

Permitted falconers are allowed to take birds from the wild into captivity for the sport of falconry. Dependent on their permit class, they may either trap a young bird within its first year of life (i.e., fledgling or hatch year age, also referred to as a passage bird), or they may take a bird from the nest and raise it in captivity as a human imprint. Wild-trapped passage birds can be released back to the wild. Hand-reared wild birds that are human imprints cannot be released back to the wild. There are also captive breeding programs where birds can be purchased for use in falconry. Captive-bred birds may be hand reared by people as human imprints, or they may be chamber raised or parent reared, meaning they are raised by the parent birds and are not human imprints. Captive-bred birds cannot be released to the wild no matter their imprint status. Falconry birds are required to have some form of individual identification, regulations around the type of identification vary by state. Leg bands are the most common form of individual bird identification. Some states provide numbered zip-tie bands to place on wild trapped birds after acquisition. Captive-bred birds have closed, complete metal leg bands placed when they are nestlings.

Falconers are required to follow all state and federal hunting regulations. They must have the appropriate hunting licenses for their quarry of interest and are limited to the same hunting seasons as gun hunters.

Veterinarians are not required to have any type of permit, or complete any sort of reporting, when providing medical care to a falconry bird. It is recommended to record the permit information, and bird identification information (i.e., band number) in the medical record for your falconry clients.

Training, Captive Management and Equipment

There is a wide range of training methods in the sport of falconry, the details of which are beyond the aim of this manuscript. The training techniques used for wild-caught passage birds do not prevent the bird from being released back to the wild. One key component of training that is important for the attending veterinarian to be aware of is weight management. Food is commonly used as reinforcement for desired behaviors. For a wild bird to be willing to engage in training for food, it needs to be mentally or physiologically hungry. As part of the initial training, the falconer will often decrease the bird's weight from its trap weight (i.e., weight at the time of trapping from the wild) to its flight weight (the optimum flying body weight), the weight that the bird is food motivated and responding as desired to the trained cues. The flight weight at the beginning of the bird's first season is often the lowest it will be in the bird's falconry career. As the bird learns and practices the desired behaviors, the weight is often increased as the falconry season goes on. The exception to this is when a bird is trapped in thin body condition, in which case the flight weight may be higher than the trap weight. When the hunting season is over, the bird's weight is increased prior to release back to the wild or in preparation for the seasonal molt. Falconers may also refer to the bird's molt weight or chamber weight. This is the weight the bird is increased to during the off season when it is not actively hunting and is molting.

As with companion birds, captive management and husbandry can have impacts on the bird's health. Weight management, housing environment, nutrition, perching, and sanitation methods can all impact the bird's health, for better or worse. The history and exam form template included below presents several questions related to captive management and husbandry that can provide important clues and insight into the health of the patient presenting to you (Fig 1).

History Form for Falconry Birds UNIVERSITY OF MINNESOTA
THE RAPTOR CENTER

Client information
Name:
Address:
Phone number:
Email address:
Falconry permit number:

Signalment and Management/Husbandry
Species:__________ **Case number**:__________ **Name**:____________ **Date**:_________
Age:_______ **Sex**: M F U **Band**: L R Closed Open **Band #**:___________
Source: Captive Bred Wild Caught
Breeder:____________________ Trap location:_________________
Hatch date:__________________ Trap date:____________________
Chamber raised or imprint Trap weight and keel score:_________
Flight weight:__________ **Molt weight**:__________ **Hx WNV vx**: Yes No
Diet:
Food types:
Food source:
Supplements or medications:
Does the bird eat wild game, what type and when:
Housing:
Indoor and/or outdoor mews:
Free lofted or tethered:
Substrate:
Ventilation:
Cleaning and disinfection methods and frequency:

Figure 1. Sample History Form for Falconry Birds.

There is a wide variety of equipment used in falconry, and a variety of management methods to keep a wild raptor in captivity. Some of this information can be found in the glossary of common terms below.

Medical Care

Falconers struggle to find access to veterinary care. It is challenging for them to find a veterinarian that is comfortable providing medical care for their raptor, and the attending veterinarian may not be familiar with the sport or common medical conditions that affect falconry birds. Because of this limited access, home care remedies and treatment are somewhat common in the falconry community. Additionally, falconers can get access to some prescription medications through online shops and farm supply stores and may start treatment on their own before connecting with a veterinarian.

Medical conditions and treatment are beyond the scope of this manuscript and presentation. Common medical conditions in falconry birds include but are not limited to the following:

Infectious disease

- Gastrointestinal and external parasites
- Aspergillosis respiratory infections
- Highly pathogenic avian influenza infection
- West Nile Virus infection
- Trichomoniasis (frounce)
- Inclusion body herpesvirus hepatitis

Traumatic conditions, sustained during hunting or in the enclosure

- Fractures
- Wounds
- Traumatic head injuries and ocular trauma

Management-related conditions

- Mid-winter anemia (functional starvation)
- Sour crop - gastrointestinal ileus where the crop contents are not passed in a timely fashion, often a secondary clinical sign related to some other systemic issue
- Pododermatitis/bumblefoot
- Carpal/wrist wounds and cere wounds

- Damaged and/or broken feathers
- Wounds associated with equipment such as mandibular wounds from hoods
- Lead toxicosis
- Visceral gout
- Amyloidosis

Age-related conditions

- Cataracts
- Osteoarthritis
- Atherosclerosis

The veterinarians at The Raptor Center at the University of Minnesota do provide remote consultation services to veterinarians.

Glossary of Common Terms

The sport of falconry includes its own vocabulary of terms. There are unique terms for equipment, behavior of the bird, training methods, and many others. The modern apprentice is a falconry website with a helpful glossary of terms (https://www.themodernapprentice.com/glossary.htm). Some common terms you may encounter when working with a falconer in a veterinary setting are provided in Table 1.

Table 1. Common falconry terms and definitions.

Term	Definition
Anklets	Leather equipment around the tarsometatarsus of the bird used as an attachment point for the jesses and leash. Similar to a collar on a dog. Can also be an attachment point for bells and telemetry
Bate	The bird flying off the handling glove or off of a perch while attached to the glove or the perch with a leash
Cast	To grab and restrain the bird for a medical examination or management procedure
Casting	The act of regurgitating the indigestible fur and feathers from a previous meal. The product that is brought up is called a pellet. Owls cannot digest bones and therefore cast bones in their pellets as well.
Chamber raised or parent reared	A captive bred bird that was raised in the breeding chamber by the parents, not a human imprint
Coping	Beak trim
Creance	A long line or tether that is used for free flight training
Deck feathers	The two central tail feathers
Eyass	A downy nesting age raptor, this can also refer to a fully grown bird that was taken from the nest as a downy chick
Feak	The action of rubbing the beak on objects to clean it
Foot	A verb meaning the bird grabbing something with the feet, can be directed towards people as a defensive behavior
Free loft	A housing technique where the bird is allowed to fly free in the enclosure without being attached to a leash
Fret marks	Also called stress bars, lines in the feather that develop during molt
Gorge	May also be referred to as allowing the bird to crop up, letting the bird eat a large meal or eat until they stop on their own
Hallux	Digit one, one of the two power talons (digit one and digit two)
Hard penned	A fully grown flight feather that is now out of blood, meaning the blood supply is no longer within the shaft of the feather and the feather is now hollow
Hawk box or giant hood	Travel container with limited light for transporting birds of prey
Hood	The leather head covering used to block vision in hawks and falcons
Imp	The act of replacing a broken flight feather with a donor feather
Imprint	A bird raised by humans that is now a human imprint or identifies as a human not a bird. There is a large spectrum of imprinting, ranging from a hard sexual imprint to a dual imprint to a soft imprint.
Intermew(s)	To keep a wild trapped bird in captivity through the molt between hunting seasons. A falconer may age the bird by saying the bird is three times intermewed, meaning they have had the bird for three molt seasons in captivity.
Jess	The leather strap that extends from the anklet around the birds tarsometatarsus to the leash to help contain the bird
Lure	An object that may or may not look like quarry that is used to recall the bird or used for lure flying, a technique to physically condition the bird
Man or manning	The process of training or acclimating a bird to human presence, a well-manned bird is comfortable around people
Mew	Bird's enclosure

Table 1 Continued.

Term	Definition
Mute	The combined urine, urate and fecal output, often called dropping in companion avian medicine
Passage	A wild trapped bird
Put over the crop	Passing food material from the crop into the proventriculus and ventriculus
Rouse	Shaking the whole body to adjust the feathers, part of preening and a comfort behavior
Stoop	Diving in flight
Tethered	Housing and management method where the bird is attached to an immovable object or perch with a leash
Telemetry	Radio-telemetry or GPS devices that are placed on the bird to track the bird when free flying. Telemetry devices can be attached to the bird via multiple methods, including a backpack mount, tail mount, leg mount, or neck mount.
Weather	To put the bird outside in open air and exposed to the elements, some falconers may have a weathering yard which is a protected outdoor area exposed to the elements.
Pellet	The indigestible material that is regurgitated/vomited by a raptor after a meal

References

1. North American Falconers' Association. Press release. Nov 16, 2010. Available at: https://cdn.ymaws.com/www.n-a-f-a.com/resource/resmgr/Docs/UNESCO_FalconryAnnouncement.pdf. Accessed August 11, 2025.

The Science of Compassion and Its Application to Veterinary Medicine

Jesse A. Fallon, DVM, PhD

Session #209

Affiliation: From Avian Conservation Center of Appalachia and Cheat Lake Animal Hospital, Morgantown, WV 26508, USA.

Abstract: Most of us begin our journey into the field of veterinary medicine because of the compassion we have for animals. We love our furry, scaled, and feathered companions and want to help them live full, healthy lives. But for many of us, somewhere along our path we encounter feelings of frustration in our diagnostic and treatment limitations, disappointment with our colleagues or management, and exasperation with non-compliant clients. Add to those personal challenges such as financial struggles, relationship stress, or emotional insecurity, combined with existential threats outside our control such as political turmoil, natural disasters, and climate change, and it seems like all we can do is try to survive. When we enter this survival mode, compassion for our patients and clients can seem less consequential and at times unimportant. But as we struggle with these tribulations, compassion can help sustain us; leaning into compassion can allow us to view the world in a more positive and productive way. Veterinary medicine thrives on compassionate care. Indeed, there is a growing body of scientific evidence that our inherent compassion for patients and clients can help us deal with the challenges of professional and personal life. In this presentation, we explore the science of compassion and apply this knowledge to clinical veterinary medicine.

What Is Compassion and How Do We Fatigue?

Compassion is an emotional response to suffering paired with a desire to help and has long been a foundational principle in the ethical practice of veterinary medicine. Indeed, the compassion we have for our non-human companions is what inspired many of us to become veterinarians in the first place! Despite the inherent nature of compassion in the practice of medicine, compassion fatigue—a form of emotional exhaustion that can occur in veterinary professionals who are repeatedly exposed to suffering and trauma in their work—seems to have become commonplace. Sometimes defined as the negative cost of caring, compassion fatigue can manifest as physical and emotional symptoms like exhaustion, depression, headaches, changes in appetite, and substance abuse.[1]

The prevalence of compassion fatigue in veterinary medicine is staggering. A comprehensive study involving over 5,000 full-time U.S. veterinarians revealed that 50% experienced high levels of burnout, and 59% reported high levels of sustained stress. Factors such as high educational debt, extensive time spent working with companion animals, and being female were associated with higher levels of compassion fatigue.[2] Other recent, albeit limited, research on the topic of compassion fatigue in veterinary medicine have found similar results.[3]

With all these challenges, what is a naturally compassionate veterinarian to do? How can we avoid burnout and fatiguing one of our most useful and heart-felt emotions? First, we should consider the neurobiologic mechanism that is associated with compassion.

The Neurobiology of Compassion: How Does it Work?

What happens in the brain when we feel compassion? Functional magnetic resonance imaging studies have demonstrated that feeling compassion engages specific neural circuits, including heightened activity in brain regions such as the anterior insula, anterior cingulate cortex, and the periaqueductal gray center —areas associated with empathy, emotional regulation, and caregiving behavior.[4,5]

Oxytocin, often called the bonding hormone, is also implicated in the physiologic basis of compassion. Elevated oxytocin levels are associated with increased trust, social bonding, and prosocial behavior. Compassion training has been found to increase oxytocin release, modulate cortisol levels, and activate parasympathetic responses that contribute to emotional balance and well-being in clinicians.[6] These findings support the idea that compassion is not merely a psychological construct but a biologically rooted human capacity that can be enhanced through deliberate practice. With an understanding of these biological mechanisms behind compassion, how can our clients and patients benefit from compassionate care?

Compassion with Clients: Building Trust

Pet owners often enter a veterinary office anxious, confused, or worried. Compassion becomes a bridge to mutual understanding in these emotionally charged moments. By listening attentively, explaining clearly, and validating the owner's emotions, veterinarians build trust and foster a collaborative approach to care. For example, delivering bad news, discussing end-of-life decisions, or setting financial boundaries all require sensitivity. A compassionate veterinarian does not shy away from these difficult conversations but approaches them with honesty and empathy. This not only supports the client emotionally but also helps prevent misunderstandings, compliance issues, and avoid resentment. Moreover, compassionate communication can often de-escalate tense situations. Veterinarians often face clients who are upset or defensive, especially when costs or outcomes are not what they hoped for. Meeting anger with understanding instead of defensiveness can transform conflict into collaboration, which is key to a sustainable relationship to practice.

Lean Into Compassion: Physiologic and Psychological Benefits

We know that compassion is an essential trait for effective delivery of patient care and client interaction; however, a growing body of scientific evidence also shows that acting compassionately can have substantial benefits for the clinicians themselves. Compassionate care—marked by empathy, attentiveness, and a desire to alleviate suffering—not only enhances interpersonal relationships in clinical settings but also contributes to lower stress, reduced burnout, and improved mental and physical health among healthcare providers.

The regions of the brain stimulated when acting with compassion are the same areas involved in reward processing and emotional regulation, suggesting that compassion can stimulate a neurologic reward response in clinicians. Acting with compassion activates the parasympathetic nervous system, promoting calmness and reducing stress.[7] For example, compassionate behavior leads to increased heart rate variability (HRV), a marker of autonomic flexibility and emotional regulation.[7] Higher HRV is associated with greater resilience to stress and reduced risk of anxiety and depression, which are common among healthcare professionals.

On a psychological level, clinicians who practice compassion report higher levels of job satisfaction and less burnout. A longitudinal study of human healthcare workers found that self-reported compassion was inversely related to emotional exhaustion and depersonalization—two key dimensions of burnout.[8] Further, a 2024 study of 868 veterinary professionals in Portugal found that higher life satisfaction was associated with lower psychological distress.[9] Crucially, this relationship was mediated by compassion satisfaction and resilience, which together explained over 65% of the effect. This underscores the importance of finding joy and fulfillment in compassionate care, as it can act as a buffer against the emotional toll of the profession. Together, these findings suggest that fostering compassion may protect clinicians from the detrimental effects of chronic stress inherent in healthcare settings.

Conclusions

Veterinary medicine is often seen through the lens of science, diagnostics, and treatment; however, behind every clinical decision lies a deep well of emotional labor, especially when caring for creatures who cannot speak for themselves. Compassion in veterinary medicine isn't just a soft skill; it is a vital part of ethical, sustainable, and effective practice and can help return joy to our work. Leaning into compassion means recognizing and valuing the emotional dimensions of animal care, fostering better outcomes for patients, clients, and veterinary professionals themselves. In a profession marked by high burnout and emotional fatigue, embracing compassion offers both a path to healing and a foundation for thriving practice.

References

1. Lutz K, Amundson N, Nishikawara R. Addressing compassion fatigue using career engagement and the hope-centered model for career development. *Can J Career Dev.* 2023 Jan 20;22(1):41-50.

2. Ouedraogo FB, Lefebvre SL, Hansen CR, Brorsen BW. Compassion satisfaction, burnout, and secondary traumatic stress among full-time veterinarians in the United States (2016-2018). *J Am Vet Med Assoc.* 2021;258(11):1259-1270.

3. McArthur ML, Andrews JR, Brand C, Hazel SJ. The prevalence of compassion fatigue among veterinary students in Australia and the associated psychological factors. *J Vet Med Educ.* 2017;44(1):9-21.

4. Lutz A, Brefczynski-Lewis J, Johnstone T, Davidson RJ. Regulation of the neural circuitry of emotion by compassion meditation: effects of meditative expertise. *PLOS One.* 2008;3(3):e1897.

5. Klimecki OM, Leiberg S, Ricard M, Singer T. Differential pattern of functional brain plasticity after compassion and empathy training. *Soc Cogn Affect Neurosci.* 2014;9(6):873-879.

6. Singer T, Klimecki OM. Empathy and compassion. *Curr Biol.* 2014;24(18):R875-878.

7. Kirby JN, Doty JR, Petrocchi N, Gilbert P. The current and future role of heart rate variability for assessing and training compassion. *Front Public Health.* 2017;5:40.

8. Fernando III AT, Consedine NS. Beyond compassion fatigue: the transactional model of physician compassion. *J Pain Symptom Manage.* 2014;48(2):289-298.

9. Peixoto MM, Cunha O. Life satisfaction, psychological distress, compassion satisfaction and resilience: when the pleasure of helping others protects veterinary staff from emotional suffering. *Vet Res Commun.* 2024;48(5):3489-3498.

Section 12

Masterclass Session I

Len Donato, VMD, DABVP (Avian Practice)
Session Chair

Choose Right, Choose Flight

Todd Driggers, DVM

Session #218

Affiliation: From Avian and Exotic Animal Clinic of Arizona, 1911 S. Lindsay Rd. Mesa, AZ 85204, USA.

Abstract: Why does an understanding of flight matter to wild animal care and veterinary teams? The goal of returning injured wild birds to nature successfully requires the team to understand the impacts of injury on routine flight and the functional ability of birds to maintain sustainable use upon release. It may enable insight into best decision practices for non-releasable wildlife regarding proper housing, spacing, aviary complexity, and environmental enrichment. From a clinical point of view, four types of alterations can be encountered in avian flight impairment: in anatomic structures, in normal motion, in coordination or aerial proprioception, and in a bird's general status. Flexibility is indispensable for the kinematics of flapping flight and is allowed by feathers, propatagium, and bone properties. With this question answered, a basic understanding of aerodynamics can inform critical clinical decisions and clarify considerations for releasable and captive birds. Aerodynamic principles are the focus of this presentation, including the characteristics common to gliding and flapping flight and the general uniqueness of the anatomy that has promoted them from dinosaurs to modern aerial geniuses.

Evolutionary Changes that Enabled Flight

Every aspect of a bird's biology impacts the evolution of flight. Evolutionary theories regarding tree gliding versus ground flapping are still debated among evolutionary scholars.[1] The necessary anatomic and physiologic changes for flight are still studied today. As it turns out, a slight change in form results in successful reproduction and genetic modifications. These changes over time have allowed birds to exploit divergent landscapes, aquatic environments, and diverse climates. The credited model original bird, the *Archaeopteryx*, was thought to be the first flighted genus and diverged into more than 10,000 species.[1] Inheritable skills and traits allow these species to occupy the habitats and exploit the niches they are created by, which is the definition of evolution. Individual species features are diverse and unique but are beyond the scope of this talk.

Light weight modifications

- Scales became feathers[2]
- Loss of teeth
- Bones got lighter in weight by hollowing and pneumatic and some skeletal fusion
- Torso got shorter
- The heart got larger and adapted to its need for altitude.
- More efficient respiratory system
- No urinary bladders or urethras

- Females' vestigial right-sided ovaries/oviduct
- Penises in Vasa parrots, ducks, and the ratites.[3]

Aerodynamics

Aerodynamics is how a bird counteracts the force of gravity, lifts off from the ground, and stays aloft while flying. The wings maintain flight by creating lift. Counterforces to lift is drag. Birds' flight is due to airflow over the leading edge (i.e., the airfoil). The wing airfoil diverts air above the wing (i.e., extrados) and below the wing (i.e., intrados). A bird's wing comprises the musculoskeletal apparatus, skin folds (i.e., patagium), and feathers. The elongated and curved shape of the extrados (i.e., upper wing surface) causes the air to move at higher velocities that change pressure gradients on each side of the wing, where low pressure effectively creates an upward pull, suction, or lift. Due to the camber (i.e., cup or curve) of the underside of the wing, a higher pressure exists can be created for takeoff. This higher pressure under the wing prevents stall due to the induced drag. In summary, high velocity on the top side of the wing creates a low pressure and an upward suction. Lower velocity on the underside of the wing has higher pressure (i.e., compression of the air), which prevents stall.

Physics principles of lift with flapping flight can be described by considering velocity represented by a vector quantity, a value with both a magnitude and a direction. Bird wings change the air's direction and speed by directing incoming air onto a new path. Higher velocity is created by the extrados (i.e., dorsal wing) curving shape. Newton's Second Law of Motion describes how force equals mass multiplied by acceleration (i.e., a change in velocity over time). In the case of a flying bird, the mass of air that is accelerated by the wings is equal to a force. The direction of the force applied to the air is equal and opposite to the force and direction of the bird. With flapping flight, lift and thrust are opposite the downstroke's direction. Minor thrust is created by the upstroke, equal and opposite the direction of the bird.

Pressure physics

When air comes in contact with the wing surface, pressure is placed on the wing surface. Atmospheric air puts pressure on every object—as measured by a barometer—and is called static pressure at the same altitude and weather conditions. This static pressure (P) is the pressure on the not-in-flight wing. Dynamic pressure ($PV^2/2$) is a pressure that is measured on a wing when a bird uses flapping flight at a velocity (V). At a given altitude, pressure is constant and obeys Bernoulli's equation of P + (Pv2/2) = constant. As a bird takes flight, the inverse relationship ensures pressure stays stable, but the distribution varies with the air velocity created by wing movements. As the flapping flight speed increases, dynamic pressure increases, and static pressure decreases. The wing shape harnesses the pressure. To summarize, the airfoil (i.e., wing) flapping compresses air under the wing, slowing its velocity and increasing velocity dorsally, creating lift. High pressure under the wing drags and prevents stalling, and low-pressure lifts. The dynamic pressure changes at the wing tip changes during downstroke, and with tip reversal of upstroke, it inverts. This is the component of the upstroke that produces more and less lift depending on the species being studied. Further kinematic studies are needed to quantify these pressures that result in the equal and opposite forces of both lift and thrust.

Wing form

Wing form and use have implications for necessary functional purpose. The aspect ratio is the wingspan from tip to tip squared, divided by the wing area. The longer the wingspan, the better the lift. Full extension results in maximal lift, and wing flexion reduces lift. Gliding birds have longer wing spans and relatively thin widths.

The body weight to wing area defines wing loading. Lightweight birds with large surface areas have low output needs (i.e., low wing loads) and are good at gliding. A low-wing-load and high-aspect-ratio bird is the general shape of most sea birds. A low wing load and a high aspect ratio allow the frigate bird to generate high lift over various conditions. The frigate is effective at taking off and landing, carrying heavy food items, and maneuvering to steal food from other flying birds. Generally, heavy-bodied birds are built for speed with high wing loads, but only to a point. Raptors are speed examples, and the peregrine falcon is the prototype of such a body shape. Higher speeds further generate lift, so wing extension in a stoop allows falcons to pull up quickly when hitting prey mid-air.[4]

High camber (i.e., the wing's curvature as seen through a cross-section) creates more lift and high drag. Birds with highly cambered (i.e., curved) wings typically have larger barrel-shaped bodies, utilize rapid wing movements for short burst escapes, and typically stall quickly. Quail and turkey are good examples of barrel bodies with high-camber wings. The male suicide birds in South Africa, known as the red-crested korhaan, have a power display that shows their form and function as a means of impressing the females. They fly nearly straight up and stall out due to the angle of attack and the drag created by the high-camber wings. Straighter, non-cambered, or flatter wings have diminished lift and are not well suited for self-powered takeoff. Straighter wings offer impressive lift-to-drag ratios during gliding and forward-cruising flight, and swifts are an excellent example of birds utilizing this anatomy. Birds with low-camber wings often nest and perch in elevated places such as under bridges on cliff faces or high elevated plateaus. Small swallows and waved albatrosses are such examples.

Wing movement

The angle of attack has to do with the way the chord of the wing (e.g., the imaginary line running from the leading edge of the wing to its trailing edge). The angle of attack can be increased if the front edge of the wing is raised to oncoming air. If the angle of attack is too high, the bird stalls. The alula regulates the airflow over the dorsal wing surface at low speeds to prevent stalling. The patagium functions may vary with different species, but its shape shifts during flight and dynamically directs the airflow above and below its airfoil. Each bird has an optimal angle of attack depending on its form and function. Every bird with its species-specific anatomy can alter angles of attack depending on its needs by muscular movements, coordination, and timing of the wing maneuver.

Drag

A force on a moving object that opposes the object's direction of travel through a fluid or air environment. A bird moving through the air is slowed down by drag. Drag has two primary factor forms: (1) the obstacle effect, where the whole bird is an obstacle to airflow and (2) the friction effect, which creates friction between the air and the wing's surface. Drag is commonly divided into three components:

- Profile drag generated by feather and body friction[5]
- Induced drag from the loss of kinetic energy in the wake (i.e., vortex), which is indissociable from the lift production
- Parasite drag, which is the drag of the body (i.e., weight)

Drag varies with lift, projection surface of the body, angles of attack, wing shape, and speed.

Profile drag or friction drag

Airflow characteristics very close to the body/wing feather surface are important aerodynamically. In this area, called the boundary layer, air viscosity is not negligible, and it decreases airflow velocity by friction. When airflow velocity is not strong enough to overcome this viscosity, airflows detach from the surface and become turbulent, which is called stalling. This phenomenon occurs mainly at low speeds and high angles of attack near landing, where circulation is inadequate. A turbulent boundary layer increases drag by friction but prevents stalling because of the higher kinetic energy of the air particles (i.e., higher pressure). The boundary layer is generally turbulent. Several features exist to enhance turbulence thereby preventing stalling, such as the surface of feathers, the differences in air transmissivity between feathers especially at the leading edge of the wing, and the comb-like leading advantages of owls' feathers.[5]

Induced drag

An excellent way to think of induced drag is the cost of accelerating air downward as lift is produced. Further stated, the downward stroke of the wing alters the overall airflow pattern around the wing, creating some airflow that opposes the direction of travel. Induced drag is highest when birds are at low speeds or hovering because the wing only intercepts the air swept by the wings. Induced drag is lowest at high speeds.

Parasitic drag

Being streamlined decreases parasitic drag. Think about the shape of a peregrine falcon in a stoop where the tuck wrist reduces the parasitic drag. High-speed bodies are generally streamlined. Examples are falcons (Falconidae) and albatross (Procellariiformes). The boundary layer is typically turbulent and several features exist to enhance turbulence thereby preventing stalling, such as the surface of feathers.

Drag factors

Below are some specific factors that affect drag.

- Automotive eddy flaps on the extrados (i.e., rows of feathers that detach from the extrados when the bird is close to stalling)
- Active and passive modifications of wing shape and airfoil (e.g., passive modification of the camber)
- Leading-edge vortex
- Defects in the wing, such as holes, decrease wing efficiency. For example, a hole in the wing would increase drag and promote turbulence, as well as reduce lift, which alters the gliding surface. Still, the importance of this effect would vary depending on hole size and position.

Gliding versus Flapping Flight

Gliding flight is harvesting wind resources of varying wind velocities or rising air. Soaring is the least energy-intensive form of gliding. Wingspans built for gliding are long, slim wings with a high aspect ratio; however, they give up maneuverability like a passenger jet. Flapping flight is a matter of active wind creation. Short, broad wings with low aspect ratios require more flapping flight and are much more energy consuming. They are also more maneuverable, like a fighter jet, and are built for speed when tucked. All flighted species use varying degrees of gliding and flapping depending on their evolutionary habitat and niche specializations.

Gliding flight

Gliding flight is accomplished by optimizing lift and diminishing drag. Harvesting air for velocity requires wind, which optimizes lift. Harvesting air for pressure requires ultraviolet vision to see rising air thermals. Vultures and many sea birds utilize this technique. This technique is employed to cover large areas on conserve metabolic energy. Gliding is also enhanced with upward wind currents from mountains or oceanic and water waves. High-velocity wind can also be harvested, providing enough power even for glide hovering.

Wingtip slots between the primaries at the wing tips are anatomic changes that enable gliding flight by decreasing the induced drag (i.e., drag induced by lift production). These slots increase the practical lift-producing part of the wing surface by vertically dispersing wingtip vortices, which could otherwise create induced drag.

Flight formations in flocks include V-shaped flock formation, which reduces profile drag by diminishing the full headwind. Low curve (i.e., low camber) straight wings create less drag. Modeling studies have suggested that the highest lift-to-drag ratio is generated by tails that are slightly forked when folded and triangular when fanned, as exhibited by frigate birds.

The lift-to-drag ratio further accomplishes effectiveness in gliding flight and prevents stalling. It is intuitive that when drag is higher than the lift, descent occurs. The higher the drag and less the lift determines the glide angle. Molting, trimmed or diseased feathers (e.g., psittacine beak and feather disease), feather destructive behavior, functional musculoskeletal defects, overweight birds, and vascular flow issues to the flight muscles are examples of causes for higher drag and less lift.

We commonly think of wings as the sole generators of lift for a flying bird. The body and tail regions generate some drag that resists a bird's forward motion, but these parts of the bird may also provide some lift on their own. Studies of small birds that fly in undulating paths have been used to explore the aerodynamics of the tail and body acting in isolation without the lift added by the wings. This is possible because these birds typically hold their wings close to their bodies during the non-flapping parts of this cycle. These studies reveal that the body and tail alone can have a lift-to-drag ratio of up to 1.2 and support up to 15% of a bird's body weight.

Flapping flight

In flapping flight, lift and thrust are generated by active motions of the wings. A wing beat is a complex, elliptic movement that shows variations among species, speeds, and maneuvers. Generally, speed increases with low aspect ratio and wing loading. Heavy birds with broad wings flex their wings to decrease the exposed wing area. For a bird that is moving forward, the velocity of the wind relative to the velocity of the bird is termed translational velocity. Adding flapping velocity and translational velocity equals the incurrent velocity.[6]

Thrust is the forward movement. The bird can maintain a steady horizontal flight speed when thrust matches the horizontal drag component. Lift supports the bird's weight and is a minor component of thrust. With flapping flight, the downstroke is the power stroke with primarily weight support (i.e., lift) and thrust, and the upstroke produces the minor of both lift and thrust. A symmetrical down and upstroke would be a hover.

Two types of upstrokes exist: (1) tip-reversal upstroke at low speed, with rotations of the primaries along their long axis and (2) feathered upstroke at medium to high speed. A major part of thrust and lift is produced during downstroke, but significant amounts of these forces are also produced during upstroke at medium to high speeds. The distal part of the wing has greater thrust than other parts. From low-speed to fast-speed flight, the dominant function of the wing beats changes from weight support (i.e., lift) gradually to thrust production.

To reduce lift production, birds may flex their wings, decrease their angle of attack, minimize wing camber, or reduce flapping velocity. Although downstroke wing motions are relatively invariant among many birds, these patterns of upstroke movements differ among species. In birds with pointed wings with a relatively high aspect ratio, such as pigeons, the mid-upstroke posture changes as a function of flight speed. During slow flight, these birds rotate their wings and hold them extended in what is known as a tip-reversal upstroke, which may represent a simple recovery stroke or may provide some lift. As flight speed increases, they shift gradually to a flatter wing posture during upstroke, with wing flexion increasing as flight speed increases. The flight posture at intermediate speed is called feathered, and in fast flight, it is called flexed. In contrast, birds with relatively rounded wings with a low aspect ratio tend to adopt a flexed wing posture during upstroke, regardless of flight speed.

In birds with high-aspect-ratio wings, flight speed determines the flapping pattern. In slow flight, wingtips trace a figure-of-eight pattern when viewed from the side (i.e., lateral view). The arc of their flapping wings—the distance they move up and down—is most significant during slow flight. In faster flight, the wingtips and wrists track an elliptical path. These features are typical of birds with high-aspect-ratio wings. In contrast, in birds with lower-aspect-ratio wings, the wingtip path is elliptical regardless of flight speed, and the wings are drawn close to the body during the upstroke.

Hovering

The primary aerodynamic challenge in hovering is that there is no forward movement to generate airflow over the wing. Instead, while hovering, a bird's pectoralis and supracoracoideus muscles must supply most of the power necessary to move air across its wings. The goal of a bird during hovering or slow flight may be better thought of as efficacy: pushing the envelope of these demanding flight styles to gain control in takeoffs and landings, escape from predators, access hard-to-reach energetically rich food, or display potential mates.

Aerodynamic Power

Of all the various forms of animal locomotion, self-powered flight is the most demanding in terms of its energy cost per unit of time or aerodynamic power. Aerodynamic power is defined for birds as the total power required to match the various forms of drag and maintain flight. Aerodynamic power varies with flight speed as an upward parabolic curve. The lowest aerodynamic power obtained at an intermediate speed is called minimum power speed. Birds should fly at this speed if their goal is to stay in the air for as long as possible for a given energy expenditure. An example might be a raptor slowly coursing over a field while it searches for small mammals.

The greatest efficiency per unit of distance traveled is obtained at the maximum range speed. Birds should fly at this speed when they are trying to maximize flight distance. An example might be a long-distance migrant who must fly for several days between locations thousands of kilometers apart.

The shape of the curve and estimates of minimum-power speed and maximum-range speed differ among bird species due to differences in morphology such as the aspect ratio of their wings and flight style. Muscle power is assessed by implants that measure the length of contraction via ultrasound methodology. Bone torque is evaluated by a strain gauge attached to the humerus. The electrical impulse is measured by pectoral implanted electrodes testing nerve conduction.

Oxygen consumption is assumed to be directly proportional to power. It varies as a U-shaped curve with velocity.

Flapping flight has an additional efficiency advantage, as it facilitates some energy storage and recovery between successive flaps. In European starlings (*Sturnus vulgaris*), for example, the furcula (wishbone) is stretched—loaded with strain energy—by the pectoralis muscle during the downstroke, and the recoil of the furcula transfers energy that helps power the next upstroke as well as breathing. The primary upstroke muscle, the supracoracoideus, is also connected to a long, thin tendon. In rock pigeons (*Columba livia*), the storage and release of the strain energy in this tendon equals up to 60% of the network of the supracoracoideus, thus substantially reducing the work and power required for upstroke.

On average, flight performance declines dramatically across all birds as body size increases. For example, large waterfowl like swans find it harder to take off and ascend rapidly during escape flight than small ducks. As described above, the tiny size of hummingbirds has allowed them to develop a unique capacity for sustained hovering. Likewise, small songbirds can fly circles around larger species like hawks, a functional ability when mobbing predators. A principal reason for this scaling pattern is that larger birds have longer wings, and their wingbeat frequencies are slower.

Dividing a constant amount of mass-specific work from the muscles by progressively longer time intervals, from having longer wings that beat less rapidly, means that the relative power for flight declines as body size increases. As species get larger, the decrease in marginal power available results in limitations on flight styles. Birds larger than a hummingbird cannot hover efficiently. Still, larger birds like loons require a running start to take off from water, and birds like the waved albatross (*Phoebastria irrorata*) of the Galápagos must take off into the wind to become airborne. Finally, the largest flight-capable birds, like condors, have only enough burst power to take off at their minimum power speed and then need to extract energy from the environment, using thermals to stay aloft.

The negative scaling of mass-specific power available from the flight muscles is an inevitable consequence of increasing body size and wingspan. Still, there are some ways in which this trend can be partially offset, at least over limited ranges of body size. As noted above, wing shape has substantial effects on flight performance. Similarly, there are physiologic differences in the flight muscles of different birds that are adapted to the power requirements of their flight styles.

Energy-Saving Strategies

Intermittent flight

The non-flapping phases of intermittent flight may consist of a flexed wing bound during which the wings are held tightly against the body, a glide during which the wings are fully extended, or some intermediate posture. During intermittent glides, the pectoralis muscle contracts, and the supracoracoideus muscle is inactive, whereas both muscles are inactive during bounds.

Mathematical models of power requirements generally suggest that flap-bounding offers savings in power during fast flight at maximum range speed or faster, particularly when lift from the body and tail contributes to weight support during the bound. In contrast, flap-gliding is estimated to be less costly at most flight speeds, particularly slow and intermediate. An ideal strategy for saving energy while flap-gliding is to gain altitude rapidly using high-power wingbeats and then to descend slowly during the glide phase. Some birds vary their non-flapping posture, shifting from intermittent glides when flying slowly to intermittent bounds when flying fast.

There appears to be a relationship between body mass and wing design that affects the use of intermittent flight. Small birds with rounded, low-aspect-ratio wings, such as the zebra finch (*Taeniopygia guttata*) (13 grams; aspect ratio=4.2), appear to use only intermittent bounds. In contrast, species of about the same body mass but with more pointed, high-aspect-ratio wings, such as the barn swallow (20 grams; aspect ratio=6.2) and northern house martin (17 grams, aspect ratio=6.5), use both bounds (or partial bounds) and glides. Birds heavier than about 300 grams do not appear to be able to use intermittent bounds; for example, pigeons only glide during pauses in wing flapping.

As in the general decline of overall flight performance with increasing mass among bird species, there is a decline in the performance of bounds as body size increases. Among passerines engaged in migratory flight and woodpeckers engaged in foraging flight, the percentage of time spent flapping increases with increasing body mass.

Gliding costs some energy, while soaring is harvesting energy from the environment to maintain flight. Several types of soaring are described below.

Slope soaring

The upward airflow direction permits the bird to maintain a glide while holding its place relative to the ground.

Thermal soaring

The bird descends in the thermal column unless the air in the thermal is rising at a faster rate relative to the ground, then the bird will rise close to the ground.

Dynamic soaring

Birds use the gradient in wind speed over the ocean's surface to travel long distances without expending much energy. The bird glides down the gradient at an angle, then turns and rises abruptly into the wind, using its momentum to gain height quickly. It then turns and glides down again, crossing the waves in large zigzags. Dynamic soaring is used most by albatrosses and other large pelagic birds.

Maneuverability and Stability

Maneuverability

Birds can use swept wings (i.e., hovering flight), slotted wings, longitudinal dihedral (i.e., between the wings and tail), and movements of the tail and wings to maintain themselves in the desired position. Whether gliding or flapping, the wing can be extended or flexed, increasing or decreasing the exposed surface area depending on the specific activity the bird desires. Complex parameter variations such as camber, twist, sweep, and dihedral characterize their wing geometry. Fighter jets and raptor wings must go from stable to a stoop, roll, or direction change. Soaring birds and airline jets have features that increase the smooth glide and stability.

Maneuvers are realized by asymmetry in the production of forces between both wings and through rolling motions of the body. For instance, a gliding bird turning to the right will either roll its body to the right, thus orienting the total lift force vector to the right, or modify its wing shape and orientation to produce more drag with its right wing.

To produce a maneuver, a bird must create force asymmetry using its wings or tail to make a rotational force (i.e., torque) about its center of mass. Any torque will result in rotation about a given axis, and most flight maneuvers include rotation about all three axes. To self-stabilize, a bird must generate a force asymmetry to correct for any perturbations—like a sudden gust of wind coming from one side—that would cause a rotation about one or more of these axes.

Maneuvering during flight is related to a measurement called wing loading, which is the bird's weight divided by the projected surface area of its wings and, by convention, the area of the body between the wings. As a general pattern, as wing loading increases, the intrinsic or preferred flight speed increases, as higher velocities of air passing the wing are required to generate the lift to keep these relatively heavy birds aloft. This relationship has an important implication for maneuvering with fixed (i.e., gliding) wings: birds with lower wing loading can turn with a tighter radius than birds with higher wing loading.

In a simple gliding turn, the turn is affected by creating a roll about the long axis of the bird's body to establish a bank angle, thus redirecting the lift force to provide a turning force. To produce this roll, a bird can manipulate one or more things that produce lift: wing surface area, angle of attack, or wing speed.

The wings can also be used to create rotation around the other two body axes. By moving the wings' center of lift forward or backward relative to the bird's center of mass, birds can create a pitching rotation to direct themselves upwards or downwards. Likewise, any left-right asymmetry will produce differences in lift and drag, causing yawing rotations.

Maneuverability influences many aspects of the ways birds use flight. For example, wing loading and maneuverability variations help account for differences in foraging behavior and prey selection in swallows, birds that often use gliding when foraging. Moreover, gliding wings are not fixed in terms of posture. For example, swifts can change the sweep of their wings, holding them straight out or angling them back towards their tail.

Stability

There are several types of stability in flying birds. These are described below.

Passive stability

When applied to flying birds, components of the bird's aerodynamic structures cause the bird to return to an original flight posture when it is disrupted by an outside force (e.g., wind) without flapping or other active movements.

Roll stability

During gliding, many birds adopt a dihedral wing posture, in which the wings form a V-shape when the bird is seen from the front or rear.

Yaw stability

Yaw stability is offered by the tail and, in some species with long legs, by the feet trailing behind the bird's body. If you think about kayaking or canoeing and the paddle on the right back side to correct for steering left, the bird does this with its tail and or feet.

Flapping motions may affect both maneuverability and stability. The contributions of flapping make it very challenging to study the act of maneuvering because the velocity of flapping wings changes in relation to the air pressure along the length of the wing.

References

1. Tobalske BW. Avian flight. In: Lovette I, Fitzpatrick J, eds. *Handbook of Bird Biology*. Hoboken, NJ: Wiley Blackwell; 2016:149-168.

2. Meehan C, Mench J. Captive parrot welfare. In: Luescher AU, ed. *Manual of Parrot Behavior*. Ames, IA: Blackwell Publishing; 2006:301-318.

3. Biewener AA. Biomechanics of avian flight. *Curr Biol.* 2022;32:R1110-R1114. doi: 10.1016/j.cub.2022.06.079.

4. Beaufrère H. A review of biomechanic and aerodynamic considerations of the avian thoracic limb. *J Avian Med Surg.* 2009;23(3):173-185. doi: 10.1647/2007-023.1.

5. Durston NE, Wan X, Liu JG, Windsor SP. Avian surface reconstruction in free flight with application to flight stability analysis of a barn owl and peregrine falcon. *J Exp Biol.* 2019;222(9):jeb185488. doi: 10.1242/jeb.185488.

6. Nadal J, Ponz C, Margalida A. Feathers for escape: the transition from juvenile to adult in red-legged partridges (*Alectoris rufa*). *Biol J Linnean Soc.* 2018;123(1):72-80. doi: 10.1093/biolinnean/blx130.

7. Tobalske BW. Biomechanics of bird flight. *J Exp Biol.* 2007;210(18):3135-3146. doi: org/10.1242/jeb.000273.

Avian Cardiopulmonary-Cerebral Resuscitation: What We Know and Don't Know

Brynn McCleery, DVM, DABVP (Avian Practice)

Session #155

Affiliation: From Red Bank Veterinary Hospital, 100 Schulz Drive, Red Bank, NJ 07701, USA.

Abstract: It is essential that practitioners working with avian species are prepared to perform cardiopulmonary-cerebral resuscitation (CPCR) if an arrest event occurs. The basics of CPCR are the same for all species; however, full CPCR efforts are not always applied to avian patients. Until taxa-specific evidence is available, recommendations for CPCR for non-domestic species must be extrapolated from what is known in humans, dogs, and cats. Reassessment Campaign on Veterinary Resuscitation guidelines currently only include recommendations for dogs and cats; however, these guidelines can be utilized for birds with alterations to account for the unique anatomy and physiology of avian patients.

Introduction

Avian patients have evolved to hide their signs of illness, and sick birds are often more critically ill than can be appreciated on visual inspection by the owner or physical examination by a veterinarian. Unfortunately, birds presenting to a veterinary clinic for care may experience cardiopulmonary arrest (CPA), and the risk of CPA is relatively high in avian patients undergoing sedation or anesthesia when compared to domestic species.[1-3] It is essential that practitioners working with avian species are prepared to perform cardiopulmonary-cerebral resuscitation (CPCR) in the event of an arrest event.

Evolution of CPCR

Like other aspects of human and veterinary medicine, CPCR techniques have evolved over time. In humans, the American Heart Association altered the compression to ventilation ratio in 2005 to minimize disruptions to chest compressions and reprioritized the airway-breathing-circulation (ABC) of CPCR to CAB in 2010.[4-6] Chest compressions were prioritized over breathing based on multiple studies showing that chest-compression-only CPCR was equivalent to CPCR performed with an artificial respirator for non-asphyxial CPA outside of a hospital.[5] In veterinary medicine, CPCR underwent a major transformation in 2012 when the Veterinary Emergency and Critical Care Society (VECCS) and American College of Veterinary Emergency and Critical Care (ACVECC) released the first evidence-based consensus veterinary CPCR guidelines, referred to as the Reassessment Campaign on Veterinary Resuscitation (RECOVER).[7] RECOVER guidelines are based on a review of the human and veterinary literature and were last updated in the summer of 2024.

CPCR in nondomestic species

Despite chickens being one of the earliest animal models for cardiopulmonary resuscitation, very few studies have been performed in recent decades to evaluate CPCR techniques in birds.[8] RECOVER guidelines currently only include recommendations for dogs and cats. However, the basics of CPCR are the same for all species. Until taxa-specific evidence becomes available, techniques and drug doses can be extrapolated from the RECOVER guidelines for nontraditional species, including birds, but some of the guidelines may need to be adapted due to the unique anatomy and physiology of avian patients.

Prevention, Preparedness, and Patient Assessment

Regardless of species, the first step for CPCR is to be prepared. Hospital and staff preparedness is essential to reduce delays in starting CPCR, which have been shown to affect outcomes in neurologic status and survival to discharge in other species.[9] Steps for preparedness should include:

- Regular audits of resuscitation equipment: Ensuring crash carts are well stocked with inventory performed at least weekly and after every code. Staff members should ensure reusable equipment is sterilized/disinfected appropriately and returned. Expiration dates of emergency drugs should also be closely monitored.

- Updating dosing charts for avian species: Dosing charts should reflect the drugs available in the crash cart. If the hospital obtains an alternative concentration of a particular drug, the dosing charts should be updated.

- Ensuring checklists and CPCR algorithms are available: These memory aids can be posted in treatment areas and/or laminated copies can be available on or inside the crash cart. (Note: a new algorithm was created for the 2024 updated RECOVER guidelines.)

- Printing data recording sheets: CPCR recording sheets should be pre-printed and placed in the crash cart or can be available electronically.

- Didactic learning and skills training for team members: CPCR review lectures and labs should be held minimally every 6 months for all staff members assisting in codes.[9]

- Preparation for sedated or anesthetized patients: Emergency doses—including the doses for any reversal agents—should be pre-calculated for avian patients undergoing sedation or anesthesia (some clinicians prefer to draw up the first dose). For shorter procedures where intubation or catheterization are not elected, equipment should be prepared in case the patient arrests and emergency intervention is required. This includes having endotracheal tubes pre-selected for emergency intubation and having catheters or needles (hypodermic versus spinal) available for intravenous (IV) or intraosseous (IO) catheterization.

- Acquiring a code status from the owner for all avian patients admitted to the hospital

Considerations for an avian crash cart

Crash carts for avian patients should contain much of the same equipment as those used for mammals with increased variation in sizes, depending on the diversity of avian species being treated. In addition to standard equipment (e.g., emergency drugs [epinephrine, ± vasopressin, atropine, ± glycopyrrolate, ± doxapram], syringes, hypodermic needles, intravenous catheters, T-sets, sterile injection caps, sterile saline flush, red rubber catheters, ± polypropylene catheters, dosing charts, data recording sheets, a stocking checklist, white medical tape, alcohol wipes for drug bottles), other equipment for avian patients should include:

- Specialized endotracheal tubes. Uncuffed and Cole endotracheal tubes of various sizes are needed due to the presence of closed tracheal rings in birds.

- Additional needles. A wider variety of hypodermic needles (ranging from 27 gauge to 18 gauge) and a selection of spinal needles is recommended for placement of intraosseous catheters.

- Equipment for initial stabilization of catheters. This equipment can include surgical glue, specialized tape, and suture.

- Equipment for emergency air sac cannulation.

Equipment that should be housed in or near the crash cart includes sterile fluid bags, warming equipment, a syringe pump, a defibrillator, various sized face masks, a bag-valve-mask (BVM) apparatus (e.g., Ambu bag), and monitor (with electrocardiography and capnography).

Identifying CPA

Previous RECOVER guidelines recommended assessment of circulation and cardiac contractility immediately after assessing airway in breathing in an unconscious patient.[10] In 2024, revised guidelines recommended suspecting CPA in any unconscious dog or cat that is found to be apneic. Circulation assessment was replaced by a shake and shout technique.[8] In these patients, starting CPCR erroneously is considered preferable to a prolonged patient assessment, which can be associated with worse patient outcomes.

Although delays in initiating CPCR in avian patients with CPA should also be avoided, the author still advises performing a very brief patient assessment on birds to assess the airway, breathing, and circulation. Avian patients often undergo respiratory arrest prior to cardiac arrest.[11] In apneic patients with a normal or mildly bradycardic heart rate, controlling the airway and providing intermittent positive pressure ventilation (IPPV) without chest compressions may be effective. Avoiding chest compressions and their possible sequelae (which are often minimal in dogs but may be more severe in birds including internal organ damage and hemorrhage) when not necessary is desirable, especially in smaller patients where these side effects may be more deleterious. In a study of CPCR techniques in chickens, side effects of compressions included cloacal prolapse, egg expulsion, and of more concern, liver fracture.[8] Another situation in which chest compressions may be avoided is in avian patients undergoing a vagally mediated dive response (a.k.a. the diving reflex). The dive response is commonly seen in aquatic vertebrate species, such as waterfowl, and can cause apnea followed by bradycardia.[9,12] Reducing stimulation of the vagus nerve by reducing anesthetic levels (when present), opening the mouth, or by temporarily discontinuing procedures that may be the cause of a vasovagal response unrelated to diving can be attempted prior to initiating full CPCR with chest compressions.

Due to these differences, the author recommends a 5-10 second maximum evaluation of the airway, breathing, and circulation of an unconscious avian patient. In most cases, cessation of breathing is the indicator that patient assessment is necessary. When assessing cardiac contractility and circulation in birds, auscultation of the heart with a stethoscope is most common. Doppler or ultrasound assessment can also be utilized. As with mammals, circulation assessment should not be based on palpation of peripheral pulses.[13] Peripheral pulses are unreliable, difficult for assessing systemic circulation, and can be inhibited by anesthetics.[9] If bradycardia is noted, avian patients should be monitored closely and interventions started because bradycardia due to the development of heart blocks often precedes cardiac arrest.[14]

Since acute feather erection has been associated with cardiac arrest or reduced cardiac blood flow in anesthetized birds and has been noted during euthanasia of multiple poultry species, patients exhibiting this change should be evaluated promptly.[15]

Basic Life Support

Basic life support (BLS) is the foundation of CPCR. If CPA is identified in an avian patient, BLS should be initiated immediately with a goal of improving venous return to the heart during the diastolic phase of the compression cycle.[c] Components of BLS include the CPCR ABC's reorganized to reflect prioritization of circulation.

Circulation

If cardiac arrest occurred, compressions should be initiated. Tissue hypoxia and ischemic injury occur rapidly during arrest.

Blood flow theories

In mammalian CPCR, there are two theories of how blood flow is produced. The first is the cardiac pump theory, where the cardiac ventricles are directly compressed during CPCR to generate cardiac output. Increased ventricular pressure opens the pulmonic and aortic valves, allowing blood flow to tissues.[9,13] In the second theory, the thoracic pump theory, the heart is not compressed directly. Instead, the widest portion of the chest is compressed leading to an increase in intrathoracic pressure, as well as compression of the aorta and vena cava leading to blood being pushed out of the thorax. Afterward, time is allowed for elastic recoil of the chest. During recoil, subatmospheric intrathoracic pressure creates a pressure gradient that pulls blood back into the thorax and lungs.[9,13] Both blood flow theories have been shown to have a positive effect on blood flow in domestic mammals.[9]

When applying these theories to birds, the thoracic pump mechanism is ineffective. Birds are adiaphragmatic, so subatmospheric intrathoracic pressure is not present. To generate blood flow, avian patients must rely on the cardiac pump theory; however, there has been debate in the literature regarding the impact of compressions on birds due to their unique anatomy.[9] Avian patients have a large sternal plate, often with a keel, that inhibits direct compression of the heart and its ventricles. Furthermore, in mammals, compressions not only provide blood flow but also may help move air through the lungs. This added benefit is not likely in birds due to their rigid lung parenchyma; however, it is likely that air is moved through the bellows-like air sacs during compressions.[8,14]

Despite these anatomic differences, compressions have been shown to be effective in avian patients. In a study comparing compression techniques in chickens, chest compressions paired with interposed abdominal compressions (IAC) increased arterial pressure up to 28% of the baseline anesthetized value when the birds were placed in dorsal recumbency. Although average arterial pressures were higher in chickens receiving sternal compressions without IAC compared to those placed in lateral recumbency, the difference was not significant.[8] In European starlings and sparrows, high rates of grossly visible internal hemorrhage were found in the coelomic cavity, pericardium, and perihepatic area after lateral rapid cardiac compressions (RCC) due to rupture of the vena cava or right atrium. Note that the RCC technique is performed for euthanasia, not CPCR.[15] Although compressions are expected to be less effective at generating blood flow in birds compared to mammals due to their unique anatomy, the limited literature shows that blood flow is generated and support the positioning of birds in dorsal recumbency during CPCR, which is a major difference compared to CPCR in most dog breeds. Updated RECOVER guidelines recommend only performing compressions in dorsal recumbency for wide-chested dogs (e.g., English and French bulldogs).[7]

Compression technique

Depth

For small animals undergoing CPCR in lateral recumbency, including cats and most dog breeds, the chest should be compressed to a depth of one-third to one-half of the total width of the chest.[7,10] Updated RECOVER guidelines recommend a reduced compression depth of one-quarter of the chest width for wide-chested dogs moved into dorsal recumbency.[7] In human infant CPCR, patients are positioned in dorsal recumbency and the chest is compressed by one-third of its depth.[16] The appropriate compression depth is unknown for avian patients. In a human CPCR study, compression depth was found to be the most important factor for maintaining targeted systolic blood pressures.[8] Due to the impedance of cardiac compressions by the sternal plate, the author currently aims to compress avian patients by one-third to one-half of the total depth of the chest to maintain adequate blood pressure during CPCR.

Hand position

Various hand positions exist for CPCR in dog, cat, and human patients of various ages. The appropriate hand position for birds is likely dependent on species and body size; the technique best utilized in ostriches and budgerigars is probably quite different! To the author's knowledge, hand positions for CPCR have not been evaluated in avian patients. For larger patients, an interlocked two-handed technique, typical of adult human and canine CPCR, may be effective. For most patients presenting to private practice (e.g., psittacine and passerine birds), an alternative one-handed technique is likely more appropriate. A finger technique, like that used in human infant CPCR, may also be utilized, but the compressor and other staff members responding to the arrest should ensure the chest is compressing adequately, and this technique may be more fatiguing.

The author utilizes a modified one-handed technique in most small or medium avian patients. Similar to the hand positions recommended for cats by the updated RECOVER guidelines, the author uses her nondominant hand to stabilize the patient's body.[7] Her dominant hand position is altered from those recommended for cats. Instead of cupping the chest with the thumb on one side and the other four fingers on the other side or using the heel of her palm for compressions, she forms a fist with her dominant hand with the thumb slightly separated, which forms a V shape that can be placed on top of the keel bone and pectoral muscles. An encircling technique is a proposed technique for cats in the updated RECOVER guidelines and can be useful in patients with higher thoracic wall compliance, like ferrets.[7,9] This technique may not be as effective for birds due to excessive coverage in small avian patients preventing access of staff members needing to intubate or catheterize the bird and risk of inadequate compressions with fingers due to a robust sternal plate.

Regardless of the hand position elected, staff should position themselves properly in comparison to the patient to generate sufficient cardiac output. Use of a step stool, lowering the table, kneeling on the table, or moving the patient to the floor may be necessary, depending on the patient size. To ensure the kinetic energy from the compressions is translated into blood movement and not into compression of padding, blankets and large warming devices beneath the patient should be removed.[9] The author finds it helpful to maintain a single thin fleece, towel or non-slip drawer liner under smaller bird to prevent excessive movement of the patient on the table during compressions. Over-compression should be avoided and, although not utilizing the thoracic pump theory for avian patients, full elastic recoil should be allowed between compressions to allow air movement through the air sacs.

Rate

The appropriate cardiac compression rate for taxa with resting heart rates significantly different than those of domestic species is unknown and requires further investigation.[9] It is likely that the appropriate compression rate for avian patients varies by species. Recommendations for avian compression rates in the literature vary widely from 60-150 bpm.[9,14] In dogs and cats, the recommended compression rate is 100-120 bpm, regardless of patient size and despite the varied resting heart rates between dogs and cats.[7,10] A metronome—physical or electronic (e.g., smartphone or tablet app)—is recommended to maintain a constant compression rate.

It takes approximately 1 minute of continuous thoracic compressions for myocardial perfusion to reach its maximum. Uninterrupted 2-minute compression cycles are associated with improved CPCR outcomes in human medicine.[9] When possible, these recommendations should be followed in avian patients. Even with an appropriately locked hand-position and good body positioning as discussed above, staff members may become more easily fatigued during avian compressions. In a study evaluating compression techniques in chickens, compressors reported muscle pain and exhaustion associated with 2-minute compression cycles.[8] If swapping of compressors is necessary midcycle, counting aloud "3-2-1-swap" is recommended to minimize missed beats.[9] RECOVER guidelines recommend that a 2- to 5-second pause is performed after every 2-minute cycle for rhythm diagnosis and palpation of pulses to assess for return of spontaneous circulation (ROSC).[7]

Other considerations

Similar to the technique used in chelonian CPCR, a wing flap technique was evaluated for avian CPCR. It was theorized that wing flapping would promote blood flow through the vascular flight muscles. This technique was not found to generate blood pressure during a pilot study and is, therefore, not recommended during avian CPCR.[9]

Open-chested CPCR has been found to produce higher cardiac output compared to closed-chested CPCR.[17] This technique is highly invasive and is not recommended by the author in avian patients due to their unique anatomy. Accessing the heart would require several bony fractures and compromise of multiple air sacs.

Airway

After initiating compressions, the second component of BLS is to secure the airway. Airway control should occur during compressions as pauses in compressions are deleterious to the effectiveness of CPCR.[6,9j] For most avian species, airway control is easier to obtain than in mammals, and their unique anatomy provides an additional option for ventilation (via air sac cannulation).

Intubation

As with mammals, intubation is the preferred method of airway control in birds. Due to their complete tracheal rings, uncuffed endotracheal tubes (ETT) should be used in birds to prevent tracheal damage. Both uncuffed standard ETT or Cole ETT can be utilized. Commercial ETT are available as small as size 1.0. The author prefers commercial tubes over homemade ETT tubes utilizing IV catheters without the stylet or trimmed red rubber catheters. The stiffer materials of homemade options may be more irritating to the tracheal mucosa, and homemade options often have a thicker wall and, therefore, reduced lumen diameter compared to commercial ETT. Some commercial tubes are excessively long, especially for smaller patients, leading to excess dead space. These tubes can often be cut shorter and the ends flame-heated to prevent sharp edges.

In many avian species, the trachea narrows as it passes down the cervical area. For this reason, ETT should only be passed a short distance. Cole ETT, which were designed for use in human pediatric patients, are a great option for avian patients as they prevent the tube from being inserted too far and help form a tighter seal on the glottis. Tube placement is typically verified visually; however, end tidal CO_2 can also be used but may be altered during CPA and/or too low to read for smaller patients.[7]

Although most birds are easy to intubate, some avian species have anatomic variations that can make intubation more challenging. Notable differences include the presence of a crista ventralis in penguins, pelicans, hornbills, kiwis, and some ducks; a deeper and more mobile glottis in chickens; and early bifurcation of the trachea in penguins.

Forced mask ventilation

In avian patients that are more difficult to intubate, such as very small birds, forced mask ventilation can be considered. The mask must be tight fitting to allow ventilation. Possible complications of forced mask ventilation include failure to ventilate, aspiration of material refluxed from the crop, and gastrointestinal bloat.[14] The flange provided with commercial masks does not usually provide an appropriately tight seal for avian patients. Commercial or homemade masks (e.g., from syringe cases) can be fitted with pieces of cut nitrile gloves with smaller holes to allow a tighter seal on the face. Self-adherent bandaging material (e.g., Vetrap) is not recommended as these materials are permeable and will allow escape of gasses, which is less concerning for CPCR than for avian anesthesia. RECOVER guidelines were updated in 2024 to recommend ventilation with a tight-fitting mask and BVM apparatus in dogs and cats that could not be intubated.[7]

Laryngeal masks

Use of supraglottic airway devices such as a laryngeal mask airway (LMA) have been anecdotally reported for avian species and could be helpful to prevent proliferative tracheal membranes from forming secondary to movement of an endotracheal tube during CPCR.[9] No formal studies have been performed evaluating LMA use in avian patients.

Air sac cannulation

Due to the ease of intubation of most avian patients, surgical options to secure an airway are not usually required. In patients where facial, oropharyngeal, or tracheal disease prohibits tracheal intubation, air sac cannulation can be performed and is considered a less traumatic option than tracheostomy in birds. The technique has been described elsewhere.[18]

In small birds where performing intubation concurrently with chest compressions may be difficult, such as budgerigars, a temporary air sac cannula can also be considered using an 18-gauge over-the-needle IV catheter placed in the caudal thoracic or abdominal air sac.[9]

Breathing

Once the airway is secured, ventilation should be initiated since hypoxia and hypercapnia both lessen the likelihood of ROSC.[9] Ventilation can be performed with a BVM apparatus or anesthetic machine with oxygen flow. Use of a mechanical ventilator is not recommended during CPCR due to expected delays in setting up the machine; however, patients can be switched to a mechanical ventilator if ROSC is achieved. The recommended positive pressure ventilation rate in dogs and cats is 10 brpm with a tidal volume of 10 ml/kg and short inspiration time of 1 second.[7,10] Both hypoventilation and hyperventilation have been shown to decrease cerebral perfusion.[9] In the literature, anecdotal ventilation rates for avian patients range from 10-50 brpm.[9,8,14]

Overinflation should be avoided to prevent volutrauma to the delicate respiratory structures of birds.[9] Birds have very thin air sacs and reduced exterior muscular and skeletal support compared to mammals, putting them at increased risk of complications from overinflation. A peak airway pressure of < 15-20 mmH_2O has been recommended for avian patients.[9] Although the author's hospital typically uses a maximum peak airway pressure of approximately 10 mmH_2O when performing IPPV for avian anesthesia, higher values may be required during CPCR to overcome chest compressions.

Interposed abdominal (coelomic) compressions

Reviews in human medicine, as well as many animal model studies, have shown that mid-abdominal compressions interposed with thoracic compressions are beneficial.[19] In pigs, IAC improves arterial pressures, generates significant ventilatory volumes alone, and creates equivalent or greater coronary perfusion when compared to thoracic compressions.[9,19] IAC in humans has been shown to improve coronary perfusion pressure, increase ROSC rates, and increase survival to hospital discharge.[8]

Chickens that underwent sternal compressions with IAC in dorsal recumbency had higher average arterial pressure (i.e., diastolic, systolic, and mean) than birds receiving sternal compressions alone or compressions in lateral recumbency. While the arterial pressure in the sternal compression with IAC group was significantly different than chickens receiving compressions in lateral recumbency, no significant differences were found compared to chickens receiving sternal compressions only while placed in dorsal recumbency.[8] Further studies are needed in avian patients to evaluate the effects of IAC during CPCR.

Advanced Life Support

The goal of advanced life support (ALS) is to reestablish electrical and mechanical activity of the heart.[13] ALS cannot be effective without good BLS, so it should be initiated only after BLS is underway. Components of ALS include applying monitoring equipment to the patient, acquiring vascular access, and administering emergency drugs and fluids, if needed.[7]

Monitoring

The first step in ALS is to apply monitoring equipment, which allows staff to assess the quality of CPCR, adjust CPCR technique, and identify ROSC.[7,9]

End-tidal CO_2

End-tidal CO_2 ($EtCO_2$) is used as an assessment of cardiac output (CO) and is proportional to pulmonary blood flow.[9] Low $EtCO_2$ in dogs (< 10 mmHg) may indicate improper esophageal intubation, ineffective CPCR technique, or hypoventilation.[13] Previously, an $EtCO_2$ goal of 12-18 mmHg was recommended for dogs and cats during CPCR; however, this was increased to 18 mmHg or higher in the updated RECOVER guidelines.[7,10] Although these guidelines may also be appropriate for avian patients, $EtCO_2$ may be falsely lower if a facemask or long ETT is used due to excess dead space or if the sampling line is long. Quick elevations in $EtCO_2$ into the 30s or mid 40s may indicate ROSC.[13] Pulses should be confirmed before compressions are discontinued if a spike in $EtCO_2$ is identified.

Electrocardiogram

An electrocardiogram (ECG) should be used to characterize cardiac arrhythmias. ECG leads can be on the propatagium of each wing and on the knee patagia, not on the bone or muscle. Needle ECG leads, atraumatic clamps, or blunted alligator clamps can be used in avian patients.[9]

Other monitoring

Pulse oximetry is expected to be nonfunctional in patients with CPA due to poor peripheral perfusion and lack of pulsatile flow.[9] Indirect blood pressure has been shown to be inaccurate in multiple species of birds when healthy and alive, so use of indirect blood pressure in birds experiencing CPA will likely be inaccurate and is often impractical due to difficulties placing the Doppler probe and cuff.[20] Similarly, blood gas measurement are considered impractical in birds during CPCR due to difficulty in obtaining appropriate blood samples.[9]

Intravenous or intraosseous access

Vascular access is the second component of ALS. Birds experiencing CPA require vascular access for effective emergency drug administration. IO catheterization is often quicker and easier to obtain in birds with CPA undergoing chest compressions. In the updated RECOVER guidelines, IO catheter placement is recommended for dogs and cats if an IV catheter cannot be placed within 2 minutes. Once the IO catheter is placed, staff should continue to place an IV catheter.[7] For birds, due to the ease of IO catheterization, the author suggests placing an IO catheter first but still attempting for IV catheterization afterward. Although IO catheterization is painful, analgesia and/or anesthesia is not necessary in emergency situations where a patient has arrested.

Sites in most avian species

- IV: basilic vein (i.e., cutaneous ulnar vein), medial metatarsal vein, jugular vein
- IO: distal ulna, proximal tibiotarsus

Alternative drug administration

In cases where vascular access cannot be obtained immediately, intratracheal (IT) administration may be an option for some drugs, including naloxone, atropine, vasopressin, and epinephrine.[13] When utilizing IT administration, drug doses may need to be increased, often doubled but up to a 10-fold increase reported for epinephrine. The drugs should be administered beyond the carina, which can be achieved using a long polypropylene or red rubber catheter and can be diluted with saline or sterile water to facilitate administration.[9,11]

Reversal agents

Once vascular access is obtained, reversal agents for opioids and anesthetics should be administered. If doses were already administered intramuscularly prior to CPA, repeat doses can be provided IV or IO. Butorphanol, the opioid used most in avian species, is unfortunately nonreversible. For any patient that receives a full mu-agonist, naloxone can be administered. The author recommends keeping flumazenil in crash carts utilized for avian patients since midazolam is commonly used to sedate birds.

Non-shockable rhythms and emergency drugs

If ROSC is not achieved with BLS and a shockable rhythm is not present, the next step is to administer emergency drugs.

Anticholinergics

Due to its quicker onset of action and cheaper cost, atropine is typically utilized over glycopyrrolate in avian patients. In the updated RECOVER guidelines, a single dose of anticholinergics is recommended after the first 2-minute round of compressions with no repeated doses since atropine persists in the body longer than vasopressors.[7] Pharmacokinetic studies of atropine in various avian species have not been performed, so it is unknown whether this recommendation applies to avian CPCR, as well.

Vasopressors

Epinephrine is the most common vasopressor used during avian CPA events; however, epinephrine is less effective in hypoxic and acidotic conditions compared to vasopressin. Vasopressin could, therefore, be considered first, interchangeably, or in combination with epinephrine during CPCR.[14] Although typically refrigerated, vasopressin is still effective for 12 months out of the refrigerator and could be kept in a crash cart. While this drug is often cost prohibitive in dogs, it may be a reasonable option for birds due to their smaller body size. No additional benefit was noted between high-dose vasopressin and low or medium doses in Sprague-Dawley rats serving as a model for CPCR, so lower doses are recommended.[14] Rats receiving low versus high doses of epinephrine had similar rates of ROSC, but high-dose epinephrine inhibited recovery of spontaneous ventilation and caused relative bradycardia post-CPR.[14] High-dose epinephrine is no longer recommended in the updated RECOVER guidelines. The updated guidelines also recommend reducing administration of vasopressors to every other 2-minute cycle after a rhythm check is performed.[7]

Arterial blood pressures measured before and after a dose of epinephrine (0.1 mg/kg IV, high dose) in chickens receiving chest compressions were not statistically different.[8]

Fluids

Fluid therapy can increase cardiac afterload and lead to pulmonary edema due to poor myocardial contractility.[13] For this reason, fluids should be administered during CPCR only to patients that are hypovolemic.

Other medications

Doxapram hydrochloride hydrate was previously commonly used during CPCR. This drug works by stimulating respiratory centers; however, it also increases oxygen consumption and carbon dioxide production. Due to these sequelae, doxapram should be avoided and only be used if mechanical ventilation is suboptimal or not possible.[14]

Shockable rhythms and defibrillation

Shockable rhythms include ventricular fibrillation and pulseless ventricular tachycardia.[7] Presentation of these rhythms is uncommon in birds; pulseless electrical activity (PEA) and asystole are more common. This is fortunate since defibrillation is not recommended on patients < 1 kg, which includes many birds seen in most veterinary clinics. Furthermore, many machines do not have ranges low enough for the recommended energy level in smaller veterinary patients (usually between 2-10 J/kg). If defibrillation is not possible due to body size, precordial thump technique can be considered.[9] In a study where ventricular fibrillation was experimentally induced in chickens, defibrillation starting at 30 joules (patient weights of 2.04 kg ± SD 0.44 kg) led to PEA in 7/10 birds; none had ROSC.[8]

Aftercare

Patient aftercare

If ROSC is achieved, birds should be moved to sternal recumbency as dorsal recumbency has been shown to decrease tidal volumes.[9] Patient monitoring should be continued with ECG and $EtCO_2$. Additional monitoring equipment can be applied including blood pressure and pulse oximetry. Complete blood cell count, a biochemistry panel, venous blood gas, packed cell volume, total solids, and blood glucose can be assessed to try and determine the cause of arrest, along with any effects of CPA that need to be addressed. Treatments of the post-CPCR bird will vary and may include transfusions, analgesic medications, mannitol, and more.[9,13]

Team aftercare

Regardless of outcome, it is important to take time to debrief the veterinary team after CPCR. In human medicine, a strengths, weaknesses, opportunities, and threats (SWOT) analysis has been shown to improve patient outcomes.[9]

Prognosis

The prognosis for a bird that received CPCR depends heavily on why CPA occurred. Similar to mammals, CPCR outcomes are better for avian patients that experience perianesthetic CPA.[21,22] While there are many things the veterinary team cannot control, timeliness of intervention, effectiveness of CPCR technique, and preparedness of the team are all factors that can be controlled and can improve prognosis.[13]

CPCR success rates for dogs and cats are variable in the literature with recovery rates of 28-60% reported;[13,22] however, survival to hospital discharge occurs in < 10% of cases (2-6%).[7,9,13] In an international multicenter retrospective study involving 709 dogs and cats that underwent CPCR, ROSC occurred in 30%; however, 60% of patients that achieved ROSC were euthanized, only 15% survived beyond 20 minutes, and only 3-5% survived to hospital discharge.[23] On the other hand, survival to discharge can be as high as 47% in dogs and cats experiencing perianesthetic arrests.[9]

Success rates reported in the literature for avian CPCR are unfortunately lower. In a retrospective study of CPCR in exotic animals (n = 146) that included 20 small birds, ROSC occurred in 9.3% of cases. CPA occurred in the hospital for 67.8% of patients (8.97% ROSC), while 32.3% were dead on arrival (12.5% ROSC with 0% survival post-CPR or to discharge). Overall survival after CPCR occurred in 2.3% of cases with only 1.2% surviving to discharge from the hospital.[24] There is only a single retrospective study evaluating CPCR prognosis specifically in birds (n = 41). ROSC occurred in 7% of cases, all of which were perianesthetic arrests. Survival after CPCR occurred in a single case (2%). While the prognosis for birds in this study is much lower than that reported for dogs and cats, it is important to note that compressions were performed in only 34% and ventilation occurred in only 51% of the avian cases, meaning that BLS was not provided to a large subset of avian patients. Furthermore, vascular access was only present in 58% of the birds.[21] It is therefore difficult to directly compare the avian cases to retrospective studies in dogs and cats that received full CPCR.

References

1. Dobbs P, Moittie S, Liptovszky M. Avian anaesthesia related mortality and the associated risk factors in a UK zoological collection. *Vet Anaesth Analg.* 2021;48:922-929.

2. Seamon AB, Hofmeister EH, Divers SJ. Outcome following inhalation anesthesia in birds at a veterinary referral hospital: 352 cases (2004-2014). *J Am Vet Med Assoc.* 2017;251:814-817.

3. Hollwarth AJ, Pestell ST, Byron-Chance DH, et al. Mortality outcomes based on ASA grade in avian patients undergoing general anesthesia. *J Exot Pet Med.* 2022;41:14-19.

4. Hazinski MF, Nadkarni VM, Hickey RW, et al. Major changes in the 2005 AHA guidelines for CPR and ECC. *Circulation.* 2005;112(24):IV 206-IV 211.

5. Zhan L, Yang LJ, Huang Y, et al. Continuous chest compression versus interrupted chest compression for cardiopulmonary resuscitation of non-asphyxial out-of-hospital cardiac arrest. *Cochrane Database Syst Rev.* 2017;3:CD010134.

6. Travers AH, Rea TD, Bobrow BJ, et al. Part 4: CPR overview: 2010 American Heart Association guidelines for cardiopulmonary resuscitation and emergency cardiovascular care. *Circulation.* 2010;122(18): S676-S684.

7. Burkitt-Creedon JM, Boller M, Fletcher DJ, et al. 2024 RECOVER guidelines: updated treatment recommendations for CPR in dogs and cats. *J Vet Emerg Crit Care.* 2024;34(S1):104-123.

8. Eisenbarth J, Cummings C, Rozanski E, et al. A proof-of-concept study evaluating cardiac compression techniques for cardiopulmonary resuscitation in laying hens (*Gallus gallus*). *J Vet Emerg Crit Care.* 2024;34(2):135-142.

9. Hall NH. Cerebro-cardiopulmonary resuscitation and postarrest care in exotic animal critical care. *Vet Clin North Am Exot Anim Pract.* 2023;26:737-750.

10. Fletcher D, Boller M, Brainard B, et al. RECOVER evidence and knowledge gap analysis on veterinary CPR. Part 7: clinical guidelines. *J Vet Emerg Crit Care.* 2012;22(S1):102-131.

11. Lichtenberger M. Shock and cardiopulmonary-cerebral resuscitation in small mammals and birds. *Vet Clin North Am Exot Anim Pract.* 2007;10(2):275-291.

12. Caputo MP, Alphin RL, Pritchett E, et al. Evaluation of the diving reflex in response to nonterminal submersion of white pekin ducks in water-based foam. *Poult Sci.* 2013;92(2):412-417.

13. Linklater A, Chih A. Cardiopulmonary resuscitation of small animals. Available at: https://www.merckvetmanual.com/emergency-medicine-and-critical-care/specific-diagnostics-and-therapy/cardiopulmonary-resuscitation-of-small-animals. Accessed April 1, 2025.

14. Pollock C. Cardiopulmonary resuscitation in exotic animals. Available at: https://lafeber.com/vet/cardiopulmonary-resuscitation-in-exotic-animals/. Accessed April 1, 2025.

15. Paul-Murphy JR, Engilis A Jr, Pascoe PJ, et al. Comparison of intraosseous pentobarbital administration and thoracic compression for euthanasia of anesthetized sparrows (*Passer domesticus*) and starlings (*Sturnus vulgaris*). *Am J Vet Res.* 2017;78(8):887-899.

16. 2020 American Heart Association guidelines for cardiopulmonary resuscitation and emergency cardiovascular care. Part 4: pediatric basic and advanced life support. Available at: https://cpr.heart.org/en/resuscitation-science/cpr-and-ecc-guidelines/pediatric-basic-and-advanced-life-support. Accessed April 16, 2025.

17. Kang J, Mann FA. Open-chest cardiopulmonary resuscitation through a transdiaphragmatic approach in dogs: a cadaveric study to describe the surgical approach and manipulations. *Am J Vet Res.* 2024;85(9):1-7.

18. Brown C, Pilny AA. Air sac cannula placement in birds. *Lab Anim.* 2006;35:23-24.

19. Babbs CF. Interposed abdominal compression CPR: a comprehensive evidence based review. *Resuscitation.* 2003;59:71-82.

20. Johnston MS, Davidowski LA, Rao S, et al. Precision of repeated, Doppler-derived indirect blood pressure measurements in conscious psittacine birds. *J Avian Med Surg.* 2011;25(2):83-90.

21. Crawford A, Ableson A, Gladden J, et al. Retrospective evaluation of cardiopulmonary arrest and resuscitation in hospitalized birds: 41 cases (2006-2019). *J Vet Emerg Crit Care.* 2022;32:491-499.

22. Hoehne SN, Hopper K, Epstein SE. Prospective evaluation of cardiopulmonary resuscitation performed in dogs and cats according to the RECOVER guidelines. Part 2: patient outcomes and CPR practice since guideline supplementation. *Front Vet Sci.* 2019;6:439.

23. Hoehne SN, Balakrishnan A, Silverstein DC, et al. Reassessment campaign on veterinary resuscitation (RECOVER) initiative small animal CPR registry report 2016-2021. *Vet Emerg Crit Care.* 2022;33(2): 143-155.

24. Onuma M, Kondo H, Ono S, et al. Retrospective investigation of cardiopulmonary resuscitation outcome in 146 exotic animals. *J Vet Med Sci.* 2017;79;1611-1614.

Section 13

Masterclass Session II

Julianne Richard, DVM, MPH, Cert AqV
Session Chair

Why Does Everything Look Like a Purple Blob? Tips For Choosing and Using Cytology in Avian Practice

Michelle Sutherland, BVM&S, BSc (Hons), Cert AVP (Zoo Med), DABVP (Avian Practice), FANZCVS (Avian Med Surg), MRCVS

Session #102

Affiliation: From The Unusual Pet Vets (Peninsula), 210 Karingal Drive, Frankston, Vic 3199.

Abstract: Cytology is a relatively cheap, rapid, and easily accessible diagnostic modality in everyday practice, but how often do you review an in-house cytology sample where you can confidently say, "Hooray, I know what this is?" This masterclass helps attendees understand the utility of cytology in avian practice, reviews basic techniques for obtaining a diagnostic cytology sample, and uses a case-based approach to guide attendees to confidently assess and interpret avian cytology samples, with reference to neoplasms. It also reviews the results of a recent study demonstrating the utility of cytology for diagnosing neoplastic disease in avian species.

Introduction

Cytology is a readily available and inexpensive technique that can provide rapid results compared to tissue histopathology. It is associated with low patient morbidity and can be useful for differentiating between an inflammatory process and neoplasia, in addition to differentiating between benign and malignant neoplastic processes.

The holy grail of a cytologic specimen is to obtain a highly cellular sample with a monolayer of well-preserved, well-stained, representative cells. Steps to successfully achieving this include using optimal collection techniques, optimal smearing techniques, and optimal staining techniques. After sample collection and slide preparation, all slides should be evaluated prior to submission to ensure that they are likely to achieve a reliable cytologic diagnosis for the case in question.

Sample Collection

Fine needle aspirates

The aim when collecting samples via fine needle aspirate (FNA) is to obtain a high cell harvest, with minimal artifactual damage to the cells, and minimal blood contamination. The precise technique and choice of equipment depend on the physical characteristics of the lesion, and whether blood contamination is a problem. The basic FNA technique uses a 25-gauge needle and a 3 ml or 5 ml syringe. The needle is inserted into the lesion and repeatedly redirected to sample a number of areas while applying a small amount of suction on the syringe. The suction is released prior to withdrawing the needle; this prevents the sample being violently sucked into the barrel of the syringe, avoiding cell rupture. When the sample has been obtained, the syringe is removed, filled with air, reattached, and then used to gently express the sample onto a clean, dry glass slide.

An alternative technique is the non-aspiration technique. This involves using a needle without a syringe attached to obtain the FNA sample. The needle is inserted and repeatedly redirected to sample different depths and directions within the lesion. The cells are detached by the cutting edge of the needle and enter the needle lumen by capillary action. After withdrawal of the needle, a syringe containing air is reattached and used to gently express the sample onto a clean, dry, glass slide. The non-aspiration technique has been shown to have no difference in cell harvest compared to the traditional aspiration technique, and is useful for sampling fragile cells (e.g. lymphoid cells).[1-3] In addition, there is greater sensitivity in placement of the needle, which is especially useful for small lesions and in small patients.

Impression smears

Impression smears often are used for the evaluation of ulcerated surface lesions; however, their use in this instance is limited as only the surface inflammatory exudate is sampled, rather than cells from deeper tissues. A better use for impression smear cytologic sampling is at the time a biopsy is taken, to allow immediate assessment of a lesion prior to fixation and processing of the tissue sample.

To obtain a good-quality impression smear, the cut surface of the excised sample is blotted a number of times to remove surface blood and serum. The dried surface is then applied to a clean, dry glass slide with gentle pressure. A number of areas can be prepared on a single slide. The preparations should be quickly air dried and then stained.

Fluid analysis

Once a body fluid (e.g., coelomic fluid) is obtained, a cytospin preparation is by far the best method of cell concentration. Few practices have access to a cytospin, so centrifugation of the preparation and sampling of the centrifuged sediment is the usual method of cell concentration. Once the slide has been prepared, it should be rapidly air dried before staining. Severe bacterial overgrowth obscures cells, as bacterial metabolic products tend to destroy cells in cytology preparations. Adding a drop of formalin to the fluid sample may be useful if the fluid cannot be examined immediately in house, although this is generally only recommended if obtaining an accurate nucleated cell count is important. An external laboratory must be advised of the presence of formalin, which excludes the use of Romanowsky stains.

Smearing Techniques

A squash preparation or a line preparation can be used to examine cytologic specimens. Generally, a squash preparation is useful for examination of mass lesions, organs, and lymph nodes, while a line preparation is useful for examining fluid or bloody aspirates.[4] Samples should be spread over the slide quickly and evenly, without excessive pressure, to avoid preparation artefacts.

Staining

Rapid Romanowsky stains (e.g., Diff Quik ® or Rapid Stain®) are generally adequate for examining cytologic specimens. It is important to ensure the stain is fresh and does not contain any precipitate, which could affect interpretation. Washed out staining can occur when cytologic preparations come into contact with formalin. Thus, it is important not to package or store cytologic and histopathologic specimens together. Evaluation of smears for cell preservation and cell yield is recommended prior to examination.

Clinicians are encouraged to prepare multiple slides from clinicopathologic samples where possible. These can be submitted to an external laboratory unstained, allowing the use of additional, more specialized stains such as Perls Prussian Blue (to detect haemosiderin deposits), or Ziehl-Neelsen (to detect acid-fast bacteria, particularly *Mycobacterium* spp.), as appropriate. Gram staining is also a readily available method for examination of microbiology samples in-house and is generally the first step in the identification of a bacterial group. It can also be used to identify some fungi.

Interpretation of Cytologic Samples

When evaluating a cytologic sample, we are essentially assessing whether a sample is inflammatory or neoplastic in origin. Recognition of basic inflammatory cells—heterophils, eosinophils, lymphocytes, macrophages, and plasma cells—is essential for interpretation of cytology samples.

Some tumors contain a large number of inflammatory cells, but these are very uncommon, even when there is tumor necrosis.

The presence of only inflammatory cells in a sample usually indicates a primary inflammatory lesion. Ulceration produces inflammation on the surface of neoplastic lesions, but even in this case the inflammatory cells typically do not extend deeply into the underlying neoplasm. Inflammatory responses can be classified as acute, chronic or granulomatous.

Neoplastic lesions can be classified according to their cell type of origin, generally falling into the categories of epithelial cell, mesenchymal cell, or round cell, although in some cases mixed cell tumors are reported.

Avian inflammation

Depending on the predominant cell type, the inflammatory response of birds can be classified as:

- Heterophilic
- Eosinophilic (rarely reported in birds)
- Mixed cell, or
- Macrophagic[5]

The cytologic features of avian inflammation are summarized in Table 1. Mixed cell inflammation (Fig 1) and chronic inflammation (Fig 2) are depicted in the figures below.

Table 1. Cytologic features of inflammation in birds.

Inflammatory response	Cytologic features
Acute inflammation	Heterophilic Inflammation • Heterophils → 80% inflammatory cells • Heterophil granules lose their rod-shaped appearance and become round/degranulated • Some monocytes present
Mixed cell inflammation (see Figure 1)	Predominantly heterophils (> 50%) • Heterophils are non-degenerate • Increased number of mononuclear leucocytes • Lymphocytes and plasma cells • Epithelioid and multi-nucleated giant cells
Chronic inflammation (see Figure 2)	Macrophages form into layers that resemble epithelium → granuloma • Epithelioid cells (clusters of macrophages with blue-grey/pink cytoplasm and distinct cytoplasmic margins) • Fibroblast proliferation • Appear large compared to normal fibrous tissue • Lymphocytes • May contain acellular, amorphous debris

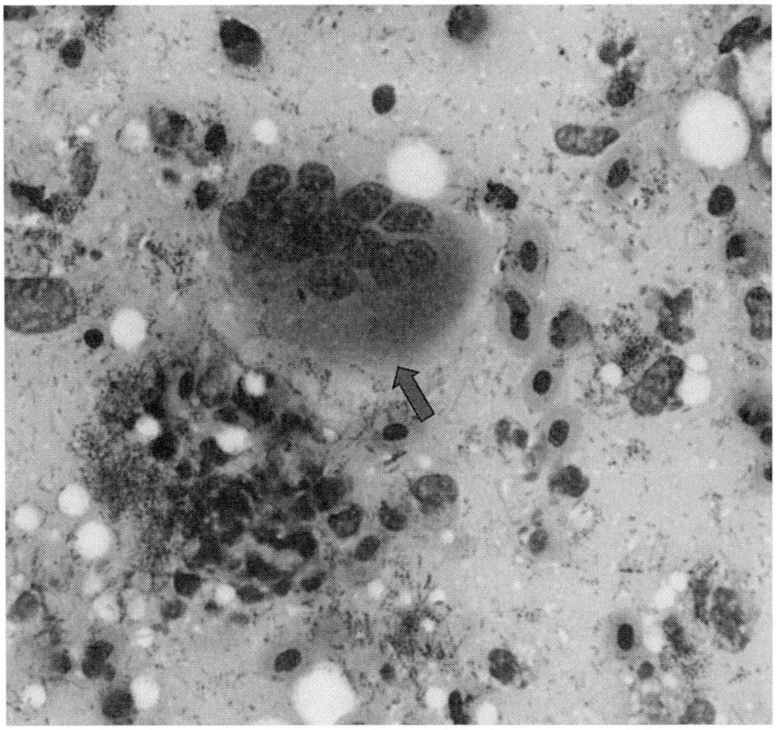

Figure 1. Multinucleated giant cells (arrow) are common features of the avian inflammatory response and can be seen in mixed cell and chronic inflammatory responses. Image courtesy of Dr Natalie Courtman.

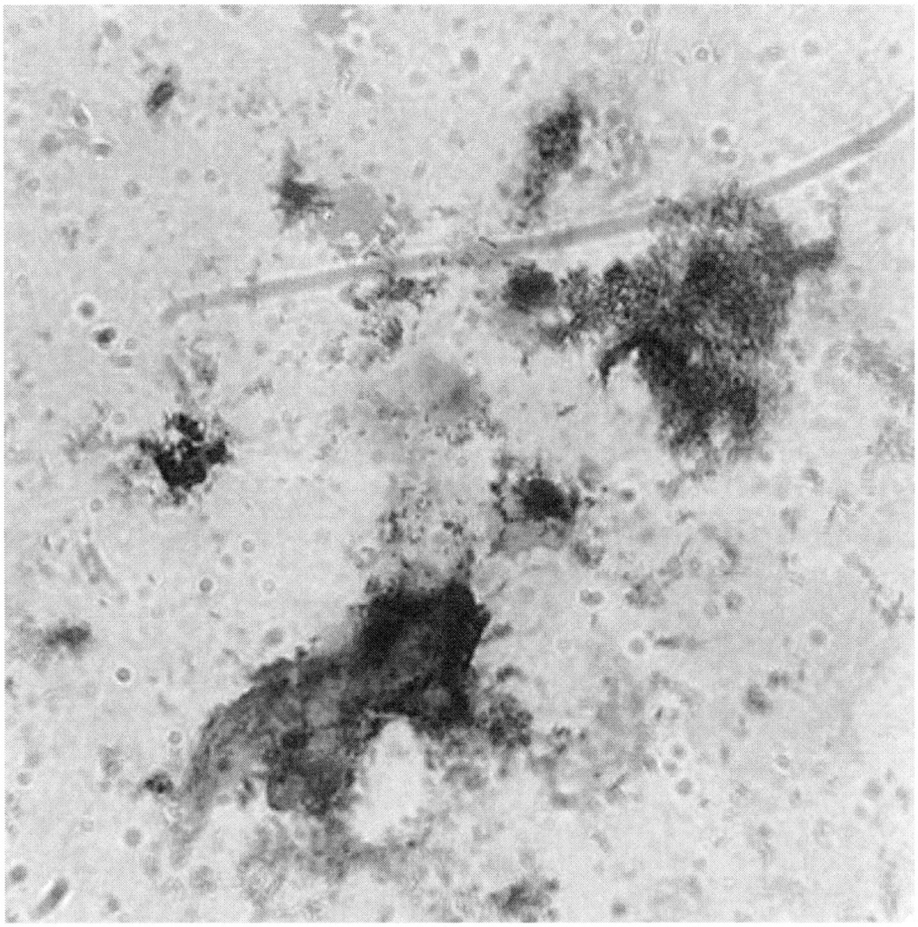

Figure 2. Chronic inflammatory lesions (e.g., bacterial abscesses) often contain a large mat of acellular, amorphous debris; any cells present are often degenerate and difficult to evaluate.

Benign neoplasia

Tissue hyperplasia is a proliferative process of tissues responding to cellular injury or chronic stimulation (e.g., glandular hyperplasia). Neoplasia is the proliferation of cells under conditions that lack the normal controls that would result in the cessation of the multiplication of normal cells.[6] Therefore, benign neoplasia consists of cells that are not malignant.

Based on cytology alone, tissue hyperplasia and benign neoplasia are indistinguishable. The cells exhibit an overall uniform appearance, and are typically immature with an increased cytoplasmic basophilia, with pale vesicular nuclei.[6] Generally, the cells are of a similar size, and have similar nuclear: cytoplasmic ratios and nuclear features. Common benign neoplastic processes that can be identified cytologically in birds include squamous cell hyperplasia (e.g., associated with hypovitaminosis A), lipomas, or lymphoid hyperplasia.[6]

Malignant neoplasia

Frequently, a correlation of clinical and cytologic findings is needed to diagnose malignant neoplasia. The number of neoplastic cells and their appearance vary with the type and location of the neoplasm, and the sampling technique. Usually, a combination of three or more features from many cells in the sample is required for the diagnosis of malignant neoplasia. It is worth remembering that a false negative diagnosis of malignant neoplasia is often less harmful than a false positive one.[7]

Epithelial cell neoplasia

Epithelial cell neoplasms are classified as carcinomas or adenocarcinomas. Cytologically, these tumors tend to yield highly cellular samples. The cells are large round, cuboidal, columnar, or polygonal cells with distinct cytoplasmic margins. Epithelial cells adhere to each other and have common adjoining borders, often producing cells that exfoliate in cellular aggregates (sheets or clusters) (Fig 3).[7] Epithelial cell tumors exhibit nuclear features of malignancy, which are described in the subsequent section on tumor evaluation. Adenocarcinomas are similar to carcinomas, but may have an acinar, tubular, or papillary cell formation (Fig 3).[7]

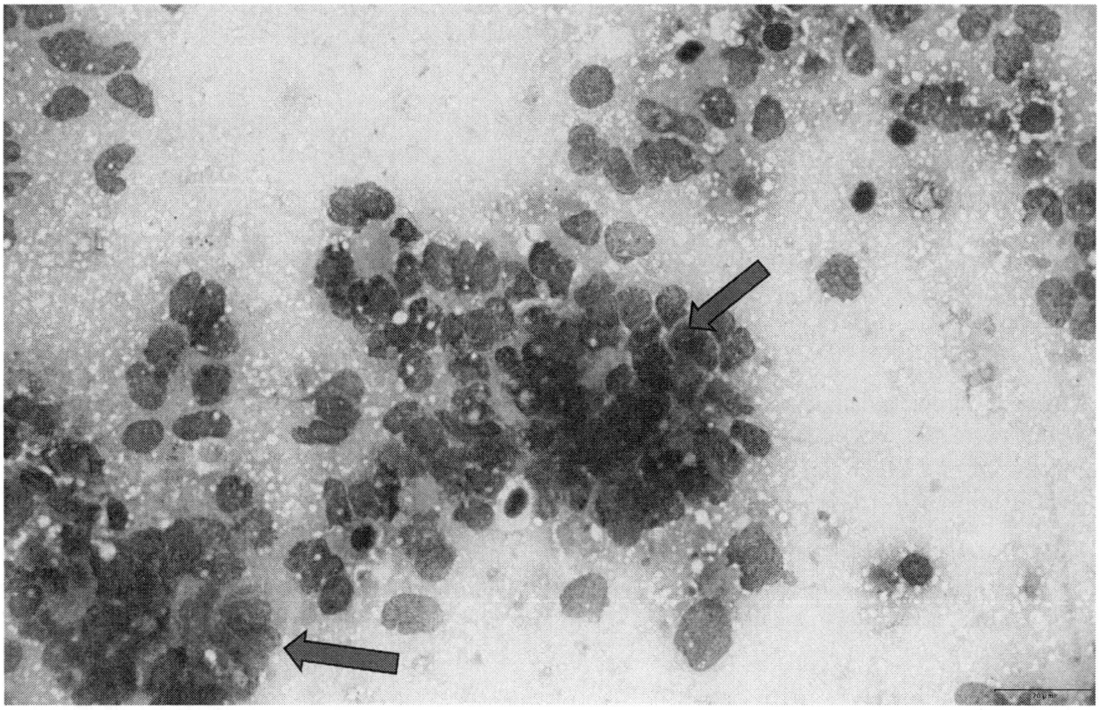

Figure 3. Cytologic appearance of an epithelial cell tumor. The cells have exfoliated in clumps with tight borders between adjoining cells. Acinar structures are noted (arrows), which is suggestive of an adenocarcinoma. Image courtesy of Dr Natalie Courtman.

Mesenchymal cell neoplasia

Mesenchymal cell neoplasms are classified as sarcomas or fibrosarcomas. These lesions typically exfoliate poorly, giving poorly cellular samples containing elongated, stellate, or spindle shaped cells with indistinct cytoplasmic margins. Mesenchymal cells generally do not occur as cellular aggregates, and more often appear as individual cells, with variably shaped nuclei that may be round, elliptical, or fusiform.[7] Fibromas and fibrosarcomas are seen as large, plump, spindle-shaped mesenchymal cells that have exfoliated singly or in aggregates within a pink amorphous background substance. These cells appear less uniform than fibroblasts, have a high nuclear: cytoplasmic ration, and contain lightly to deeply basophilic cytoplasm with indistinct margins (Fig 4).

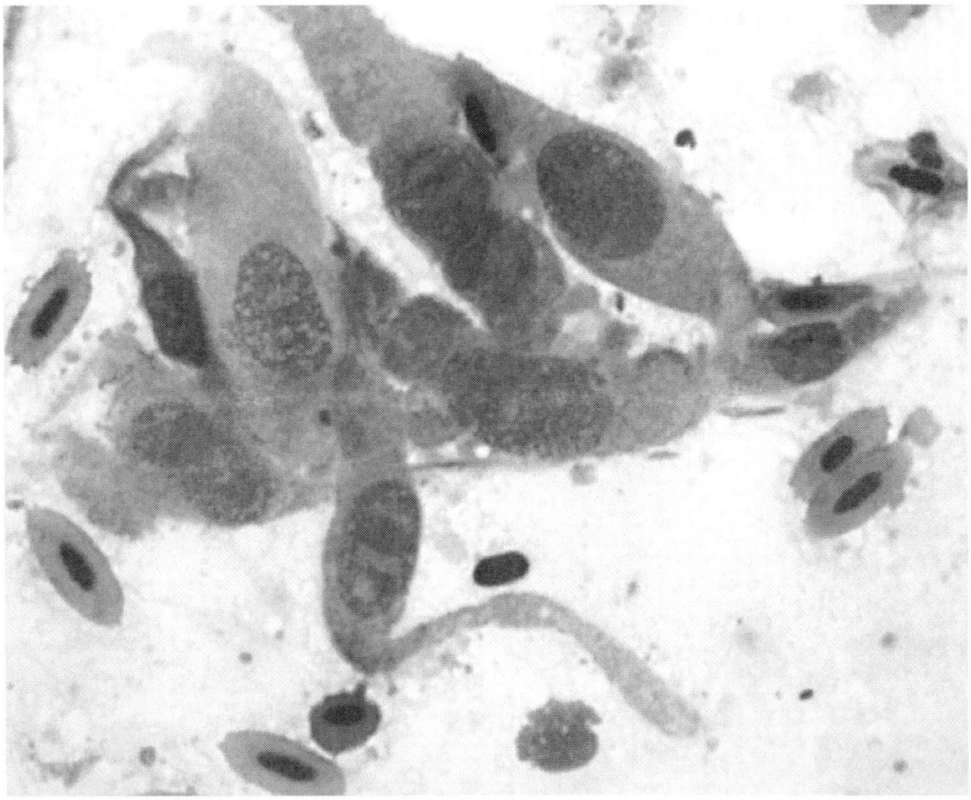

Figure 4. Cytologic appearance of a mesenchymal cell tumor. Large, plump spindle-shaped mesenchymal cells are seen in an aggregate within a collagenous background. Neoplastic mesenchymal cells appear less uniform than fibroblasts, have a high nuclear: cytoplasm ratio, and basophilic cytoplasm. The cells seen here demonstrate nuclear and cellular pleomorphism. Image courtesy of Dr Natalie Courtman.

Round cell neoplasia

Round cell neoplasia occurs as a result of abnormal development of cells that have no normal structural interaction. The cells are round or oval, have distinct cytoplasmic margins, and are smaller than epithelial cells. Typically, round cells exfoliate as individual cells, with round nuclei. Lymphoma is the most common type of round cell tumor seen in birds, since histiocytic and mast cell tumors are rarely reported.[7] Cytologically, lymphoma is seen as a predominance of medium and large lymphocytes that represent 60-90% of the cell population. The neoplastic lymphocytes are often immature. Lymphoma should be suspected if there is a monotypic, homogenous, or heterogenous population of lymphocytes devoid of inflammatory cells.[7] Neoplastic lymphocytes show nuclear and cytoplasmic pleomorphism, cytoplasmic basophilia, and mitotic figures (Fig 5). Abnormal lymphocytes should thus be evaluated for cytoplasmic volume, basophilic intensity, positioning of the cytoplasm around the nucleus, and the presence/absence of vacuolisation and granules.[7]

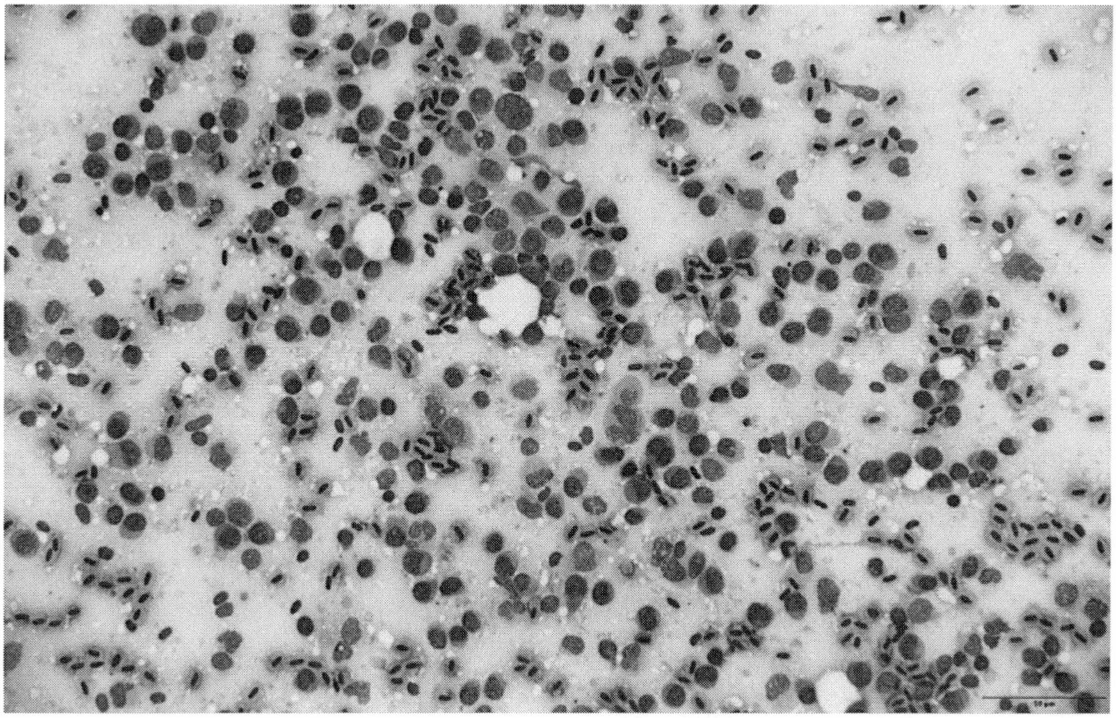

Figure 5. Cytologic appearance of round cell neoplasia (lymphoma). The cells exfoliate as individual cells, with round nuclei. Image courtesy of Dr Natalie Courtman.

Tumor Evaluation: Criteria of Malignancy

When evaluating a cytologic sample for neoplasia, the general cellular features, cytoplasmic criteria, and nuclear criteria of malignancy should be assessed (Table 2).

Table 2. Criteria of malignancy in cytologic samples.

Criteria	Description
General cellular features	• Presence of a monomorphic cell population in absence of inflammation • Pleomorphism of non-inflammatory cells • Increased cellularity from tissues that usually exfoliate poorly • Appearance of cells that are foreign to the tissue being sampled (may indicate metastasis)
Nuclear criteria of malignancy	• Hypertrophy • Anisokaryosis • Variable nuclear: cytoplasmic (N:C) ratios • Nuclear pleomorphism • Abnormal mitoses • Abnormal chromatin patterns • Multinucleation • Irregular cell membrane • Abnormal nucleoli
Cytoplasmic criteria of malignancy	• Basophilia • Vacuolisation • Variations in staining quality • Inclusions • Small cytoplasmic volume (high N:C ratio) • Variable cytoplasmic margins

Avian Cytology versus Histopathology

A recent study evaluated the accuracy of cytology for detecting neoplasia in birds and for determining whether a neoplastic process was benign or malignant, using histopathology as the gold standard.[8] Over a 3.5-year period, tissue samples from 141 companion and aviary birds with clinically suspected, naturally occurring solid neoplasms were collected via surgical biopsy (n = 53) or at necropsy examination (n = 88) from a population of birds presenting to an avian veterinary clinic in Melbourne, Australia. Neoplastic processes were identified in 73.7% (n = 104) of samples 83.7% (n = 87) interpreted as malignant neoplasia and 16.3% (n = 17) interpreted as benign neoplasia.[8] The overall estimated prevalence of neoplasia in study population (n = 5125) was 2.35% (95% CI 2-2.97).[8]

Cytologic examination was performed on 41.8% (59/141) of the suspected tumors. In 45 of these cases impression smears were examined, in 14 cases FNAs were examined, and both an FNA and impression smear were examined from the same tissue in 26 cases. In four cases, multiple sites from the same bird were examined cytologically. The study showed that there was complete agreement between the cytologic and histopathologic diagnoses in 72.8% (43/59) of cases examined.[8]

Cytology correctly identified 87.5% of these cases (35/40) as malignant neoplasms and 55.6% (5/9) as benign processes. Overall, the study showed that there was no significant difference between the use of cytology and histopathology for the detection of avian neoplasms ($P = 1.00$), nor was there any significant difference between the use of cytology and histopathology for the detection of malignant avian neoplasms ($P = 0.185$).[8]

The study also evaluated cell yields obtained in the cytologic specimens. Of the 59 cases evaluated cytologically, 26 (45.6%) were interpreted as having high cellularity, 15 (26.3%) had moderate cellularity, and 17 (29.8%) as having low cellularity. The two remaining cases were excluded from statistical analyses due to inadequate cellularity. In both of the excluded cases, FNA and impression smears had been collected and were later histopathologically diagnosed as lipomas.[8]

When considering the overall diagnostic cell yield obtained for cytologic examination by tissue type, the study found that samples obtained from the kidneys, gastrointestinal, and reproductive tracts exfoliated well via impression smear and FNA, giving samples with a high cell yield for cytologic interpretation, while samples from the liver gave a moderate cell yield, and samples from the integument yielded samples of low cellularity.[8] The authors concluded that these findings may be a consequence of tumor types commonly identified in the viscera, as carcinomas and lymphomas tend to exfoliate readily, and hence cytology may be a more useful tool for initial detection of these neoplasms.[8] This presents a challenge to us as clinicians, since in practice we are more likely to sample lesions from integument by cytology rather than biopsy at initial presentation. In this study, of the 19 samples taken from integument 90% of the samples that returned low cell yields were lipomas or xanthomas, lesions that typically exfoliate poorly. In addition, both lipomas and xanthomas comprise mainly lipid, and lipids stained with alcohol based Romanowsky stains dissolve, so slides from these lesions may be devoid of cells.[9]

The extent of agreement between cytologic and histopathologic diagnoses in the study was categorized as complete agreement, partial agreement, or no agreement.

Complete agreement was defined as comparable cytologic interpretation and histopathologic diagnoses across the categories of non-neoplastic (e.g., inflammatory), benign neoplasia, and malignant neoplasia. Cases were classified as being in partial agreement when the cytologic interpretation gave useful insight as to the basic disease process (e.g., neoplastic, inflammatory) without a specific diagnosis being given. Cases were classified as disagreeing when the cytologic interpretation did not correspond to the histopathologic diagnosis (e.g., interpreted as neoplastic instead of inflammatory).

There was a complete or partial agreement between the histopathologic diagnosis and the cytologic interpretation of malignant neoplasia cases in 70.6% (12/17) of samples associated with the reproductive tract, 69.2% (9/13) of samples affecting the integument, 72.7% (8/11) of samples affecting the gastrointestinal tract, and 66.6% (2/3) of samples affecting the kidneys.[8] The authors concluded that cytology is a reliable tool for differentiating between both neoplastic and inflammatory processes and benign and malignant neoplastic processes in birds.

The accuracies of FNAs and impression smears for examining avian tumors were also compared as part of this study. Overall, the best cell preservation was obtained by performing impression smears from tissues, with 62.2% (n = 28) returning high cellularity for cytologic examination, compared to 53.8% (n = 14) when samples were obtained by FNA.[8]

A limitation of this study is that for the visceral tissues examined, many samples were obtained at necropsy, and thus it not possible to comment on the utility of cytology for identification of neoplastic processes in these tissues in live birds. This is especially relevant when considering that gaining access to internal organs may be limited by the unique avian anatomy. For example, the kidneys are recessed in synsacrum, making ultrasound-guided percutaneous FNA challenging. In addition, biopsy samples obtained under general anaesthesia or at necropsy may yield different results compared to those obtained from a conscious or sedated live bird due to limitations in obtaining a representative tissue sample, which may affect the sensitivity of cytologic interpretation, so it is important to interpret the findings of this study with these conditions in mind.

Limitations of Cytology

Cytology should be considered a guide. In general, cytology has an inferior ability to identify tissue architecture compared to tissue histopathology, and the characteristics of the cells obtained may not be sufficient in many instances to yield a definitive diagnosis or indicate the probable behaviour of the lesion. In addition, obtaining diagnostic samples from poorly exfoliating lesions may be challenging. In such cases, examination of the overall architecture of the tissue may be required, for which cytology is not appropriate. If complex, expensive, or life-threatening therapy is being considered, then the diagnosis should if possible be confirmed by histopathology.

Conclusions

Cytology is a readily available and inexpensive technique that can provide rapid results compared to tissue histopathology. Cytology is associated with low patient morbidity, is useful for differentiating between an inflammatory process and neoplasia, and may help differentiate between benign and malignant neoplasms. Limitations of cytology include an inferior ability to identify tissue architecture and difficulty in obtaining a diagnostic specimen from lesions that exfoliate poorly. Despite these limitations, cytology can add significant value to in-house, patient side diagnostic information, and its use is encouraged to provide both personal challenge and fulfillment in avian practice.

Acknowledgments: The author thanks Drs Natalie Courtman, Richard Ploeg, and Kathy Luk for their assistance with cytologic and histopathologic interpretation, and some of the images used in the proceedings and presentation.

References

1. Mair S, Dunbar F, Becker PJ, et al. Fine needle cytology-is aspiration suction necessary? A study of 100 masses in various sites. *Acta Cytol.* 1989;33:809-813.

2. Storch IM, Sussman DA, Jorda M, et al. Evaluation of fine needle aspiration vs. fine needle capillary sampling on specimen quality and diagnostic accuracy in endoscopic ultrasound-guided biopsy. *Acta Cytol.* 2007;51:837-842.

3. Whitlock J, Taeymans O, Monti P. A comparison of cytological quality between fine-needle aspiration and non-aspiration techniques for obtaining ultrasound-guided samples from canine and feline lymph nodes. *Vet Rec*. 2021;188:e25.

4. Liffman R, Courtman N. Fine needle aspiration of abdominal organs: a review of current recommendations for achieving a diagnostic sample. *J Small Anim Pract*. 2017;58:599-609.

5. Campbell TW. The cytology of inflammation. In: Campbell T, ed. *Exotic Animal Hematology and Cytology.* Ames, IA: John Wiley & Sons, Inc; 2015:235-266.

6. Campbell TW. The cytology of hyperplasia/benign neoplasia. In: Campbell TW, ed. *Exotic Animal Hematology and Cytology*. Ames, IA: John Wiley & Sons, Inc; 2015:267-275.

7. Campbell TW. The Cytology of malignant neoplasia. In: Campbell TW, ed. *Exotic Animal Hematology and Cytology.* Ames, IA: John Wiley & Sons, Inc; 2015:277-307.

8. Sutherland M, Luk K, Courtman N, et al. . Survey of the clinical, cytological, and histopathological features associated with neoplasms in captive avian species in Melbourne, Australia. *J Avian Med Surg*. 2023;37: 243-265.

9. Raskin R. Skin and subcutaneous tissues. In: Raskin RE, Meyer D, Boes KM, eds. *Canine and Feline Cytology: A Color Atlas and Interpretation Guide*. 3rd Ed. St. Louis, MO: Elsevier; 2016:34-90.

Gross Lesion Recognition for the Avian Clinician

Michael M. Garner, DVM, DACVP

Session #109

Affiliation: From Northwest ZooPath, 654 West Main Street, Monroe, WA 98272, USA.

Abstract: Gross lesion recognition, like other imaging modalities, is a bit of an art form. Vast zones of grey may confront the investigator when considering what the lesion is, what it could be, and what to do with it. And nobody is good at all of it. A seasoned livestock pathologist could possibly clear an entire day's necropsies without the help of a single histology slide. A beekeeper would recognize immediately the maggot stages that parasitize his colony. A dog-cat pathologist in a busy private laboratory or a companion animal clinician may not have a clue about gross lesions in livestock or bees. Historically, our attempts at accurately diagnosing disease by gross lesion recognition have been a humbling experience, and that has been the impetus for all further diagnostic specialties. Now that we have put gross pathology in its proper place, why even bother with such an imperfect science? Obviously, gross lesions, or lack thereof, are the first visual indication of what may be wrong with the patient. These are the lumps, the effusions, the asymmetrical oddities, the discolorations, the odiferous clues to disease that (hopefully) stimulate a scientific thought process culminating in a list of differential diagnoses. The purpose of this masterclass is to present images of common and not-so-common avian gross lesions in live and necropsy specimens and, in a participatory manner, establish a differential diagnosis and means for establishing a definitive diagnosis.

Recommended Reading

1. Campbell TW, Ellis C. *Avian and Exotic Animal Hematology and Cytology.* Ames, IA: Iowa State University Press; 2007.

2. Garner MM. An overview of biopsy, cytology and necropsy techniques for clinical veterinarians. In: Aguilar RF, Hernandez SM, Divers SJ, Perpinan D, eds. *Atlas de Medicina de Animales Exoticos.* 2 da Buenos Aires: Inter-Medica S.A. I.C.I.; 2010:437-443.

3. Latimer KS, Rakich PM. Necropsy examination. In: Ritchie BW, Harrison GJ, Harrison LR, eds. *Avian Medicine: Principals and Application.* Lake Worth, FL: Wingers Publishing; 1994:356-376.

Index

THE FIRST LAB DEDICATED EXCLUSIVELY TO BIRDS, REPTILES, & EXOTIC SMALL MAMMALS

WHY SHOULD YOUR HOSPITAL GIVE US A TRY..?

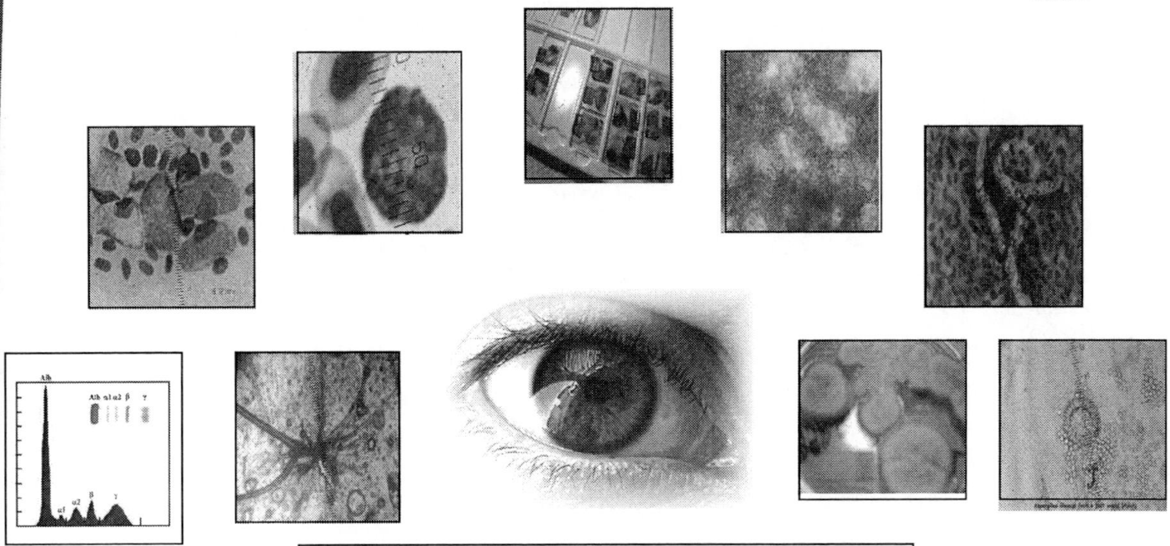

OUR TRAINED EYE.

At Avian & Exotic Lab, we live and breathe in the realm of blood picture analysis, the intricate differences between the species & the cell evaluation for toxicity & subtle changes that can prove invaluable to you, your staff, and, ultimately, your patient. We search blood fields looking for anything that will help you accurately determine the state of your clients' health.

We apply the highest level of hands-on care to every patient sample sent to us for testing.

AT AVIAN & EXOTIC, YOU WILL NOT BE "LOST IN THE CROWD"!

Author Index

Author Index

Association of Avian Veterinarians
PO Box 9
Teaneck, NJ 07666
USA

CURRAN ASSOCIATES INC.
proceedings.com

ISBN 979-8-3313-2911-2

11th Asia-Pacific Regional Conference of the International Society for Terrain-Vehicle Systems

Online
26-28 September 2022

ISTVS

istvs.org since 1962